Android 应用程序开发与实践

主　编　于海宁　马　超　张宏国
副主编　方　舟　曲家兴

哈爾濱工業大學出版社
HITP HARBIN INSTITUTE OF TECHNOLOGY PRESS

内 容 简 介

本书以 Android 应用程序开发为主题，以 Android 官方 API 文档为基础，详细且深入地描述了 Android 应用程序开发所涉及的各个环节，从基础组件开发、定位服务与百度地图、网络通信技术、数据持久化、传感器开发等多个角度进行介绍，为读者提高程序设计能力提供了很大的参考与帮助。

全书共分 14 章，主要为 Android 简介，Hello Android 应用程序，Java 语言基础知识，Android 用户界面设计，活动、意图与广播，Android 后台服务，Android 数据存储，位置服务与地图应用，Android 多线程，Android 网络通信开发，Android 近距离通信开发，Android 传感器开发，以及两个典型的 Android 实践案例等内容。

本书讲述由浅入深，由 Android 的基础知识到实际开发应用，再到开发实践，结构清晰、语言简练，书中范例难度适中，非常适合作为大学本科 Android 程序设计相关课程的理论课教材。

图书在版编目（CIP）数据

Android 应用程序开发与实践/于海宁，马超，张宏国主编. —哈尔滨：哈尔滨工业大学出版社，2020. 7（2024. 6 重印）

ISBN 978 - 7 - 5603 - 8953 - 0

Ⅰ. ①A… Ⅱ. ①于… ②马… ③张… Ⅲ. ①移动终端-应用程序-程序设计 Ⅳ. ①TN929. 53

中国版本图书馆 CIP 数据核字（2020）第 135199 号

策划编辑 张凤涛
责任编辑 周一曈
装帧设计 博鑫设计
出版发行 哈尔滨工业大学出版社
社　　址 哈尔滨市南岗区复华四道街 10 号　邮编 150006
传　　真 0451 - 86414749
网　　址 http://hitpress. hit. edu. cn
印　　刷 哈尔滨博奇印刷有限公司
开　　本 787mm × 1092mm　1/16　印张 25. 75　字数 575 千字
版　　次 2020 年 7 月第 1 版　2024 年 6 月第 2 次印刷
书　　号 ISBN 978 - 7 - 5603 - 8953 - 0
定　　价 198. 00 元

前　言

Android(安卓)是由 Google 公司和开放手机联盟领导及开发的一款基于 Linux 内核的开放源代码的操作系统,主要应用在移动设备上,如智能手机、平板电脑、数字电视和可穿戴设备等。截至 2015 年的第四季度,Android 平台手机的全球市场份额已经达到 80.75%,苹果的 iOS 系统占据了 17.75% 的份额,除 iOS 和 Android 外,还有 Windows Phone 和黑莓,不过二者分别都只占据了 1.1% 和 0.2% 的份额,其他系统共占据了 0.2% 的份额。随着 Windows Phone 的成绩持续不振、黑莓放弃自家平台而转投 Android 阵营,Windows Phone 和黑莓的市场占有率很可能会持续下滑。目前,全球知名手机品牌多采用安卓系统,如三星、华为、联想、小米、HTC 等,而采用 iOS 系统的手机仅有苹果一家。

Android 操作系统的出现为学习 Java 编程语言的读者提供了新的学习方向,巨大的市场需求也提供了更多的就业机会,同时急需更多的开发者来提供更加丰富的应用。多数学习开发的读者在熟悉了语法知识之后,都迫不及待地想编写一款属于自己的软件,这是值得肯定的学习编程的积极态度。但是,如果所选项目过大、过于复杂,往往很难实现功能。因此,Android 应用程序开发的初学者应选择功能单一、结构简单的案例项目。

本书从学习 Android 的实用性以及技术热点出发,充分考虑 Android 初学者在进行移动设备应用程序开发时所需要掌握的基础知识,其内容共分为 14 章。第 1 章介绍 Android 应用程序开发的基础知识;第 2 章介绍如何在 Android 中创建一个最简单的 Hello World 应用程序;第 3 章介绍 Java 语言的基础知识;第 4 章介绍 Android 用户界面设计基础;第 5 章介绍 Android 中的活动组件、意图对象以及广播接收器组件;第 6 章介绍 Android 后台 Service 组件;第 7 章介绍数据库与存储技术;第 8 章介绍位置服务与地图应用;第 9 章介绍 Android 多线程;第 10 章介绍 Android 网络通信开发;第 11 章介绍近距离通信开发;第 12 章介绍传感器开发;第 13 章介绍理财日记本案例开发;第 14 章介绍围住神经猫游戏案例的开发。

本书在编写过程中参考了 Android 6.0 官方 API 文档,按照知识的逻辑关系来分章,循序渐进,突出重点,对知识点的讲解与介绍尽可能做到全面、准确,并给出知识点的适用场合;对于重点、难点,给出易于理解的案例项目,按步骤讲解实现方式。全书所有章节讲解知识的方式统一、结构清晰,方便读者快速查询相关问题。在介绍章节内容时,根据不同知识点的具体情况,介绍知识点的分类、周边信息,并总结功能实现的步骤。

本书的撰写工作由哈尔滨工业大学的于海宁老师、哈尔滨理工大学的马超老师和张宏国老师、黑龙江省网络空间研究中心的方舟老师和曲家兴老师共同完成。于海宁老师负责撰写了第 1 章、第 2 章、第 4 章和第 9 章的内容，共计 15 万字；马超老师负责撰写了第 5 章、第 12 章和第 13 章的内容，共计 15 万字；张宏国老师撰写了第 3 章、第 6 章和第 11 章的内容，共计 12 万字；方舟老师负责撰写了第 7 章和第 10 章，共计 11 万字；曲家兴老师负责撰写了第 8 章和第 14 章，共计 4.5 万字。全书由于海宁老师统稿完成。

由于作者学术与经验的欠缺，在本书的结构、知识点与难点的选择和解析过程中难免会存在一定的不足，希望广大读者不吝赐教。相关技术问题可以发送邮件到 machao8396@163.com 进行交流，作者会尽量给予答复。

编　者

2020 年 3 月

目　　录

第 1 章　Android 简介

Android 一词的本义是“机器人”，同时也是 Google 于 2007 年 11 月宣布的基于 Linux 平台的开源手机操作系统的名称。2008 年 9 月，来自德国的移动运营商 T-Mobile 公司在美国纽约举办了一场大型的新品发布会，并利用这个机会向所有智能手机爱好者们隆重介绍了全世界第一款基于 Android 平台的智能手机，T-Mobile G1。

T-Mobile G1 由来自中国台湾的宏达国际电子股份有限公司（High Tech Computer，HTC）定制，其内部研发代号为 Dream（中文含义为梦想）。Android 平台由 Google 发起的“开放手持设备联盟”开发，因此大家也习惯称 T-Mobile G1 或 HTC Dream 为 Google 手机。目前，Android 逐渐扩展到平板电脑及其他领域，如电视、数码相机、游戏机和各种智能穿戴设备等。

1.1　Android 发展史

2003 年 10 月，Andy Rubin 等创建了 Android 公司，并组建了 Android 团队。2005 年 8 月，Google 低调收购了成立仅 22 个月的高科技企业 Android 及其团队，Andy Rubin 成为 Google 公司工程部副总裁，继续负责 Android 项目。

2007 年 11 月，Google 公司正式向外界展示了这款名为 Android 的操作系统，并且在这一天宣布建立一个全球性的联盟组织——开放手持设备联盟（Open Handset Alliance）。该组织最初由 34 家手机制造商、软件开发商、电信运营商和芯片制造商共同组成。开放手持设备联盟负责研发改良 Android 系统，这一联盟将支持 Google 发布的手机操作系统和应用软件，Google 以 Apache 免费开源许可证的授权方式发布了 Android 的源代码。

2008 年 5 月，在 Google I/O 大会上，Google 提出了 Android HAL 架构图。同年 8 月，Android 获得了美国联邦通信委员会（FCC）的批准。同年 9 月，Google 正式发布了 Android 1.0 系统，这也是 Android 系统最早的版本。

2010 年 2 月，Linux 内核开发者 Greg Kroah-Hartman 将 Android 的驱动程序从 Linux 内核“状态树”（“staging tree”）上除去。从此，Android 与 Linux 开发主流分道扬镳。

2010 年 10 月，Google 宣布 Android 系统达到了第一个里程碑，即电子市场上获得官方数字认证的 Android 应用数量已经达到了 10 万个，Android 系统的应用增长非常迅速。

2011 年 1 月，Google 称每日的 Android 设备新用户数量达到了 30 万部。到 2011 年 7

月,这个数字增长到55万部,而Android系统设备的用户总数达到了1.35亿,Android系统已经成为智能手机领域占有量最高的系统。

2011年8月,Android手机已占据全球智能机市场48%的份额,并在亚太地区市场占据统治地位,终结了Symbian(塞班)系统的霸主地位,跃居全球第一。

2011年9月,Android系统的应用数目已经达到了48万,而在智能手机市场,Android系统的占有率已经达到了43%,继续排在移动操作系统首位。

2012年1月,Google Android Market已有10万开发者推出超过40万活跃的应用,大多数的应用程序为免费。Android Market应用程序商店目录在新年首周周末突破40万基准,距离突破30万应用仅4个月。在2011年早期,Android Market从20万增加到30万应用也花了4个月。

Android在正式发行之前,最开始拥有两个内部测试版本,并且以著名的机器人名称来对其进行命名,它们分别是铁臂阿童木(Astro)和发条机器人(Bender)。后来由于涉及版权问题,因此Google将其命名规则变更为用甜点作为它们系统版本代号的命名方法。甜点命名法开始于Android 1.5发布,作为每个版本代表的甜点尺寸越变越大,然后按照26个字母排序:纸杯蛋糕(Cupcake, Android 1.5)、甜甜圈(Donut, Android 1.6)、松饼(Éclair, Android 2.0/2.1)、冻酸奶(Froyo, Android 2.2)、姜饼(Gingerbread, Android 2.3)、蜂巢(Honeycomb, Android 3.0/3.1/3.2)、冰激凌三明治(Ice Cream Sandwich, Android 4.0)、果冻豆(Jelly Bean, Android 4.1/4.2)、奇巧巧克力(KitKat, Android 4.4)、棒棒糖(Lollipop, Android 5.0)、棉花糖(Marshmallow, Android 6.0)、牛轧糖(Nougat, Android 7.0)、奥利奥(Oreo, Android 8.0)、馅饼(Pie, Android 9.0)。Android版本信息见表1-1。

表1-1 Android版本信息

Android版本	发布日期	代号
Android 1.0	2008年9月	Astro(铁臂阿童木)
Android 1.1	2009年2月	Bender(发条机器人)
Android 1.5	2009年4月	Cupcake(纸杯蛋糕)
Android 1.6	2009年9月	Donut(甜甜圈)
Android 2.0/2.1	2009年10月	Éclair(松饼)
Android 2.2	2010年5月	Froyo(冻酸奶)
Android 2.3	2010年12月	Gingerbread(姜饼)
Android 3.0/3.1/3.2	2011年2月	Honeycomb(蜂巢)
Android 4.0	2011年10月	Ice Cream Sandwich(冰激凌三明治)
Android 4.1	2012年6月	Jelly Bean(果冻豆)
Android 4.2	2012年10月	Jelly Bean(果冻豆)

续表 1-1

Android 版本	发布日期	代号
Android 4.4	2013 年 9 月	KitKat(奇巧巧克力)
Android 5.0	2014 年 10 月	Lollipop(棒棒糖)/ Android L
Android 6.0	2015 年 5 月	Marshmallow(棉花糖)/ Android M
Android 7.0	2016 年 5 月	Nougat(牛轧糖)/ Android N
Android 8.0	2017 年 3 月	Oreo(奥利奥)/ Android O
Android 9.0	2018 年 8 月	Pie(馅饼)/ Android P

2018 年 8 月,Google 发布了目前最新的版本——Android 9.0 的正式版本,其主要的更新包括:对"刘海屏"设备进行了适配;加入了 WiFi Round - Trip - Time 技术;支持 HDR VP9 的设备上播放来自于 YouTube 和 Play Movies 的 HDR 视频;多媒体 APIs 也重新编写了,增加了可用性;进一步优化了系统的效率,等等。

1.2　Android 平台架构及特性

Android 系统采用层次化的架构(图 1-1),包括四个功能层,自下向上依次为 Linux 内核(Linux Kernel)层、函数库和运行时(Libraries 和 Android Runtime)层、应用程序框架(Application Framework)层、应用程序(Applications)层。

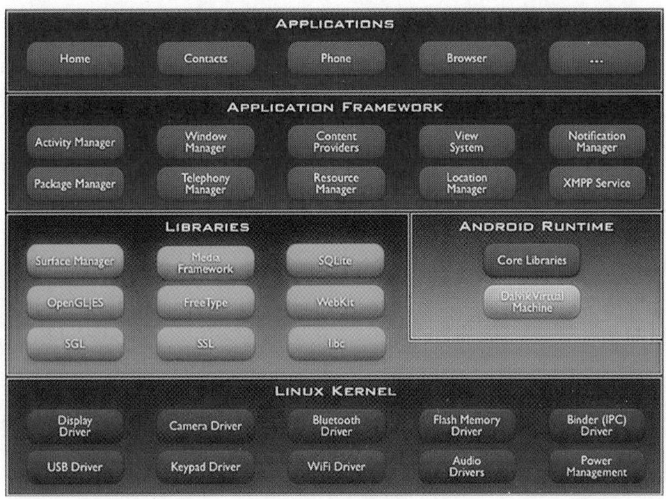

图 1-1　Android 平台架构

(1)Linux 内核层。

Android 系统以 Linux 操作系统内核为基础,借助 Linux 内核服务实现硬件设备驱动、进程和内存管理、网络协议栈、电源管理以及无线通信等核心功能。自 Android 4.0 版本之后，开始采用更新的 Linux 3.X 内核，目前 Android 6.0 版本采用的是 Linux 3.4 内核。

与此同时，Android 内核对 Linux 内核进行了增强，增加了一些面向移动计算的特有功能。例如，可以根据需要杀死进程来释放所需要内存的低内存管理器(Low Memory Keller)，为进程之间提供共享内存资源，同时为内核提供回收和管理内存的匿名共享内存(Ashmem)机制，以及类似于 COM 和 CORBA 分布式组件架构的轻量级的进程间通信 Binder 机制。这些内核的增强使 Android 在继承 Linux 内核安全机制的同时进一步提升了内存管理、进程间通信等方面的安全性。

(2)函数库和运行时层。

函数库大部分由 C/C ++ 编写，所提供的功能通过 Android 应用程序框架为开发者所使用，典型的功能包括：专门为基于嵌入式 Linux 的设备定制的系统 C 库，支持多种常用音频、视频以及静态图像文件的媒体库，2D/3D 图形引擎，Web 浏览器的软件引擎，安全套接层，SQLite 数据库引擎等。除此之外，还有 Android NDK(Native Development Kit)，即Android原生库，其直接使用 Android 系统资源，并采用 C 或 C ++ 语言编写程序的接口，能自动将生成的动态库和 Java 应用程序一起打包成应用程序包文件，即.apk 文件，但其安全性和兼容性可能无法保障。

Android 运行时包含核心库和 Dalvik 虚拟机。核心库提供了大多数 Java 语言所需要调用的功能函数，并提供 Android 的核心 API，如 android.os、android.net、android.media 等；Dalvik 虚拟机是基于 Apache 的 Java 虚拟机(JVM)，并被改进以适用于低内存、低处理器速度的移动设备环境。Dalvik 是基于寄存器的，一般认为基于寄存器的实现使用等长指令，在效率速度上较传统基于栈的 JVM 更有优势，并且 Dalvik 允许在有限的内存中同时高效地运行多个虚拟机的实例，并且每一个 Dalvik 应用作为一个独立的 Linux 进程执行，都拥有一个独立的 Dalvik 虚拟机实例。

(3)应用程序框架层。

应用程序框架层提供开发 Android 应用程序所需的一系列类库，使开发人员可以进行快速的应用程序开发，方便重用组件，也可以通过继承实现个性化的扩展。具体包括的模块如下。

①活动管理器(Activity Mananger)。管理应用程序的生命周期，并提供常用的导航回退功能，为所有程序的窗口提供交互的接口。

②窗口管理器(Window Manager)。对所有开启的窗口程序进行管理。

③内容提供器(Content Provider)。提供一个应用程序访问另一个应用程序数据(如联系人数据库)的功能，或者实现应用程序之间的数据共享。

④视图系统(View System)。构建应用程序的基本组件，包括列表(Lists)、网格

(Grids)、文本框(Text Boxes)、按钮(Buttons),还有可嵌入的 Web 浏览器。

⑤通知管理器(Notification Manager)。使应用程序可以在状态栏中显示自定义的提示信息。

⑥包管理器(Package Manager)。对应用程序进行管理,提供的功能有安装应用程序、卸载应用程序、查询相关权限信息等。

⑦资源管理器(Resource Manager)。提供对非代码资源的访问,如本地化字符串、图片、音频、布局文件等。

⑧位置管理器(Location Manager)。提供位置服务。

⑨电话管理器(Telephony Manager)。管理所有的移动设备功能。

⑩XMPP 服务。是 Google 在线即时交流软件中一个通用的进程,提供后台推送服务。

(4)应用程序层。

Android 平台的应用程序层上包括各类与用户直接交互的应用程序,或由 Java 语言编写的运行于后台的服务程序,如与 Android 系统一起发布的核心应用程序,典型的包括 Email 客户端、SMS 短消息程序、电话拨号及联系人管理程序、浏览器、日历、地图等,以及第三方人员开发的其他应用程序。

1.3 开发环境的搭建

目前,主流的 Android 开发环境有两类:一类是 Eclipse + ADT + Android SDK 的组合方式,另一类是 Android Studio + Android SDK 的组合方式。前者是国内最为普及的开发环境,是学校和软件公司的主流选择,但目前 Google 停止了对 ADT 插件的更新,因此后者将逐渐取代前者成为今后的唯一选择。本书对这两种开发环境的搭建过程均进行了详细的介绍,但是考虑到学生在实际学习过程中对 Eclipse 更为熟悉以及未来就业之后的便捷,本书后续章节的示例均在 Eclipse + ADT + Android SDK 开发环境下予以介绍。

1.3.1 Eclipse + ADT

本书讲解的 Eclipse + ADT + Android SDK 开发环境安装以 Windows 10 为平台,安装的软件为 JDK 1.8、Eclipse Mars.2 (4.5.2) Release for Windows (64 bit)、ADT 23.0.7 和 Android SDKr 24.3.4。

1. 安装 JDK

安装 Eclipse 集成开发环境之前,首先需要安装 JRE,在 Windows 10 上安装 JDK 1.8(包含 JRE)的方法如下。

(1)打开下载网址。

网址为 http://www. oracle. com/technetwork/java/javase/downloads/jdk8 – downloads – 2133151. html。下载页面如图 1 – 2 所示,根据自己所拥有的计算机配置选择下载的软件版本,此处本书选择的是 Windows x64,即 jdk – 8u77 – windows – x64. exe。

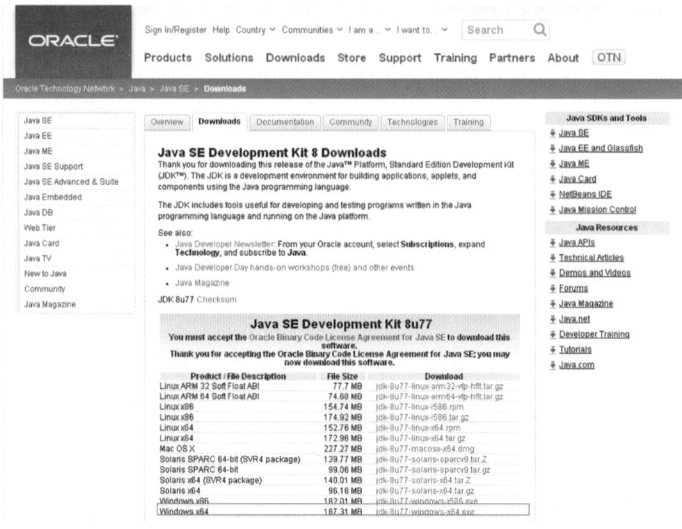

图 1 – 2　下载页面

(2)下载之后,双击文件“jdk – 8u77 – windows – x64. exe”,打开安装程序,依次选择默认选项即可,过程简单,此处不再赘述。

安装完成之后,需要确认一下是否安装成功,在 Windows 10 平台中,同时按下“win + R”,在打开的对话框中输入“cmd”并按下“确定”,在打开的 CMD 窗口中输入“java – version”命令,如果显示如图 1 – 3 所示的提示信息,则说明安装成功。

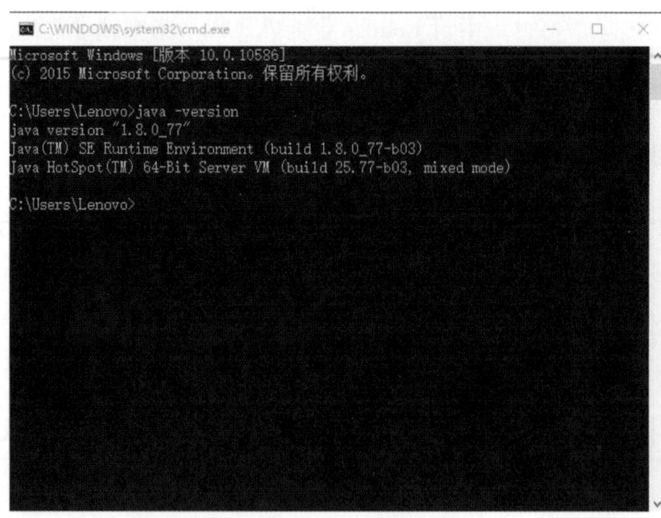

图 1 – 3　提示信息

如果经过上述步骤发现安装失败,则需要进行路径配置。在桌面上右击“此电脑(或计算机)”图标,在弹出的快捷菜单中选择“属性”命令,在弹出的对话框中选择“高级系统设置”选项卡,在弹出的对话框中选择“环境变量”按钮,弹出“环境变量”对话框,如图 1 - 4 所示。

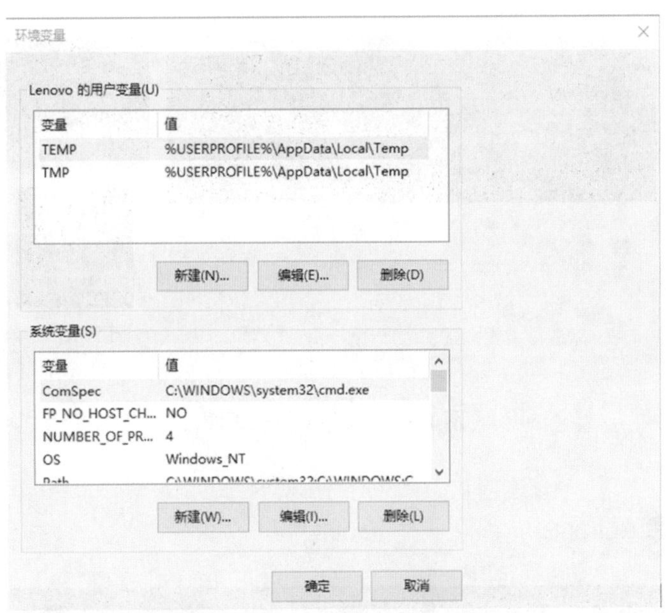

图 1 - 4 “环境变量”对话框

接下来,设置环境变量的步骤如下。

①“新建”一个“系统变量”,在“变量名”中输入“JAVA_HOME”,在“变量值”中输入 JDK1.8 的安装目录,本书为 C:\Program Files\Java\jdk1.8.0_77。

②“新建”一个“系统变量”,在“变量名”中输入“classpath”,在“变量值”中输入.;% JAVA_HOME% \lib\rt.jar;% JAVA_HOME% \lib\tools.jar。

③“编辑”一个“变量名”为“Path”的系统变量,在“变量值”最前面添加% JAVA_HOME% \bin。

2. 安装 Eclipse

(1)打开 Eclipse 的官方下载网址 http://www.eclipse.org/downloads/,Eclipse 下载页面如图 1 - 5 所示,此处选择 Eclipse Installer for Windows 64bit。

(2)来到选择镜像页面,如图 1 - 6 所示,选择最近的镜像即可,此处推荐的下载镜像为“China - University of Science and Technology of China (http)”。

(3)下载完成之后,双击文件“eclipse - inst - win64.exe”,打开安装程序,选择版本类型 Eclipse IDE for Java Developers,然后依次选择默认选项即可,此过程不再赘述。本书将 Eclipse安装到目录 E:\Eclipse 中。打开目录,双击“eclipse.exe”图标,Eclipse 能自动找到先前安装的 JDK 路径。

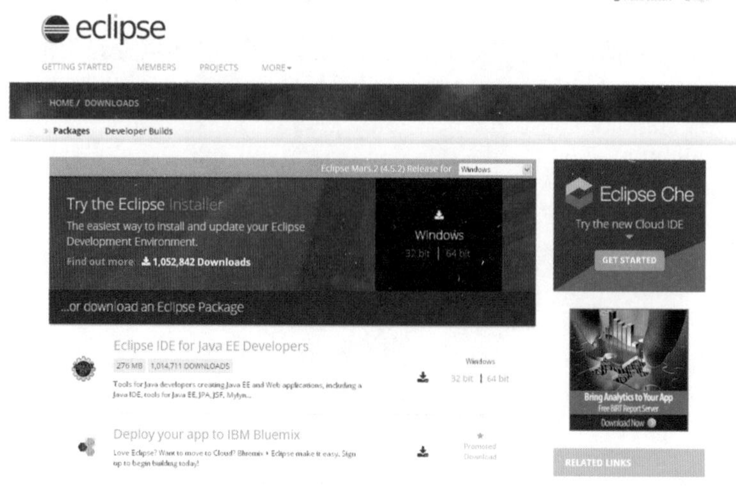

图 1－5　Eclipse 下载页面

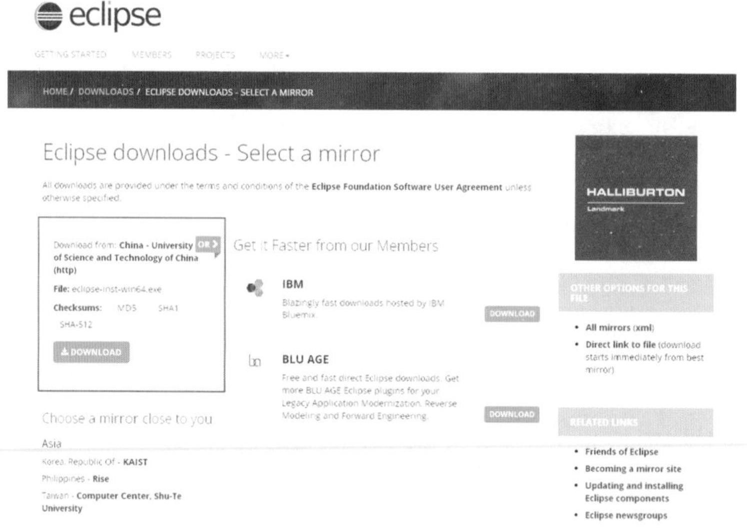

图 1－6　选择镜像页面

3. 安装 ADT

Android 开发工具(Android Development Tools, ADT)是一款支持 Eclispe IDE 的插件,它扩展了 Eclipse 的能力,使其能够快速地建立新的 Android 工程,创建应用程序 UI,添加基于 Android 框架 API 的包。使用 Android SDK 工具调试应用程序,是为了发布应用程序而导出签名(或未签名)的. apk 文件。

目前使用的是 Android 官方网站 ADT 23. 0. 7 版本,下载地址为“http://pan. baidu. com/s/1boH9obP”,该版本的使用约束包括:

(1)Java 7 或更高;

(2)Eclipse Indigo (Version 3.7.2)或更高;

(3)SDK Tools r24.1.2 或更高。

在 Eclipse 中安装 ADT 的过程:启动 Eclipse,选择 Help→Install New Software,打开Eclipse的插件安装界面,点击“Add”,本地手动安装 ADT 如图 1-7 所示。点击“Archive”按钮,选择 ADT 插件压缩包在本地磁盘中的位置。在 ADT 插件安装前,会提示用户对需要安装的插件(即 Developer Tools)的具体内容(如 Android DDMS、Android Development Tools、Android Hierarchy Viewer 等)进行选择和确认,全选即可,然后点击“Next”按钮完成剩余安装步骤,安装成功后需要按照提示重启 Eclipse 以使 ADT 生效。

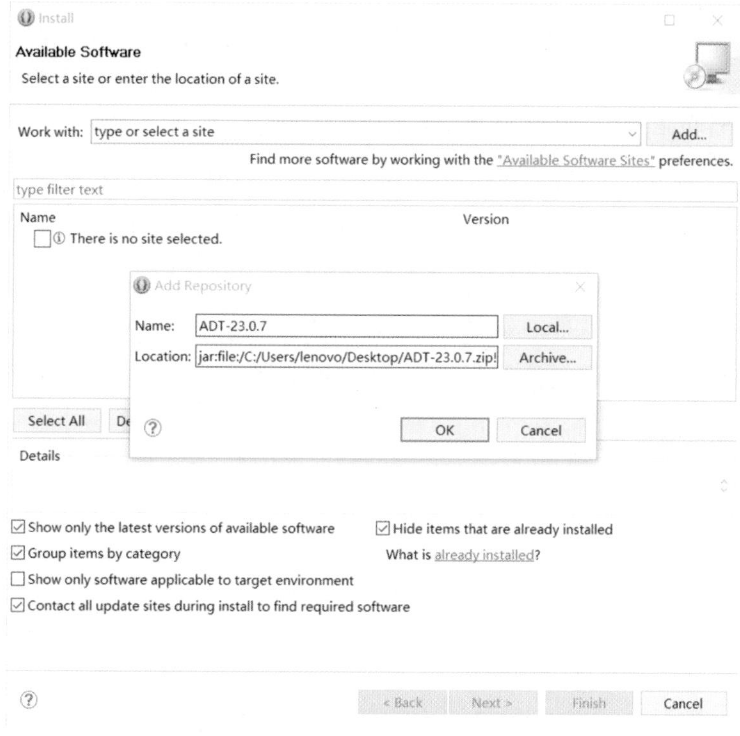

图 1-7 本地手动安装 ADT

注意:由于在大陆地区无法访问 Android 官方网站,因此在“Location”中添加网址“http://dl-sll.google.com/Android/eclipse/”,进行在线安装时,需要借助 VPN 软件。

4. 安装 Android SDK Manager

下载并安装 Android SDK Manager,本书提供的版本是 installer_r24.3.4-windows.exe,其下载地址为“http://pan.baidu.com/s/1nvkvfBz”。下载 Android 6.0 以及相应工具,如图 1-8 所示,安装成功后可以自行选择需要下载的文件,点击右下角按钮开始下载/安装(需要借助 VPN 软件)。

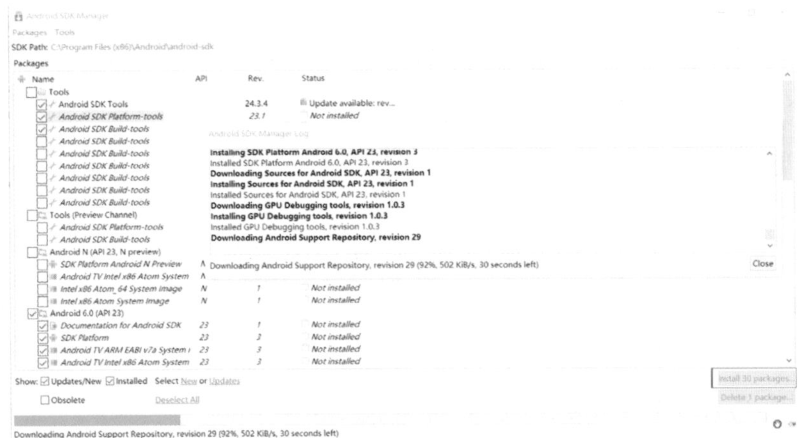

图 1-8　下载 Android **6.0** 以及相应工具

下一步需要在 Eclipse 中设置 Android SDK Manager 的保存路径。选择 Windows→Preferences，打开 Eclipse 的配置界面，配置 Android SDK 如图 1-9 所示。点击 SDK Location 后面的 Browse 按钮，选择 Android SDK Manager 的保存路径，最后点击“Apply”。

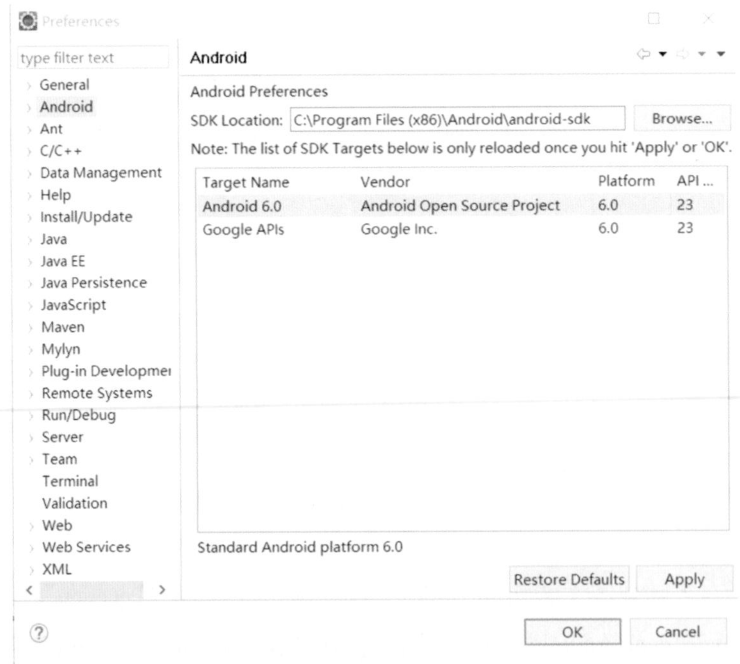

图 1-9　配置 Android SDK

注意：如果读者没有 VPN 软件，可以通过下载地址“http://pan.baidu.com/s/1i45RH1j”下载压缩包“android-sdk.zip”。下载到本地之后直接解压缩即可，无需安装，android-sdk 文件夹中包含图 1-8 中 Android SDK Tools、Android SDK Platform-tools、Androd SDK Build - tools 以及 Android 6.0(API 23)和 Extras 目录下的全部文件。

如果一直在使用 Eclipse + ADT,应该注意到 Android Studio 现在是 Android 官方的集成开发环境。因此,建议熟悉并逐渐向 Android Studio 迁移,以此来接收所有的最新版本的集成开发环境更新。

1.3.2　Android Studio

Android Studio 是 Google 于 2013 年 I/O 大会针对 Android 开发推出的新的开发工具,目前很多开源项目都已经在采用了。Google 的更新速度也很快,截至 2019 年 8 月,Android Studio 3.5 已经发布了。

与 Eclipse 相比,Android Studio 的优势体现在以下几个方面:

①Google 专门为 Android"量身订做"了一款基于 IntelliJ IDEA 改造的 IDE;

②在启动速度、响应速度、内存占用等方面优于 Eclipse;

③UI 更漂亮,其自带 Darcula 主题的炫酷黑界面堪称高大上;

④提供智能保存,开发者不用每次都按 Ctrl + S;

⑤支持 Gradle,在配置、编译、打包方面做得更好;

⑥具有更加智能的 UI 编辑器,自带多设备的实时预览;

⑦内置终端,可以方便地进行命令行操作;

⑧更完善的插件系统,如可以直接搜索下载 Git、Markdown、Gradle 等插件;

⑨自带如 GitHub、Git、SVN 等流行的版本控制系统。

下面开始下载并安装 Android Studio,具体步骤如下。

(1)打开浏览器,在地址栏输入"http://developer.android.com/sdk/index.html",进入 Android Studio 下载页面,如图 1-10 所示。

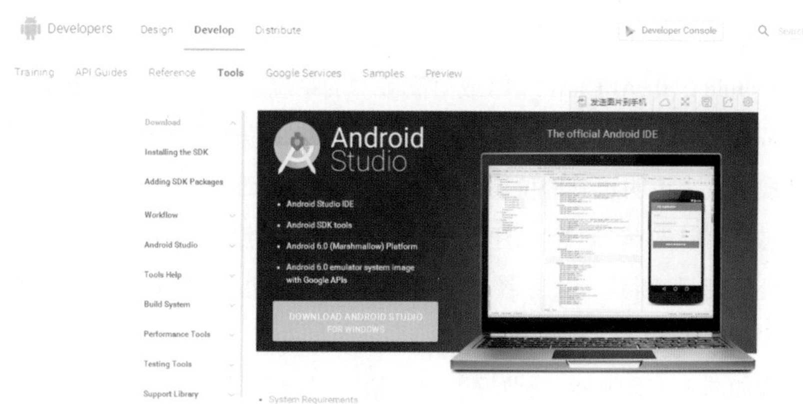

图 1-10　Android Studio 下载页面

目前，大陆地区暂时无法访问 Google 的 Android 开发者官网，作者提供了 Android Studio 集成开发包（android - studio - bundle - 141. 2456560 - windows. exe）在百度云网盘上的下载地址 http://pan. baidu. com/s/1cLUCgA。此外，读者也可以使用 VPN 软件实现对 Android 开发者官网的访问。

（2）根据系统情况，选择需要安装的软件版本进行下载（图 1 - 11），推荐使用集成开发包，本书选择的是 android - studio - bundle - 141. 2456560 - windows. exe，它提供了所有开发 Android App 需要用到的工具，包括 Android Studio IDE 和 Android SDK。

All Android Studio Packages

Select a specific Android Studio package for your platform. Also see the Android Studio release notes.

Platform	Package	Size	SHA-1 Checksum
Windows	android-studio-bundle-141.2456560-windows.exe (Recommended)	1209163328 bytes	6ffe608b1dd39041a578019eb3fedb5ee62ba545
	android-studio-ide-141.2456560-windows.exe (No SDK tools included)	351419656 bytes	8d016b90bf04ebac6ce548b1976b0c8a4f46b5f9
	android-studio-ide-141.2456560-windows.zip	375635150 bytes	64882fb967f960f2142de239200104cdc9b4c75b
Mac OS X	android-studio-ide-141.2456560-mac.dmg	367456698 bytes	d0807423985757195ad5ae4717d580deeba1dbd8
Linux	android-studio-ide-141.2456560-linux.zip	380943097 bytes	b8460a2197abe26979d88e3b01b3c8bfd80a37db

图 1 - 11　选择需要安装的软件版本

（3）在 Windows 平台上安装 Android Studio。

在安装 Android Studio 之前需要确认是否已经安装了 JDK6 或以上版本（只有 JRE 是不够的）。当采用 Android 5.0（API level 21）或更高版本的系统进行开发时，需要安装 JDK7。

在有些 Windows 平台中，执行脚本无法发现 JDK 的安装位置。如果遇到这一问题，则需要通过设置环境变量来说明正确的位置。在 Windows 10 桌面上，鼠标移动到“此电脑”图标，右击打开上下文菜单，选择“属性”，打开控制面板→所有控制面板项→系统，依次选择高级系统设置→环境变量，新建一个系统变量 JAVA_HOME 指向 JDK 文件夹，如 C:\Program Files\Java\jdk1. 7. 0_67。

安装向导页面如图 1 - 12 所示，按照安装向导依次点击 Next 按钮，本书将 Android Studio 默认安装到目录 C:\Program Files\Android\Android Studio 下。Android Studio 使用的个人工具和其他 SDK 包被安装到了独立的目录（如 C:\Users\Lenovo\AppData\Local\Android\sdk）中。如果需要直接使用工具，可以在 CMD 窗口下直接访问安装位置。

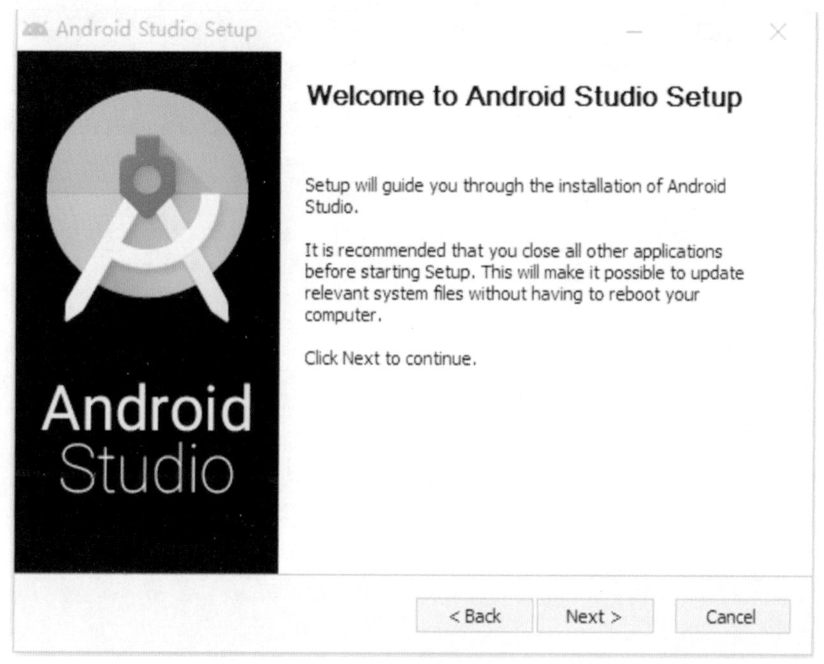

图 1－12　安装向导页面

通过上述步骤，已经完成了 Android Studio 的下载与安装。下面来打开 Android Studio，在首次打开 Android Studio 时，需要选择 Android Studio 的设置以及下载依赖组件，设置页面如图 1－13 所示。

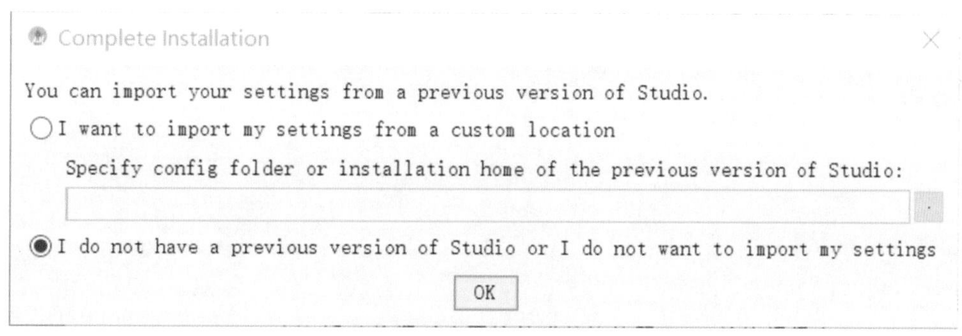

图 1－13　设置页面

如图 1－14 所示为下载组件页面。在这一过程中，需要借助 VPN 软件。此外，建议暂时关闭反病毒软件或其他使用 SDK 的执行程序。

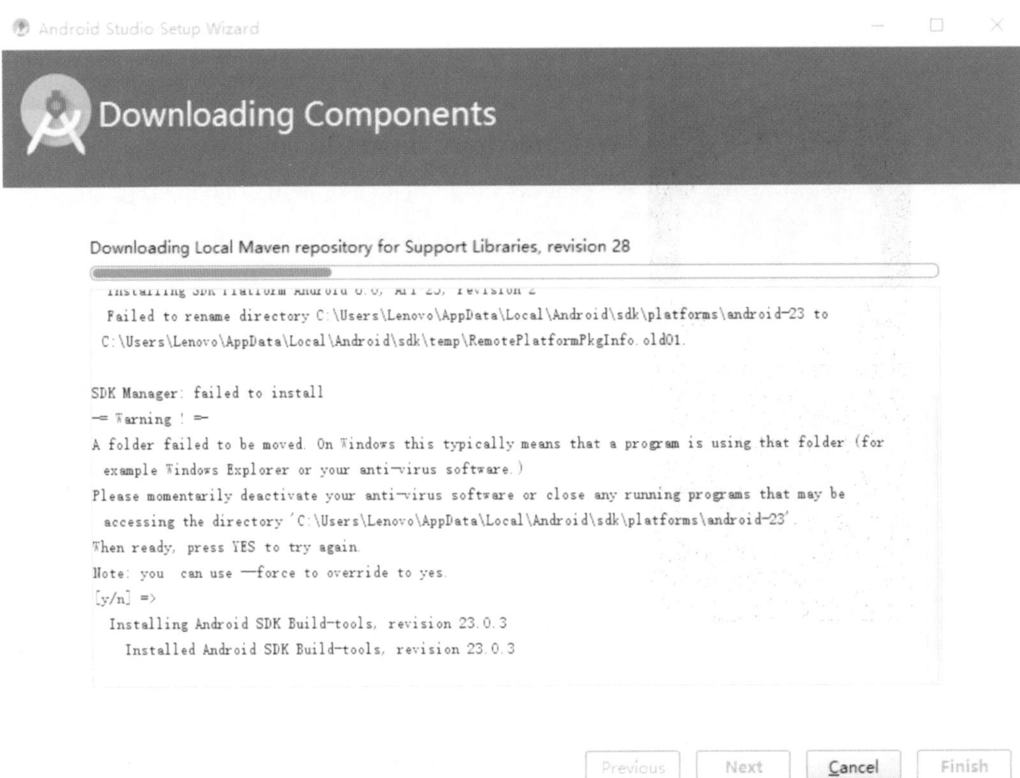
Android Studio Setup Wizard
Downloading Components
Downloading Local Maven repository for Support Libraries, revision 28
Failed to rename directory C:\Users\Lenovo\AppData\Local\Android\sdk\platforms\android-23 to
C:\Users\Lenovo\AppData\Local\Android\sdk\temp\RemotePlatformPkgInfo.old01.
SDK Manager: failed to install
-= Warning ! =-
A folder failed to be moved. On Windows this typically means that a program is using that folder (for
example Windows Explorer or your anti-virus software.)
Please momentarily deactivate your anti-virus software or close any running programs that may be
accessing the directory 'C:\Users\Lenovo\AppData\Local\Android\sdk\platforms\android-23'.
When ready, press YES to try again.
Note: you can use --force to override to yes.
[y/n] =>
Installing Android SDK Build-tools, revision 23.0.3
Installed Android SDK Build-tools, revision 23.0.3
Previous
Next
Cancel
Finish

图 1 – 14　下载组件页面

第 2 章　Hello Android 应用程序

对于一门应用程序开发技术的初学者来说，HelloWorld 程序是最基本、最简单的程序，也通常是初学者编写的第一个程序，它被用来确定程序开发环境，以及运行环境是否已经安装成功。本章分别在 Eclispe + ADT 和 Android Studio 环境中创建 HelloWorld 程序，以此来帮助 Android 移动设备应用程序的初学者熟悉并掌握 Android 应用程序的开发环境。

2.1　Eclipse + ADT 环境

2.1.1　创建工程

下面通过创建一个 HelloWorld 工程来熟悉 Eclipse + ADT 开发环境。打开 Eclipse，选择 File→New→Project 命令，打开如图 2－1 所示创建工程的向导页面，选择“Android Application Project”。

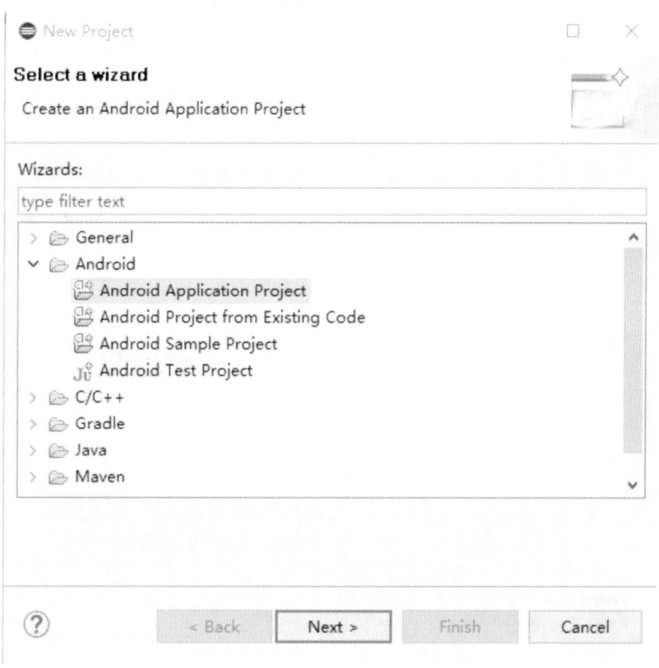

图 2－1　创建工程的向导页面

在点击“Next”按钮后，进入如图 2 – 2 所示的填写工程信息页面。在此页面中，需要填写应用程序名称、工程名称、应用程序包名，并且可以设置兼容的最小 SDK 版本等信息。

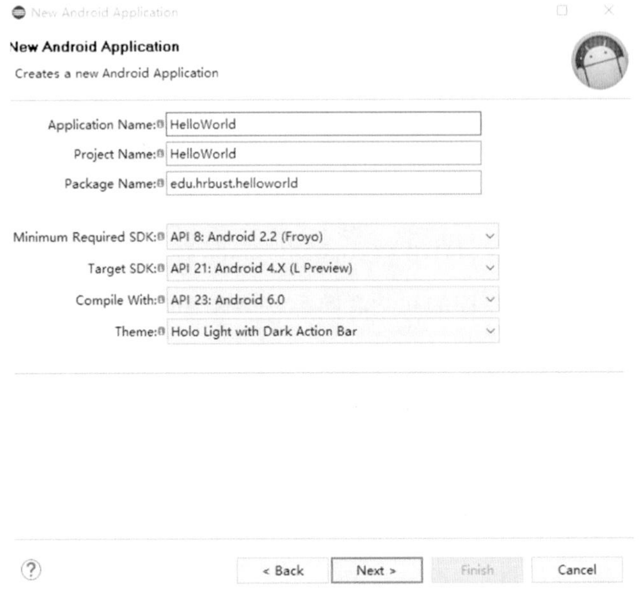

图 2 – 2　填写工程信息页面

在设置完相关信息之后，继续点击“Next”按钮后，进入如图 2 – 3 所示的选择活动模板页面。图 2 – 3 中的页面给出了在创建活动时可以选择的活动模板，这里选择一个空的活动模板。然后继续点击“Next”按钮，直到最后点击“Finish”按钮完成工程的创建。

New Android Application
Create Activity
Select whether to create an activity, and if so, what kind of activity.
Create Activity
Blank Activity
Blank Activity with Fragment
Empty Activity
Fullscreen Activity
Master/Detail Flow
Navigation Drawer Activity
Tabbed Activity
Blank Activity
Creates a new blank activity with an action bar.
< Back
Next >
Finish
Cancel

图 2 – 3　选择活动模板页面

2.1.2　目录结构

在上一节创建完成的 Android 应用程序的工程界面左侧是 HelloWorld 应用程序的目录结构，如图 2－4 所示，可以看到以下的目录和文件。

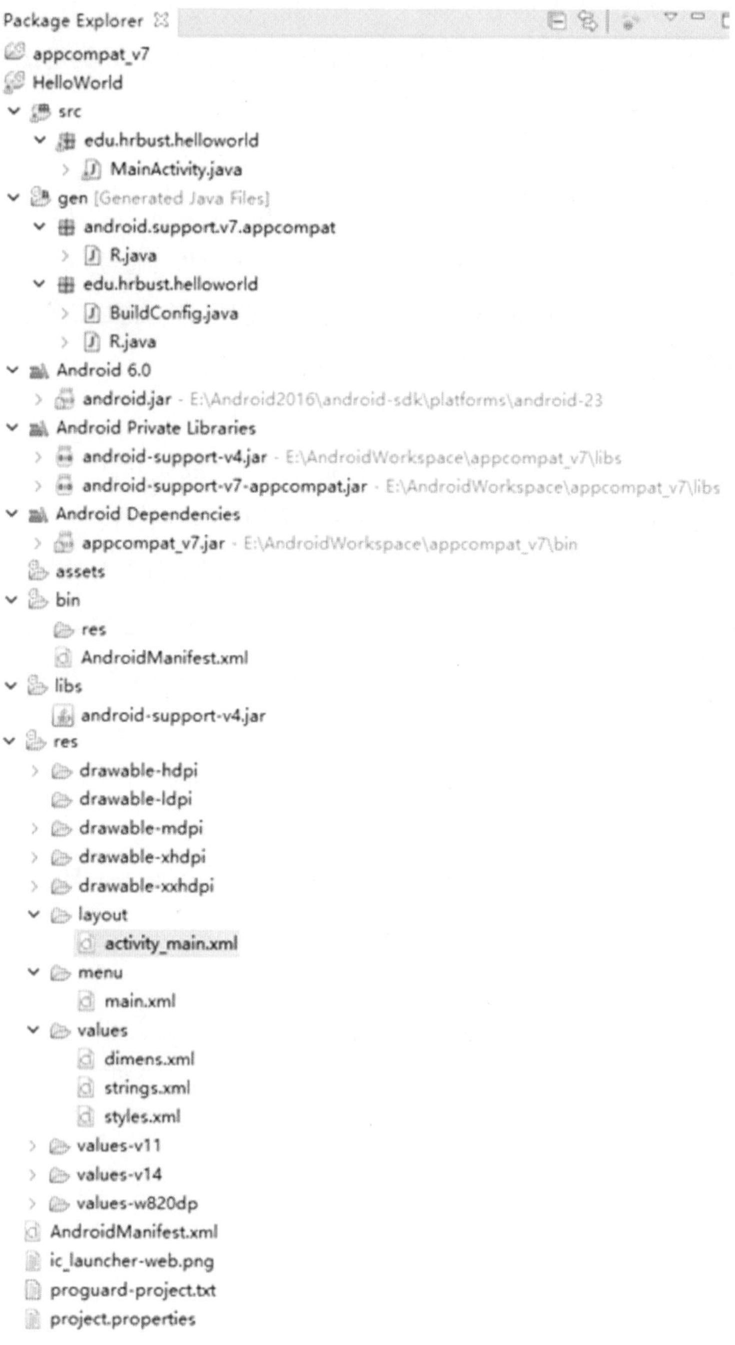

图 2－4　HelloWorld **应用程序的目录结构**

(1)src 目录——程序文件。里面保存了程序员直接编写的程序文件。与一般的 Java 项目一样,src 目录下保存的是项目的所有包及源文件(. java)。

(2)gen 目录。存放编译器自动生成的一些 java 代码,. java 格式的文件是在建立项目时自动生成的,这个文件是只读模式,不能更改。这个目录中最关键的文件就是 R. java,R 类中包含很多静态类,静态类的名字都与 res 中的一个名字对应,就像是一个资源字典大全,其中包含了用户界面、图像、字符串等对应各个资源的标识符,R 类定义了该项目所有资源的索引。

(3) Android 6. 0、Android Private Libraries、Android Dependencies 目录。这三个目录是库。Android 6.0 文件夹下包含 android. jar 文件,这是一个 java 归档文件,其中包含构建应用程序所需的所有 Android SDK 库(如 Views、Controls)和 APIs,通过 android. jar 将自己的应用程序绑定到 Android SDK 和 Android Emulator,允许使用所有 Android 的库和包,且使应用程序在适当的环境中调试。

(4)assets 目录。除了提供 res 目录存放资源文件外,android 在 assets 目录也可以存放资源文件。assets 目录下的资源文件不会在 R. java 自动生成 id,所以读取 asset 目录下的文件必须指定文件的路径,可以通过 AssetManager 类来访问这些文件。

(5)bin 目录。该目录是编译之后的文件以及一些中间文件的存放目录,ADT 先将工程编译成 Android 虚拟机文件 classes. dex,最后将该 classes. dex 封装成 apk 包。

(6)libs 目录。该目录用于存放第三方库(新建工程时,默认会生成该目录,没有的话手动创建即可)。

(7)res 目录。存放项目中的资源文件,该目录中有资源添加时,R. java 会自动记录下来。

(8)AndroidManifest. xml 文件——配置文件。配置文件 AndroidManifest. xml 是一个控制文件,里面包含了该项目中所使用的 Activity、Service、Recevier。AndroidManifest. xml 文件的详细细节见表 2-1。

表 2-1　AndroidManifest. xml 文件的详细细节

参数	说明
manifest	根节点,描述了 package 中所有的内容
xmlns:android	包含命名空间的声明
package	声明应用程序包
uses-sdk	该应用程序所使用的 sdk 相关版本
application	包含 package 中 application 级别组件声明的根节点。此元素也可包含 application 的一些全局和默认的属性,如标签、icon、主题、必要的权限等。一个 manifest 能包含零个或一个此元素(不能大于一个)

续表 2－1

参数	说明
android:icon	应用程序图标
android:label	应用程序名字
activity	用来与用户交互的主要工具。Activity 是用户打开一个应用程序的初始页面，大部分被使用到的其他页面也由不同的 < activity > 实现，并声明在另外的 < activity > 标记中。注意，每一个 activity 必须有一个 < activity > 标记对应，而无论它给外部使用或是只用于自己的 package 中。如果一个 activity 没有对应的标记，则不能运行它。另外，为支持运行环境查找 activity，可包含一个或多个 < intent－filter > 元素来描述 activity 所支持的操作
intent－filter	声明了指定的一组组件支持的 Intent 值，从而形成了 Intent Filter。除能在此元素下指定不同类型的值外，属性也能放在这里来描述一个操作所需的唯一的标签、icon 和其他信息
action	组件支持的 Intent action
Category	组件支持的 Intent category，这里指定了应用程序默认启动的 activity

在上面的目录中，res 目录下一般有以下几个子目录。

①drawable－hdpi、drawable－ldpi、drawable－mdpi、drawable－xhdpi、drawable－xxhdpi 目录。存放应用程序可以使用的图片文件，子目录根据图片质量分别保存。

②layout 目录。屏幕布局目录，可以在该文件内放置不同的布局结构和控件来满足项目界面的需要，也可以新建布局文件。

③menu 目录。存放定义了应用程序菜单资源的 XML 文件。

④values、values－v11、values－v14、values－w820dp 目录。存放定义了多种类型资源的 XML 文件，如软件上需要显示的各种字体，还可以存放不同类型的数据，如 dimens. xml、strings. xml、styles. xml。

2.1.3　创建虚拟机

选择 Window→Android Virtual Device Manger 命令，打开如图 2－5 所示的创建模拟器页面开始创建模拟器。

在图 2－5 中点击“Create”按钮，打开如图 2－6 所示配置模拟器参数页面，开始对模拟器的名称、设备、目标等参数进行配置。最后，点击“OK”按钮完成模拟器的创建。

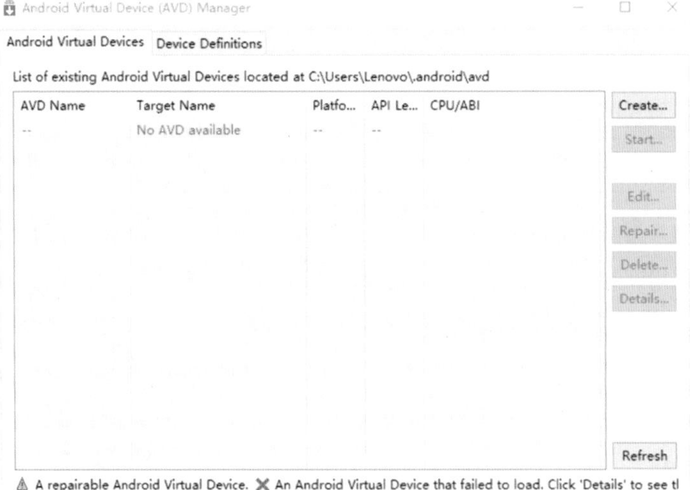

图 2-5　创建模拟器页面

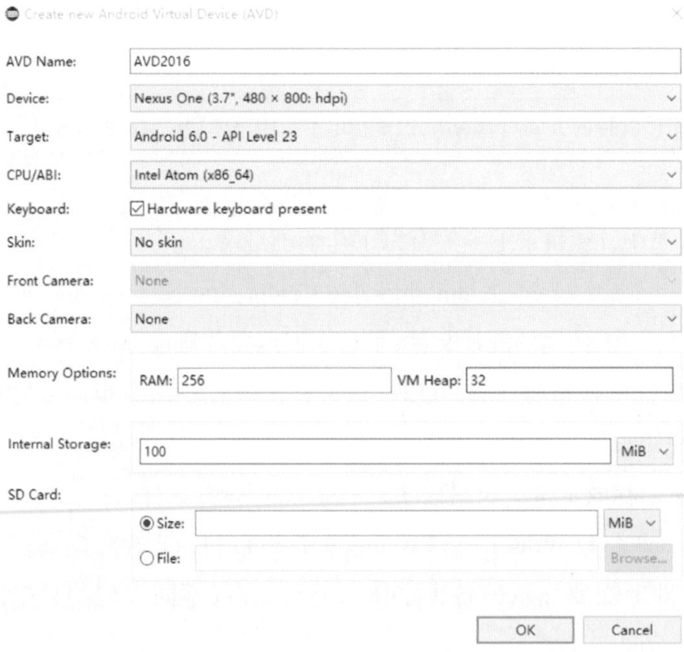

图 2-6　配置模拟器参数页面

2.2　Android Studio 环境

1.3.2 节中介绍了 Android Studio 开发环境的搭建，下面继续介绍如何在 Android Studio 环境下创建 Hello Android 应用程序。

2.2.1　创建工程

如图 2 -7 所示为创建工程的向导页面，在这个页面中可以新建一个项目，即选择“Start a new Android Studio project”，也可以导入本地项目或者 GitHub 上的项目等，页面的左侧可以查看最近打开的项目等。

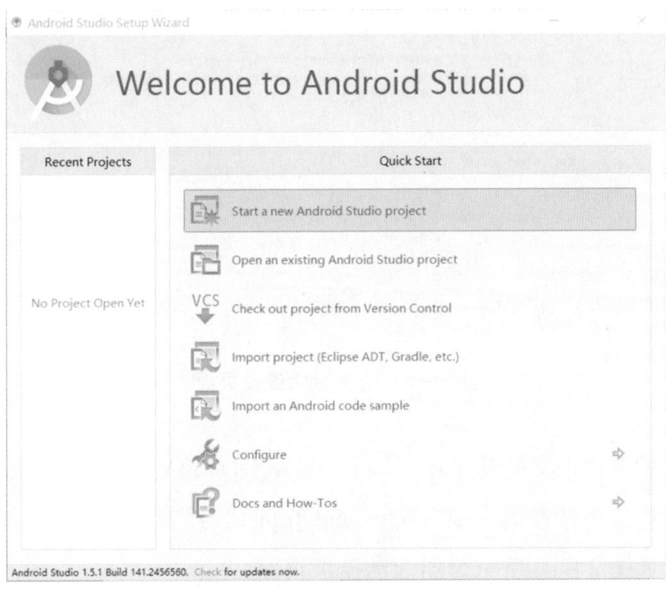

图 2 -7　创建工程的向导页面

这里直接新建一个项目，进入如图 2 -8 所示的新建工程页面，然后在页面中填入项目名称、应用程序包名和项目路径等，然后点击“Next”按钮，进入如图 2 -9 所示的选择目标设备页面。

Create New Project
New Project
Android Studio
Configure your new project
Application name: HelloWorld
Company Domain: hrbust.edu.cn
Package name: cn.edu.hrbust.helloworld　Edit
Project location: C:\Users\Lenovo\AndroidStudioProjects\HelloWorld

图 2 -8　新建工程页面

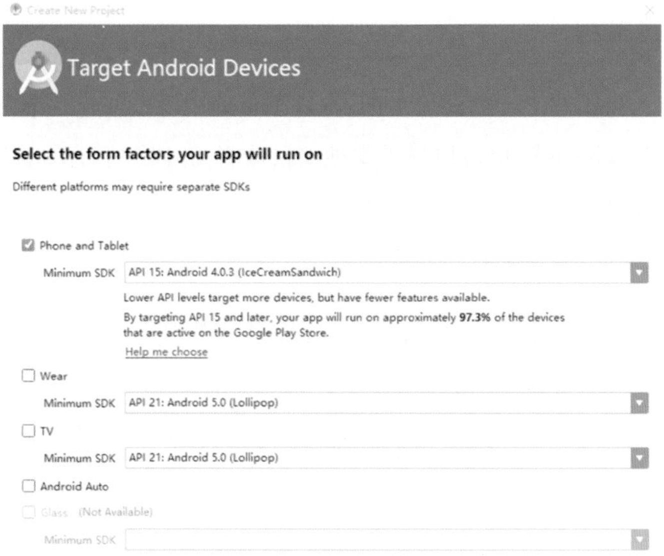

图 2 – 9　选择目标设备页面

如图 2 – 9 所示页面中支持从 Phone and Tablet、Wear、TV、Android Auto、Glass 等设备中选择一项目标设备，这里选择第一项“Phone and Tablet”，接着需要选好最小 SDK，然后点击“Next”按钮，进入如图 2 – 10 所示添加活动模板页面。

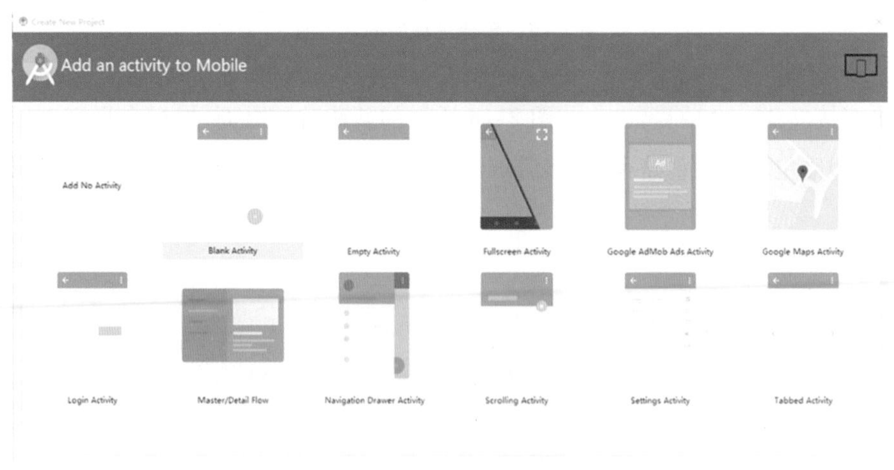

图 2 – 10　添加活动模板页面

如图 2 – 10 所示页面中支持从众多的活动模板中选择一项，可以看出 Android Studio 环境提供的模板展现形式较 Eclipse + ADT 环境要直观一些，这里选择一个 Blank Activity 模板，然后点击“Next”按钮，进入如图 2 – 11 所示的自定义活动信息页面。

如图 2 – 11 所示页面中支持自定义活动的名称、布局文件的名称、活动的标题等信息，还可以选择是否采用碎片。信息填写完成后，点击”Finish”按钮，此时会出现如图 2 – 12 所示的进度条,这里需要下载 Gradle,只在第一次时会下载(这里需要借助 VPN 软件,下载过程稍慢,需要耐心等待)。

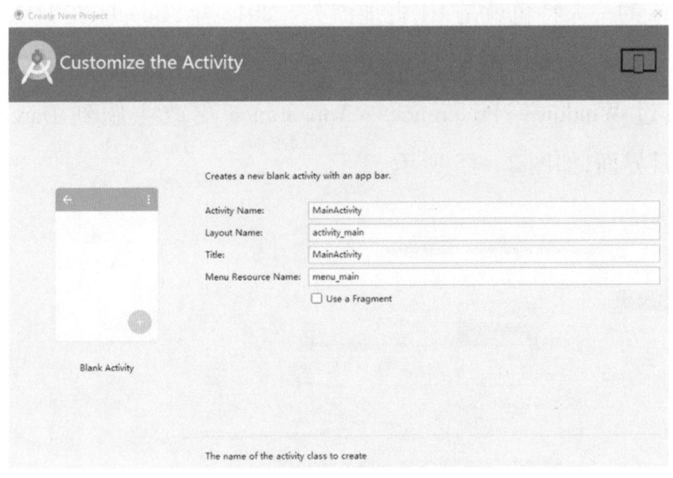

图 2－11　自定义活动信息页面

图 2－12　进度条

下载成功之后,便看到如图 2－13 所示 HelloWorld 项目的完整界面。至此,一个 HelloWorld 的 Studio 项目就完成了,图片中也可以看到默认是一个白色主题。

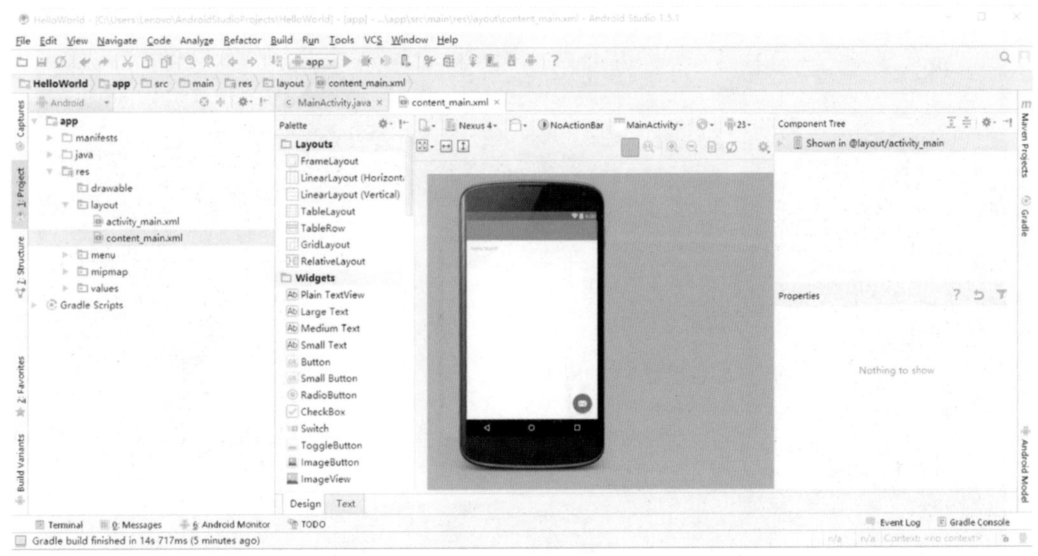

图 2－13　HelloWorld 项目的完整界面

2.2.2 个性化设置

Android Studio 的一大特色就是 UI 更加漂亮。2013 年,Google 在 I/O 大会上演示了黑色主题 Darcula,其展现出的炫酷黑界面可以说是十分“高大上”。主题个性化设置如图 2－14所示,通过 Window→Preference→Appearance 更改主题到 Darcula,就可以获得 Darcula主题的项目界面,如图 2－15 所示。

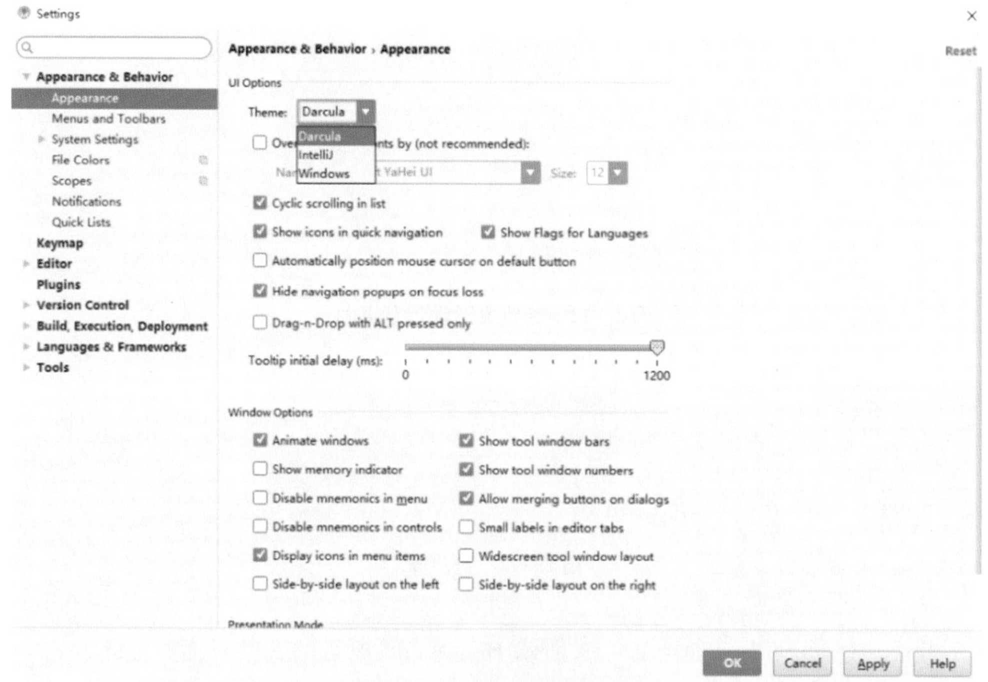

图 2－14　主题个性化设置

图 2－15　Darcula 主题的项目界面

如图 2－15 所示，项目界面的左侧显示了在 Android Studio 环境下创建 HelloWorld 工程之后得到的默认的目录结构，可以发现其与 Eclipse 的目录结构有些区别，Studio 一个窗口只能有一个项目，而 Eclipse 则可以同时存在很多项目，如果读者看着不习惯，可以点击左上角进行切换，将“Android” 模式切换到“project”模式，切换后的“project”模式的目录结构如图 2－16 所示。

图 2－16　切换后的“project”模式的目录结构

在目录结构方面，Android Studio 和 Eclipse 存在以下区别。

（1）Studio 中有 Project 和 Module 的概念，前面说到 Studio 中一个窗口只能有一个项目，即 Project，代表一个 workspace，但是一个 Project 可以包含多个 Module，如读者项目引用的 Android Library、Java Library 等，这些都可以看作一个 Module。

（2）上述目录中将 java 代码和资源文件（图片、布局文件等）全部归结为 src，在 src 目录下有一个 main 的分组，同时划分出 java 和 res 两个文件夹，java 文件夹则相当于 Eclipse 下的 src 文件夹，res 目录结构则一样。

在调整完应用程序的目录结构之后，读者也许还需要调整字体的大小或者样式。在 Darcular 主题中，字体的默认大小是 12，但是对于一般读者来说，字体可能偏小。为此，可以到 Window→Preferences→Settings 页面中搜索 Font，找到 Colors&Fonts 下的 Font 选项，能看到默认字体大小是 12，但是无法修改，需要先保存才可以修改，点击 Save As 输入一个名字，如 MyDarcular，然后就可以修改字体大小和字体样式了。编辑区字体偏好设置界面如图 2－17所示。

点击确定之后再回到项目页面发现中央编辑区的字体是变大了，但是 Studio 默认的一些字体大小如左侧的目录结构区域的字体大小却没有变化，看起来很不协调。为解决这一问题，还需要到 Window→Preferences→Appearance 中修改，Studio 默认的字体偏好设置界面如图 2－18 所示。

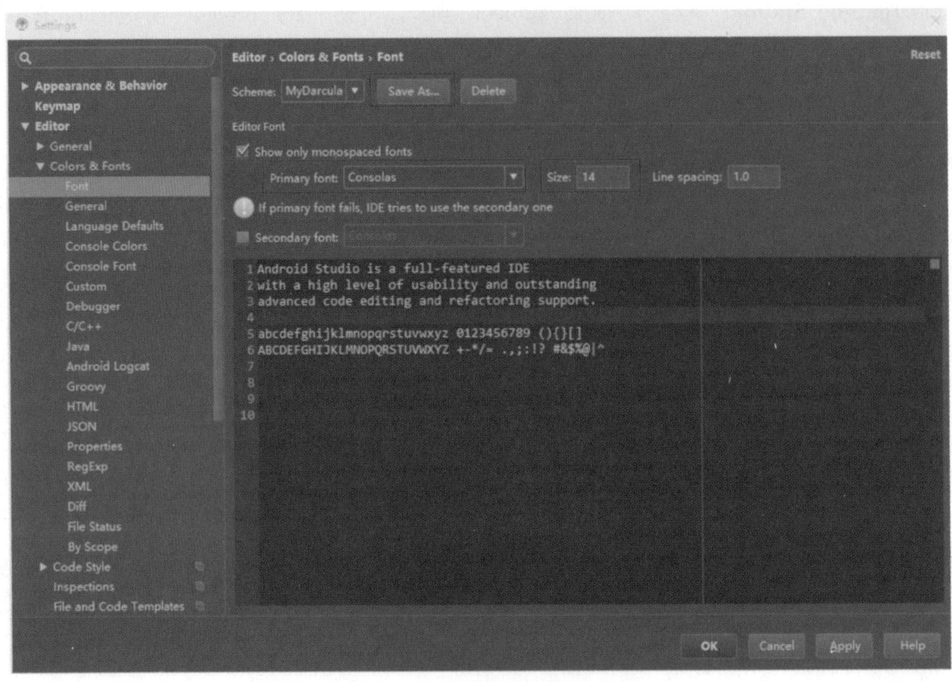

图 2－17　编辑区字体偏好设置界面

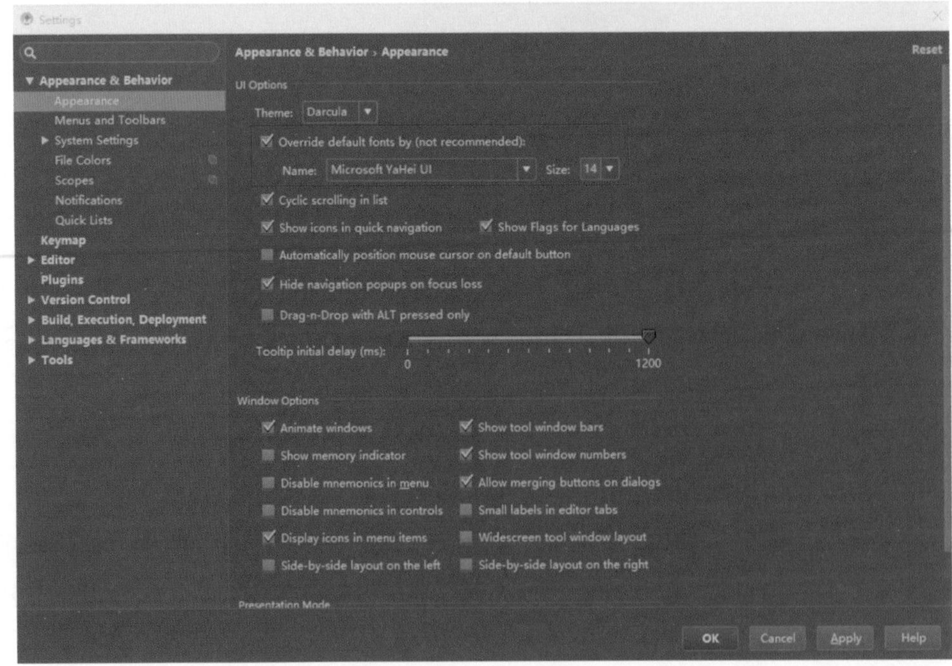

图 2－18　Studio 默认的字体偏好设置界面

2.2.3　运行应用程序

在 Android Studio 环境下运行程序时，与在 Eclipse 中比较像，点击菜单栏的绿色箭头直接运行。Android Studio 默认安装会启动模拟器。如果想让应用程序安装到真机上，可以按照如图 2－19 所示的配置执行设备方式进入配置界面，在下拉菜单中选择 Edit Configurations，在配置界面中选择提示或者是 USB 设备。配置完成后，下次执行应用程序就会看到如图 2－20 所示的选择执行设备界面。

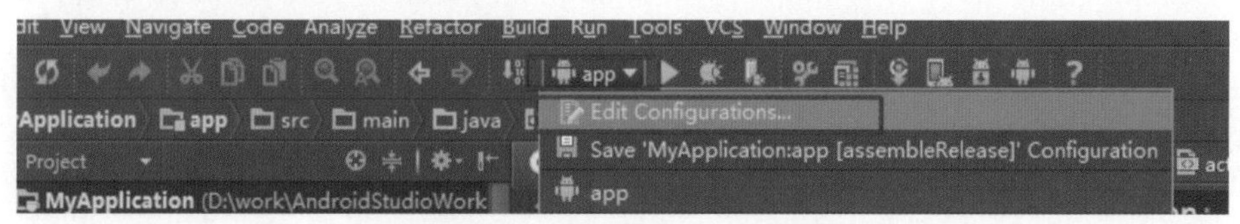

图 2－19　配置执行设备方式

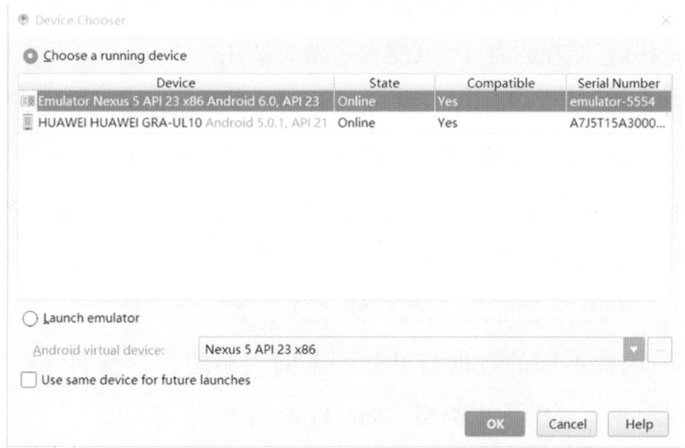

图 2－20　选择执行设备界面

HelloWorld 应用程序在真机和模拟器上的执行结果如图 2－21 所示。

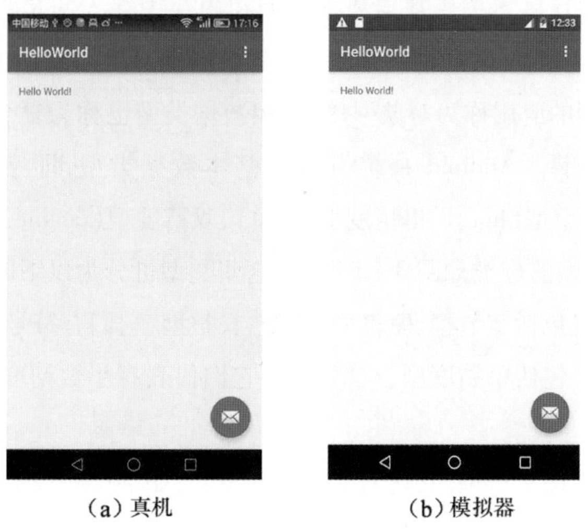

（a）真机　　　　（b）模拟器

图 2－21　HelloWorld 应用程序在真机和模拟器上的执行结果

第3章　Java语言基础知识

3.1　语法基础

3.1.1　数据类型

程序在执行过程中,需要对数据进行运算,还需要储存数据。这些数据可能是由使用者输入的,可能是从文件中取得的,甚至是从网络上得到的。在程序运行的过程中,这些数据通过变量(Variable)存储在内存中,以便程序随时调用。

数据存储在内存的一块空间中,为取得数据,必须知道这块内存空间的位置。然而若使用内存地址编号,则相当不方便,所以通常用一个变量来表示,变量(Variable)是一个数据存储空间的表示。将数据指定给变量,就是将数据存储至对应的内存空间;调用变量,就是将对应的内存空间的数据取出来使用。

一个变量代表一个内存空间,数据就存储在这个空间中。然而由于数据在存储时所需要的容量各不相同,因此不同的数据需要分配不同大小的内存空间来存储。在Java语言中,对于不同的数据,用不同的数据类型(Data Type)来区分。

Android系统使用的数据类型可以分为两大类:基本数据类型和引用数据类型。基本数据类型是由程序设计语言系统定义,不可再划分的数据类型。基本数据类型的数据所占内存的大小与软/硬件环境无关。基本数据类型在内存中存入的是数据值本身;引用数据类型在内存中存入的是指向该数据的地址,不是数据本身,它往往由多个基本数据组成。因此,对引用数据类型的应用称为对象引用,引用数据类型也称为复合数据类型,在有些程序设计语言中称为指针。Android系统有八个基本数据类型,即字节型(byte)、短整型(short)、整型(int)、长整型(long)、单精度型(float)、双精度型(double)、字符型(char)、布尔型(boolean)。数据类型的分类如图3-1所示,这些类型可分为以下四组。

(1)整数型。该组包括字节型、短整型、整型、长整型,它们有符号整数。

(2)实数型。该组包括单精度型、双精度型,它们代表有小数精度求的数字。实数型又称浮点型。

(3)字符型。该组即字符型,它代表字符集的符号,如字母和数字。

(4)布尔型。该组即布尔型,它是一种特殊的类型,表示真/假值。

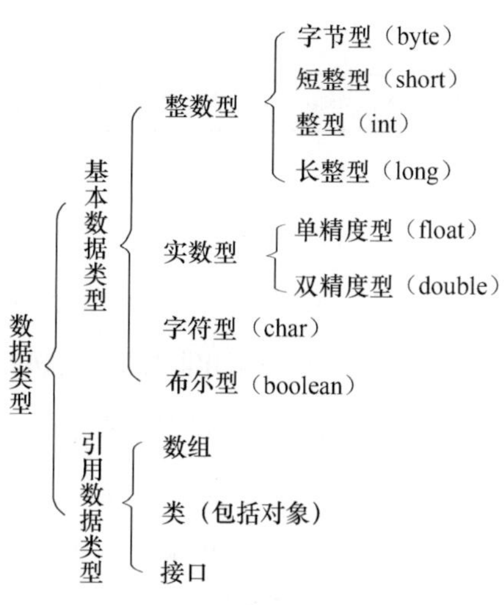

图 3－1 数据类型的分类

每一种具体数据类型都对应着唯一的类型关键字、类型长度和值域范围，Android 系统的基本数据类型见表 3－1。

表 3－1 Android 系统的基本数据类型

类型	数据类型关键字	适用于	类型长度	值域范围
字节型	byte	非常小的整数	1	－128～127 内的整数
短整型	short	较小的整数	2	$-2^{15} \sim 2^{15}-1$ 内的整数
整型	int	一般整数	4	$-2^{31} \sim 2^{31}-1$ 内的整数
长整型	long	非常大的整数	4	$-2^{31} \sim 2^{31}-1$ 内的整数
单精度型	float	一般实数	4	$-3.402\,823\times10^{38} \sim 3.402\,823\times10^{38}$ 内的数
双精度型	double	非常大的实数	8	$-1.797\,7\times10^{308} \sim 1.797\,7\times10^{308}$ 内的数
字符型	char	单个字符	1	
布尔型	boolean	判断	1	true 和 false

3.1.2 变量与常量

在程序中，每一个数据都有一个名字，并且在内存中占据一定的存储单元。在程序运行过程中，数据值不能改变的量称为常量，数据值可以改变的量称为变量。

在 Android 系统中，所有常量及变量在使用前必须先声明其值的数据类型，也就是要遵守“先声明后使用”的原则。声明变量的作用有两点：一是确定该变量的标识符（变量名），以便系统为其指定存储地址和识别它，这便是“按名访问”原则；二是为该变量指定数据类型，以便系统为其分配足够的存储单元。

声明变最的格式如下：

数据类型　　变量名 1,变量名 2 ,……　;

例如：

```
int  a;                //a 的值在程序运行过程中可能发生变化,将其声明为变量
int  x,  y,  sum;      //同时声明多个变量,变量之间用逗号分隔
```

在 Android 系统中,常量的声明与变量的声明非常类似,例如：

```
final int Day =24;            //Day 的值整个程序中保持不变,将其声明为常量
final double PI =3. 14159;    //声明圆周率常数。
```

从上面的示例中可以看出,常量声明的前面多了一个关键字 final,并赋了一个固定值。

在习惯上,常量用大写字母表示,变量用小写字母表示,以示区别。

3.1.3　对变量赋值

在程序中经常需要对变量赋值,在 Android 程序中用赋值号(=)表示。所谓赋值,就是把赋值号右边的数据或运算结果赋给左边的变量,其一般格式如下：

变量 = 表达式;

例如：

```
int x  =  5;          //指定 x 为整型变量,并赋初值 5
char c  ='a';         //指定 a 为字符型变量,并赋初值 'a'
```

如果同时对多个相同类型的变量赋值,可以用逗号分隔,例如：

```
int x =5,y =8,sum;    //将 x + y 的运算结果赋给变量 sum
sum =x + y;
```

在 Android 程序中,经常会用到形如“x =x + a”的赋值运算,例如：

```
int x =5;
x =x +2;
```

其中,右边 x 的值是 5,加 2 后,把运算结果 7 赋给左边的变量 x,所以 x 的值是 7。

3.1.4　关键字

所谓关键字,就是 Android 系统中已经规定了特定意义的单词,它们用来表示一种数据类型,或者表示程序的结构等。注意,不可以把这些单词用作常量名或变量名。

Android 系统中规定的关键字有 abstract、boolean、break、byte、case、catch、char、class、continue、default、do、double、else、extends、false、final、finally、float、for、if、implements、import、instanceof、int、interface、long、native、new、null、package、private、protected、public、return、short、static、 super、switch、synchronized、this、throw、throws、transient、true、try、void、volatile、while 等。

3.1.5　转义符

Android 系统中提供了一些特殊的字符常量，这些特殊字符称为转义符。通过转义符，可以在字符串中插入一些无法直接输入的字符，如换行符、引号等。每个转义符都以反斜杠(\)为标志。例如，"\n"代表一个换行符，这里的"n"不再代表字母 n，而是作为"换行"符号使用。常用的以"\"开头的转义符见表 3－2。

表 3－2　常用的以"\"开头的转义符

转义符	含义
\b	退格
\f	走纸换页
\n	换行
\r	回车
\t	横向跳格(Ctrl＋I)
\'	单引号
\"	双引号
\\	反斜杠

3.2　基本数据类型应用示例

3.2.1　整型与浮点型

1. 整形

当用变量表示整数时，通常将变量声明为整型。

【例 3－1】用整型变量计算两个数的和。

```
/* 计算两个数的和 */
用整形运算方法 */package com.ex02_01;

import android.app.Activity:
import android.os.Bundle;
import android.widget.TextView);
public class Ex02_01Activity extends Activity{
```

```
8    {
9    public void onCreate(Bundle savedInstanceState)
10   {
11   super. onCreate(savedInstanceState))
12   //setContentView(R. layout. activity_main);
13   int x,y,sum;
14   x = 3;
15   y = 5;
16   sum = x + y;
17   TextView txt = new TextView(this);
18   txt. setText("x - -3;" + "\n y -5;" + "\n x + y = " + sum);
19   setContentView(txt);
20       }
21   }
```

程序运行结果:

```
x =3
y =5
x + y =8
```

语句说明:在程序的第 13 行声明了三个整型变量,即 x、y、sum,这三个变量与存储器中相应类型的存储单元对应。在程序运行到第 14 行语句时,数值 3 存放到被编译器命名为 x 的内存单元中。在程序运行到第 15 行语句时,数值 5 存放到被编译器命名为 y 的内存单元中。在程序运行到第 16 行语句时,将内存单元 x 和内存单元 y 中的值相加并将结果放到变量 sum 中,进行加法运算的内存单元如图 3 -2 所示。

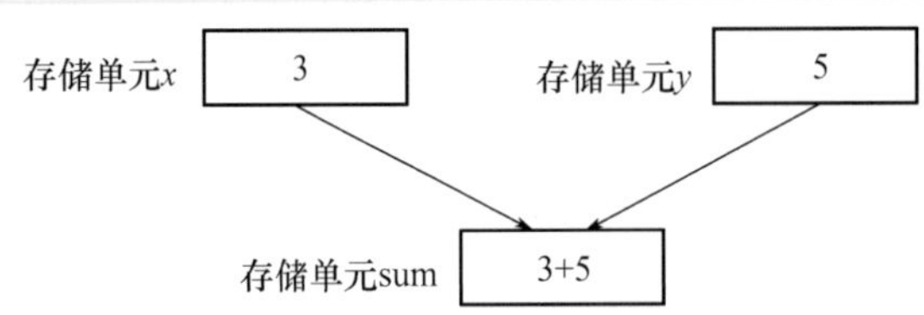

图 3 -2　进行加法运算的内存单元

2. 浮点型

【例 3 -2】用双精度浮点型变量计算一个圆的面积。

```
1    /* 计算圆的面积 */
2    package com. ex02 02;
```

```
import android. app. Activity;
import android. os. Bundle;
import android. widget. TextView;
public class Ex02 02Activity extends Activity {

    public void onCreate( Bundle savedInstanceState) {

        super. onCreate (savedInstanceState);
        //setContentView (R. layout. activity main);
        double pi, r, s;
        r = 10.8;
        pi = 3.14159;
        s = pi * r * r;
        TextView txt = new TextView (this);
        txt. setText( "圆的面积为:" + s);
        setContentView (txt);
            }
        }
```

运行结果:

圆的面积为 366.435 057 6

3.2.2　字符型

1. 字符变量

在 Android 系统中,存储字符的数据类型是 char,一个字符变量每次只能存放一个字符。一个字符变量在内存中占用两个字节(16 位)的存储空间,也就是说,一个字符变量可以存放两个字节长度的字符。例如,一个汉字是两字节长度,一个字符变最可以存放一个汉字。但当存放的字符是 8 位时,如一个字母是一个字节长度(8 位),则一个字符变量只能存放一个字母。

【例 3－3】演示字符型变量的用法。

```
/* 字符型变量的用法 */
package com. ex02 03;

```

```
import android. app. Activity;
import android. os. Bundle
import android. widget. TextView;
  public class Ex02 03Activity extends Activity {

      public void onCreate (Bundle savedInstanceState) {

          super. onCreate( savedInstanceState) ;
          //setContentView (R. layout. activity_main) ;
          char chl, ch2,ch3;
          chl = 88;
          ch2 = 'Y';
        ch3 = '汉';
        TextView txt = new TextView (this) ;
          txt. setText( "ch1,ch2,ch3: " + ch1 + ". "ch2 +"." +ch3) ;
          setContentView (txt) ;
                }
          }
```

程序运行结果:

ch1,ch2,ch3: X,Y,汉

2. 字符串

用双引号括起来的多个(也可以是一个或空)字符常量称为字符串。字符串是程序设计中最常用的数据类型。

例如,"我对 Android 很痴迷! n"和"a + b = "等都是字符串。

字符串与字符有以下区别:字符是由单引号括起来的单个字符,而字符串是由双引号括起来的,且可以是零个或多个字符。例如,'abc' 是不合法的,而" "是合法的,表示空字符串。

在 Android 系统中,用 String 来定义字符串,String 是 Android 系统的一个类,其使用方法见例 3 -4。

在 Android 系统中,还经常使用 Charsequence 定义字符串类型,其用法基本与 String 相同。

【例 3 -4】字符串用法示例。

```
 /* 字符串的用法 */
package com. ex02 04
```

```
import android. app. Activity
import android. os. Bundle;
import android. widget. TextView;
public class Ex02 04Activity extends Activity {

        public void onCreate( Bundle savedInstanceState)

        super. onCreate ( savedInstanceState) ;
        //setContentview( R. layout. activity_main) ;
        String str;
        str =   "This is a String!" ;
        TextView txt  =  new TextView ( this) ;
        txt. setText ( str) ;
        setContentView ( txt) ;
            }
        }
```

程序运行结果:

This is a String!

3.2.3　布尔型

Java 中表示逻辑值的基本类型称为布尔型,它只有两个值,即 true 和 false,且它们不对应于任何整数值。例如:

boolean b = true;

布尔型是所有诸如 a < b 这样的关系运算的返回类型,它对管理控制语句的条件表达式也是必需的。

【例 3 –5】布尔型变量的用法示例。

```
/ * 布尔型变量的用法 * /
package com. ex02 05;

import android. app. Activity;
import android. os. Bundle;
import android. widget.  TextView;
```

```
public class Ex02_05Activity extends Activity {

        public void onCreate( Bundle savedInstanceState) {

                super. onCreate (savedInstanceState);
                //setContentView (R. layout. activity main);
                boolean a, b, c;
                a = true;
                b = false;
                c = 10 <8;
                TextView txt = new TextView (this);
        txt. setText( "a is " +a+ " \n b is " +b+ " \n10 < 8 is" + c);
              setContentview( txt);
          }
    }
```

程序运行结果:

a is true

b is false

10 < 8 is false

3.2.4 数据类型的转换

在 Java 语言中,对于已经定义了类型的变量,允许转换其类型。变量的数据类型转换分为"自动类型转换"和"强制类型转换"两种。

1. 自动类型转换

在程序中已经对变量定义了一种数据类型,若想以另一种数据类型表示,则要符合以下两个条件:

(1)转换前的数据类型与转换后的数据类型兼容;

(2)转换后的数据类型比转换前的数据类型表示的范围大。

对于基本数据类型按精度从"低"到"高"的顺序如下:

byte → short → int → long → float → double

低 ————————→ 高

当把级别低的变量赋给级别高的值时,系统自动进行数据类型转换。

例如:

```
int x = 10;
float y;
y = x;
```

这时,y 的值为 10.0。

2. 强制类型转换

强制类型转换是指把级别商的变量的值赋给级别低的变量,其转换格式如下:

(类型名)要转换的值或变量;

例如:

设有

```
int a;
double b = 3.14;
```

则

```
a = (int)b;
```

结果 a = 3,b 仍然是 double 类型,b 的值仍然是 3.14。

从该示例可以看出,采用强制类型转换将高类型数据转换成低类型数据时,可能会降低数据的精确度。

3.3　程序控制语句

3.3.1　语句的分类

语句组成了一个执行程序的基本单元,它类似于自然语言的句子。Java 语言的语句可以分为以下几类。

1. 表达式语句

```
x = 3;
y = 5;
sum = x + y;
```

在一个表达式的最后加上一个分号就构成了一个语句,分号是语句不可缺少的部分。

2. 复合语句

用{　}把一些语句括起来就构成了复合语句。

```
{
String str = "我对 Android 很痴迷!"
TextView txt = new TextView (this);
txt. setText(str);
}
```

3. 控制语句

制语句用于控制程序流程及执行的先后顺序,主要有顺序控制语句、条件控制语句和循环控制语句三种类型。

4. 包语句和引用语句

包语句和引用语句是 Android 系统提供的一种类名管理机制。在介绍类和接口之后,将详细讲解包语句和引用语句的用法。

3.3.2　顺序控制语句

顺序控制是指计算机在执行这种结构的程序时,从第一条语句开始,按从上到下的顺序依次执行程序中的每一条语句。顺序控制是程序最基本的结构,包含选择控制语句和循环控制语句的程序,在总体执行上也是按顺序结构执行的。

【**例 3-6**】交换两个变量的值。

在编写程序时,有时需要把两个变量的值互换,交换值的运算需要用到一个中间变量。

例如,要将 a 与 b 的值互换,可用下面这样一段程序:

```
int a,b,temp; ← 设 temp 为中间变量
temp = a; ← 第一步:把 a 的值放到中间变量 temp 中
a = b; ← 第二步:把 b 的值放到变量 a 中,这时变量 a 中存放的是 b 的值
b = temp; ← 第三步:把 temp 中原 a 的值放到变量 b 中,这时变量 b 中得到的是原 a 的值
```

其中,temp 是中间变量,它仅起过渡作用。a、b 两数的交换过程如图 3-3 所示。

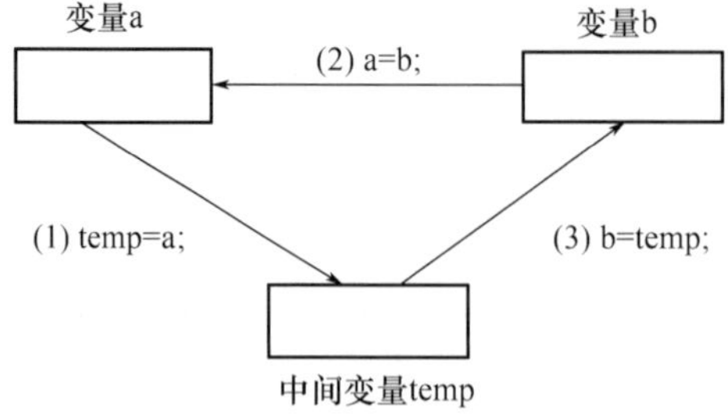

图 3-3　a、b 两数的交换过程

程序代码如下：

```
/ * 交换两变量的值 * /
package com. ex02_06;

import android. app. Activity;
import android. os. Bundle;
import android. widget. TextView;
public class Ex02 - 06Activity extends Activity {

public void onCreate( Bundle savedInstanceState) {

    super. onCreate ( savedInstanceState) ;
    //setContentView ( R. layout. activity_main) ;
    int a = 3, b = 5, temp;
    temp = a;
    a = b;
    b = temp;
    TextView txt = new TextView (this) ;
    txt. setText ("a = " + a + "\t b =" + b);
    setContentView ( txt) ;
        }
    }
```

程序运行结果：

a = 5

b = 3

3.3.3 if 语句

1. 单分支选择结构

if 语句用于实现选择结构，它判断给定的条件是否满足，并根据判断结果决定执行某个分支的程序段。对于单分支选择语句，其语法格式如下：

```
if (条件表达式)
 {
  若干语句;
 }
```

条件表达式两边的括号必不可少

这个语法的意思是:当条件表达式给定的条件成立时(true),执行其中的语句块;当条件不成立(false)时,则跳过这部分语句,直接执行后续语句。

单分支的 if 条件语句如图 3 –4 所示。

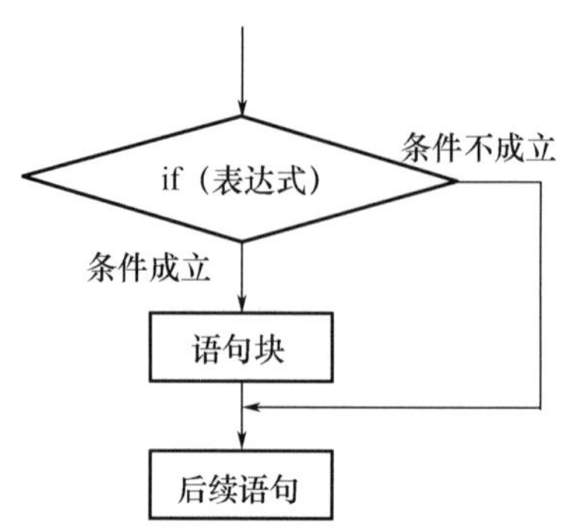

图 3 –4　单分支的 if 条件语句

【例 3 –7】设有任意三个数 a、b、c,按从小到大的顺序依次输出这三个数。首先将 a 与 b 比较,如果 a < b,本身就是从小到大排列的,则保持原顺序不变;但如果 a > b,则需要交换 a、b 两个变量的值。依此类推,再将 a 与 c 比较、b 与 c 比较,最后得到从小到大排序的结果。按从小到大的顺序排列输出三个数,如图3 –5 所示。

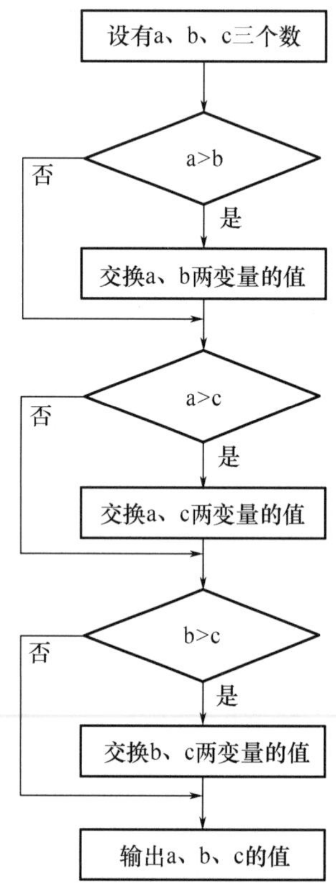

图 3 –5　按从小到大的顺序排列输出三个数

程序代码如下:

```
/* 对任意 3 个整数,按从小到大的版序排列 */
 package com. ex02_07;

import android. app. Activity;
```

```
import android. os. Bundle;
import android. widget. TextView;
public class Ex02_07Activity extends Activity {

        public void onCreate(Bundle savedInstanceState) {

        super. onCreate(savedInstanceState)
      //setContentView (R. layout. activity_main);
  int a = 9, b = 5 , c = 7;
  if(a > b){ t = a; a = b; b = t;}
  if(a > c){ t = a; a = c; c = t;}
  if(b > c){ t = b; b = c; c = t;)
  TextView txt = new TextView (this);
  txt. setText ("a = " + a +"\t b =" + b + "\t c = " + c);
  setContentView (txt);
           }
  }
```

程序结果如下：

a = 5

b = 7

c = 9

2. 双分支选择结构

有时需要在条件表达式不成立时执行不同的语句，可以使用双分支选择结构的条件语句，即 if – else 语句。双分支选择结构的语法格式如下：

```
if(表达式)
{  语句块 1;}
else
{  语句块 2;}
```

该语法的意思是，当条件表达式成立时(true)，执行语句块 1；否则(else)，执行语句块 2。双分支选择结构的条件语句如图 3 –6 所示。

if – else 语句的扩充格式是 if – else if。一个 if 语句可以有任意个 if – else if 部分，但只能有一个 else。

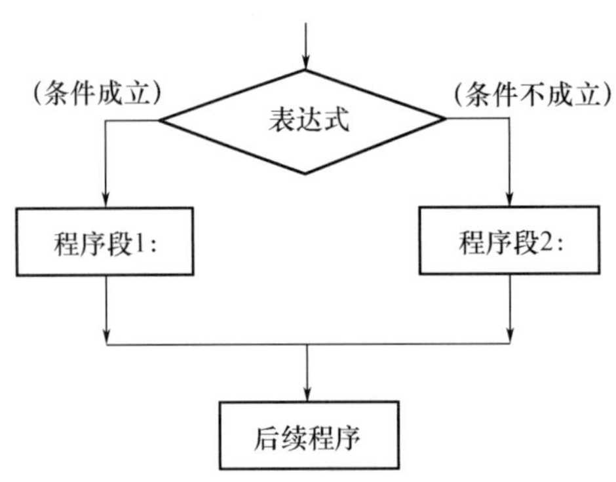

图 3-6 双分支选择结构的条件语句

3.3.4 switch 语句

switch 语句是一个多分支选择语句,又称开关语句,它可以根据一个整型表达式有条件地选择一个语句执行。if 语句只有两个分支可以选择,而实际问题中常常需要用到多分支的选择,当然也可以用嵌套 if 语句来处理,但如果分支较多,则嵌套的 if 语句层数太多,会造成程序冗长且执行效率降低。

switch 的语法结构形式如下:

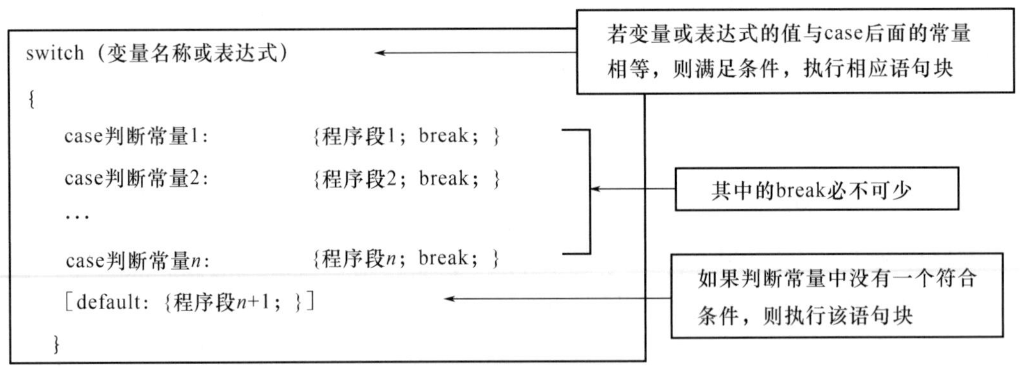

switch 语句首先计算条件表达式的值,如果表达式的值和某个 case 后面的判断常量相同,则执行该 case 里的若干条语句,直到 break 语句为止;如果没有一个判断常量相同,则执行 default 后面的若干条语句。default 语句块可以没有。在 case 语句块中,break 是必不可少的,break 表示终止 switch,跳转到 switch 的后续语句继续运行。

switch 语句的流程如图 3-7 所示。

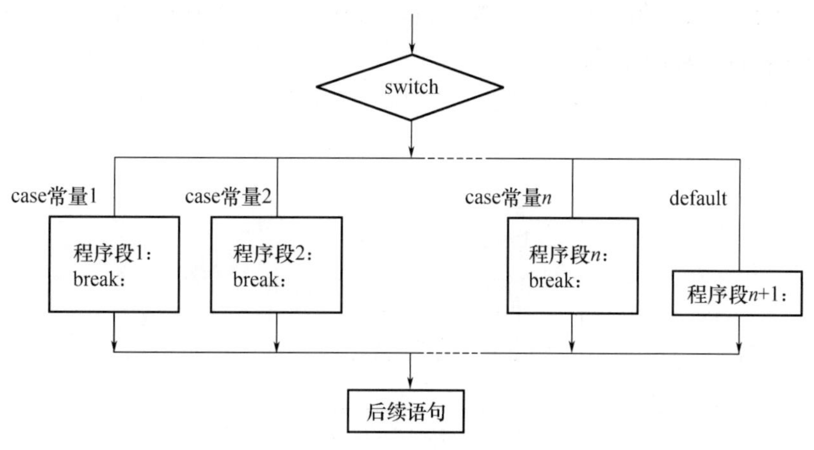

图 3－7　switch 语句的流程

3.3.5　循环语句

在程序设计构成中，经常需要将一些功能按一定的要求重复执行多次执行多次，这一过程称为循环。

循环结构是程序设计中一种很重要的结构，其特点是在给定条件成立时，反复执行某程序段，直到条件不成立为止。给定的条件称为循环条件，反复执行的程序段称为循环体。

1. for 循环语句

for 循环语句的语法结构如下：

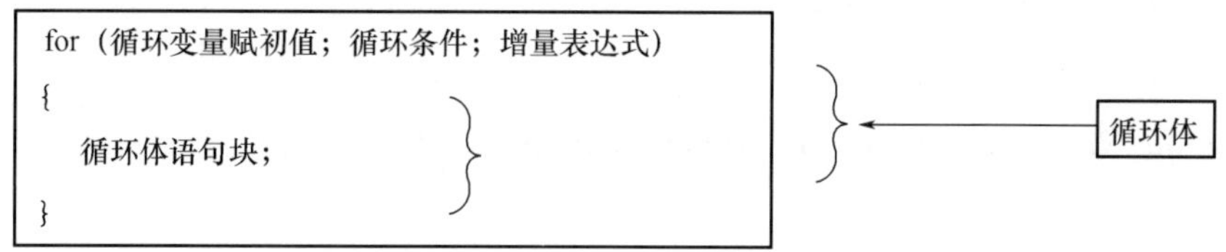


```
for (循环变量赋初值; 循环条件; 增量表达式)
{
    循环体语句块;
}
```

在 for 语句中，其语法成分如下：

（1）循环变量赋初值是初始循环的表达式，它在循环开始的时候就被执行一次；

（2）循环条件决定什么时候终止循环，这个表达式在每次循环的过程中被计算一次，当表达式的计算结果为 false 时，这个循环结束；

（3）增量表达式是每循环一次循环变量增加多少（即步长长）的表达式；

（4）循环体是被重复执行的程序段。

for 语句的执行过程是：首先执行循环变量赋初值，完成必要的初始化工作；再判断循环条件，若循环条件满足，则进入循环体中执行循环体的语句；执行完循环体后，接着执行 for 语句中的增量表达式，以便改变循环条件，这一轮循环就结束了。第二轮循环又从判断循环条件开始,若循环条件仍能满足,则继续循环;否则,跳出整个 for 语句,执行后续语句。循环语句的执行过程如图 3 – 8 所示。

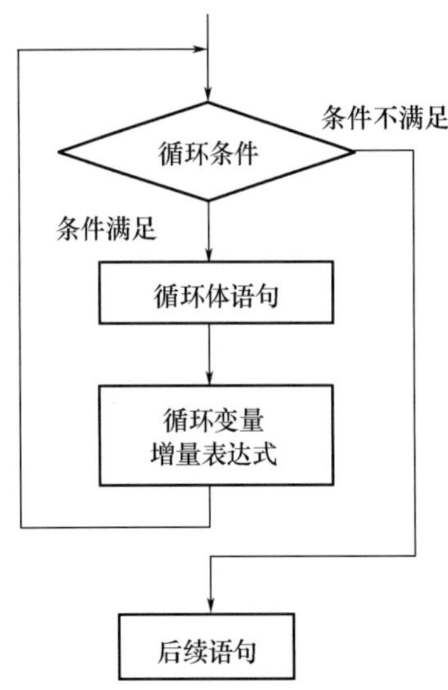

图 3 – 8　循环语句的执行过程

【例 3 – 8】 求从 1 加到 100 的和。

```
/* 利用循环求和 */
package com. ex02_08;

import android. app. Activity
import android. os. Bundle;
import android. widget. TextView;
public class Ex02_08Activity extends Activity {

public void onCreate( Bundle savedInstanceState) {

super. onCreate (savedInstanceState);
//setContentview (R. layout. activity_main);
int sum = 0;
for (int i - 1; i < = 100; i + + )
{
    sum = sum + i;
}
TextView txt = new TextView (this);
```

```
txt. setText ("1 + 2 + 3 + … + 100 =" + sum);
setContentView (txt);
        }
}
```

程序结果如下：

1 +2 +3 +… +100 =5050

在程序中，i 是改变条件表达式的循环变量。在开始循环之初，循环量 i =1，sum =0，这时 i<100，满足循环条件，因此可以进入循环体，执行累加语句 sum + i =1 +0 =1，将结果再放回到变量 sum 中，完成第一次循环。接着，循环变量自加 1(i + +)，此时 i =2，再与循环条件比较……如此反复，sum = sum + i 一直累加，直到运行了 100 次，i =101，循环条件 i < = 100 不再满足，循环结束。

2. while 循环结构

Android 系统有两种 while 循环语句，即 while 语句与 do-while 语句。while 与 do-while 循环结构的流程图如图 3 –9 所示。

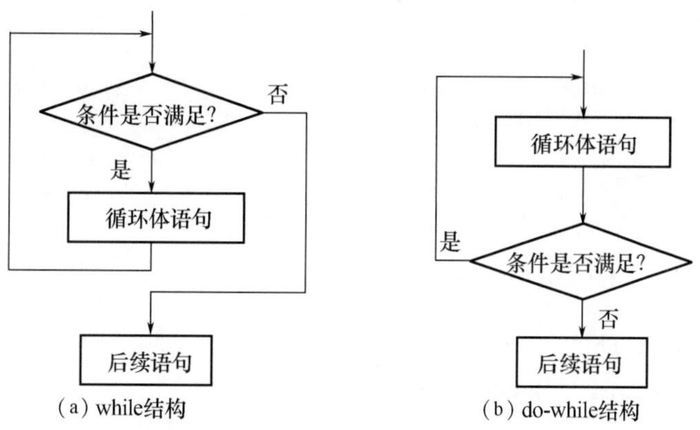

图 3 –9　while 与 do – while 循环结构的流程图

(1) while 语句。

while 语句的基本语法结构如下：

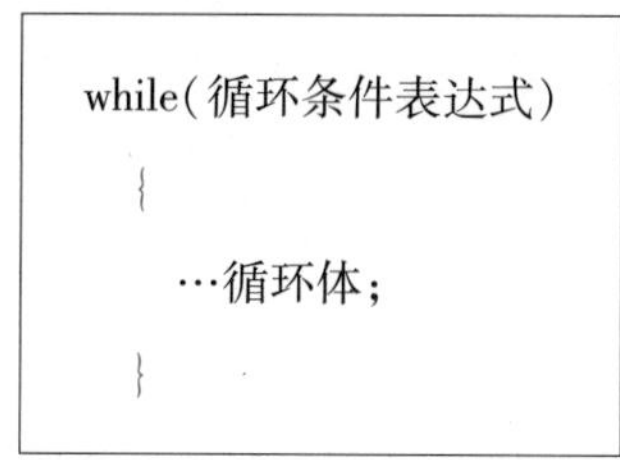

```
while(循环条件表达式)
    {
      …循环体;
    }
```

首先，while 语句执行条件表达式，返回一个 boolean 值(true 或者 false)。如果条件表达式返回 true，则执行大括号中的循环体语句，然后继续测试条件表达式并执行循环体代码，直到条件表达式返回 false 为止。

(2) do-while 语句。

do-while 语句的语法结构如下：

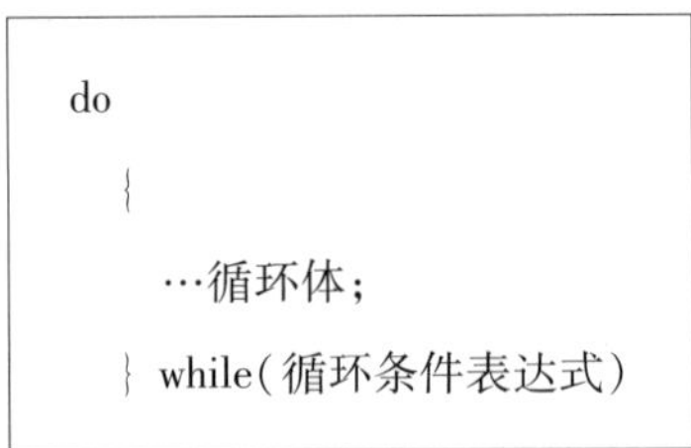

```
do
  {
    …循环体;
  } while(循环条件表达式)
```

do-while 语句与 while 语句的区别在于，语句先执行循环中的语句再计算条件表达式，所以 do-while 语句的循环体至少被执行一次。

【例 3 - 9】计算 1! + 2! + 3! + … + 10!。

算法分析：这是一个多项式求和问题。每一项都是计算阶乘，可以利用循环结构来处理，其代码如下：

```
/* while 循环 */
package com. ex02_09;

import android. app. Activity
import android. os. Bundle;
import android. widget. TextView;
public class Ex02 09Activity extends Activity {

  public void onCreate( Bundle savedInstanceState) {

        super. onCreate( savedInstanceState);
        //setContentView (R. layout. activity_main);
      int sum = 0,I = 1,p = 1;
      do
      {
        p = p * i;
        sum = sum + p
        i + +
      } while (i < = 10);
 TextView txt = new TextView (this);
```

```
txt. setText("1! + 2! + 3! + … +10! = " + sum);
setContentView (txt);
    }
}
```

程序结果如下:

1! + 2! + 3! + … +10! = 4037913

3. 循环嵌套

循环可以嵌套,在一个循环体内包含另一个完整的循环,称为循环嵌套。循环嵌套运行时,外循环每执行一次,内层循环都要执行一个周期。

【例 3-10】应用循环嵌套,编写一个按 9 行 9 列排列输出的九九乘法表程序。

算法分析:用双重循环控制九九乘法表按 9 行 9 列排列输出,用外循环变量 i 控制行数,i 从 1 到 9 取值。内循环变量 j 控制列数,由于 i * j = j * i,因此内循环变量 j 没有必要从 1 到 9 取值,只需从 1 到 i 取值就够了。外循环变量 i 每执行一次,内循环变量 j 执行 i 次,其代码如下:

```
/* 循环嵌套应用 */
package com. ex02_10;

import android. app. Activity
import android. os. Bundle;
import android. widget. TextView
public class Ex02_10Activity extends Activity {

public void onCreate(Bundle savedInstanceState) {

super. onCreate (savedInstanceState);
TextView txt - new TextView (this);
int i,j
for (i=1; i <= 9; i++)
    {
        for (j=1; j <= i; j++)
            {
            System. out. print (i+"x" +j+" =" +i*j+"\t");
            }
```

```
            System. out. println( );
        }
    setContentView (txt);
    }
}
```

注意:循环可以嵌套,但不能交叉。

3.3.6 跳转语句

Android 程序设计中主要应用了两种跳转语句,即 break 语句和 continue 语句。

1. break 语句

break 语句有两种作用:一是用来退出 switch 结构,跳出 switch 结构继续执行后续语句;二是用来中止循环。

在循环体中使用 break 语句强行退出循环时,忽略循环体中的任何其他语句和循环的条件测试,终止整个循环,程序跳到循环后面的语句继续运行。

【例 3 –11】使用 break 语句跳出循环,其代码如下:

```
/* 使用 break 语句跳出循环 */
package com. ex02 –11;
import android. app. Activity
import android. os. Bundle;
import android. widget. TextView
public cass Ex02 11Activity extends Activity
{
    public void onCreate (Bundle savedInstanceState)
        {
            super. onCreate (savedInstanceState);
            //setContentView (R. layout. activity_main);
            TextView txt = new TextView (this)i
            String output = "";
            for (int i –0; i<100; i+ +)
              {
              if (i=10) {break; }
                output = output  +"i  =  "  + i "\n";
                }
```

```
19 output = output + " \n 循环 10 次后,跳出循环!;
20 txt. setText ( output) ;
21 setContentView ( txt) ;
22         }
23 }
```

程序结果如下:

i = 0

i = 1

i = 2

i = 3

i = 4

i = 5

i = 6

i = 7

i = 8

i = 9

循环 10 次后,跳出循环!

语句说明:循环变量 i 的取值从 0 开始,当 i = 10 时,满足第 16 行 if 语句的条件,运行 break 语句,跳出循环,转向执行第 19 行的语句(注意,最后一次执行循环体时,第 17 行的语句没被执行)。

2. continue 语句

continue 语句用来终止本次循环,其功能是终止当前正在进行的本轮循环,即跳过后面剩余的语句,转而执行循环的第一条语句,计算和判断循环条件,决定是否进入下一轮循环。

【例 3 - 12】应用 continue 语句输出用" * "排列的三角形。

```
/ * 应用 continue 语句输出用" * "排列的三角形 * /
package com. ex02_12;
import android. app. Activity
import android. os. Bundle
import android. widget. TextView;
public class Ex02_12Activity extends Activity
{
 public void onCreate ( Bundle savedInstanceState)
  {
```

```
10     super. onCreate( savedInstanceState) ;
11     //setContentView (R. layout. activity_main) ;
12     TextView txt = new TextView (this) ;
13   String output = "" ;
14   for (int i =0; i < 5; i + +) {
15            for (int j =0; j <5; j + +) {
16   if (j >i) {
17     continue
18            }
19   output = output + " * " + " " ;
20   }
21   output = output + "\n" ;
22   }
23   txt. append (output) ;
24   setContentview (txt) ;
25      }
26   }
```

程序结果如下:

```
*
*  *
*  *  *
*  *  *  *
*  *  *  *  *
```

语句说明:本例第 17 行的 continue 语句终止了内循环,而跳转到第 14 行继续执行外循环的下一轮循环。

3.4　类 与 对 象

类与对象是 Android 的核心和本质,它们是 Java 语言的基础,编写一个 Java 程序,在某种程度上来说就是定义类和创建对象。定义类和建立对象是 Java 编程的主要任务。

3.4.1　类的定义

类是组成 Java 程序的基本要素,本节将介绍如何创建一个类。

1. 类的一般形式

类由类声明和类体组成，而类体又由成员变量和成员方法组成，其形式如下：

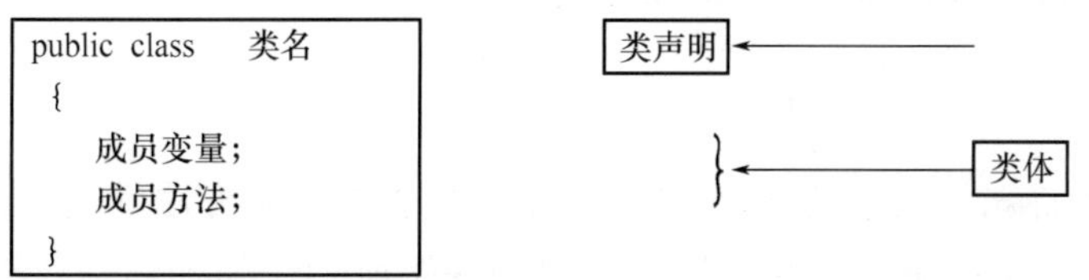

2. 类的继承性

继承性是面向对象的程序中两个类之间的一种关系，即一个类可以从另一个类（即它的父类）中继承状态和行为。被继承的类（父类）也可以称为超类，继承父类的类称为子类。继承为组织和构造程序提供了一个强大而自然的机理。

面向对象系统允许一个类建立在其他类之上。例如，山地自行车、赛车及双人自行车都是自行车，那么在面向对象技术中，山地自行车、赛车及双人自行车就是自行车类的子类，自行车类是山地自行车、赛车及双人自行车的父类。

子类声明的一般形式如下：

```
class 类名[extends 父类名][implements 接口列表]
 {
 …
 }
```

对各组成部分的具体说明如下。

(1)类的关键字 class。

在类声明中，class 是声明类的关键字，表示类声明的开始，类声明后面跟着类名，按习惯类名要用大写字母开头，并且类名不能用阿拉伯数字开头。给类命名时，最好取一个容易识别且有意义的名字，避免 A、B、C 之类的类名。

(2)声明父类。

extends 为声明该类的父类，表明该类是其父类的子类。一个子类可以从它的父类继承变量和方法。

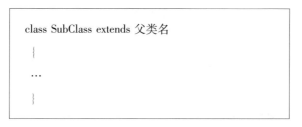

创建子类的格式如下：

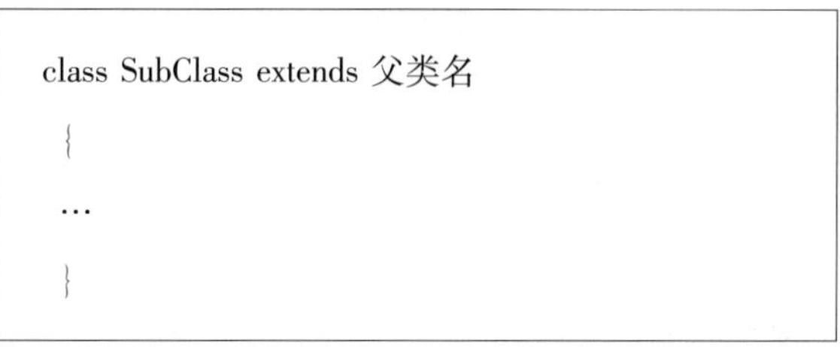

```
class SubClass extends 父类名
 {
 …
 }
```

(3)实现接口。

为在类声明中实现接口,要使用关键字 implements,并且在其后面给出接口名。要实现有多个接口时,各接口名以逗号分隔,其形式如下:

```
implements 接口 1,接口 2,…
```

接口是一种特殊的抽象类,这种抽象类中只包含常量和方法的定义,而没有变量和方法的实现。一个类可以实现多个接口,以某种程度实现“多继承”。

【**例 3 -13**】创建一个 Activity 的子类,其代码如下:

```

package com. ex02_13;

import android. app. Activity
import android. os. Bundle;
import android. widget. TextView
public class Ex02_13Activity extends Activity
{
  int x, y, sum;
    public void onCreate (Bundle savedInstanceState)
    {
        super. onCreate (savedInstanceState)

        x = 3;
        y = 5;
        sum = x + y;
        TextView txt = new TextView(this);
        txt. setText( "x = 3;" + "\n y = 5;" + "\n x + y =" + sum);
        setContantView (txt);
    }
}
```

3.4.2 对象

类是一个抽象的概念,而对象是类的具体化。类与对象的关系相当于普通数据类型与其变量的关系。声明一个类只是定义了一种新的数据类型,类通过实例化创建了对象,才真正创建了这种数据类型的物理实体。

1. 对象的创建

创建对象的一般格式如下：

```
类名 对象名 = new 类名([参数列表]);
```

该表达式隐含了两个部分:对象的声明、实例化和初始化。

(1)对象的声明。

声明对象的一般形式如下：

```
类名 对象名
```

声明对象并不为对象分配内存空间,而只是分配一个引用空间,即对象的引针,是32 位的地址空间,它的值指向一个中间的数据结构,其存储有关数据类型的信息以及当前对象所在堆维的地址,而对于对象所在的实际的内存地址是不可操作的,这就保证了安全性。

(2)实例化和初始化。

实例化是为对象分配内存空间和进行初始化的的过程,其一般形式如下：

```
对象名 = new 构造方法([参数列表]);
```

运算符 new 为对象分配内存空间,它调用对象的构造方法,返回引用。一个类的不同对象分别占据不同的内存空间,在执行类的构造方法进行初始化时,可以根据参数类型或个数调用相应的构造方法,进行不同的初始化,实现方法重构。

2. 对象的使用

类是不能直接使用的,我们使用的是类通过实例化成为的对象,而对象是通过访问对象变量或调用对象方法来使用的。

通过运算符“.”可以实现对对象的变量访问和方法的调用。变量和方法可以通过设定访问权限来限制其他对象对它的访问。

(1)访问对象的变量。

对象创建之后,对象就有了自己的变量。对象通过使用运算符“.”实现对自己变量的访问。

访问对象成员变量的格式如下：

```
对象名.方法名([参数列表]);
```

例如,设有一个 A 类,其结构如下：

```
class A
{int x; }
```

如果要对其变量 x 赋值,则先创建并实例化类 A 的对象 a,然后再通过对象给变量 x：

A a new = A();

a.x = 5

(2)调用对象的方法。

对象通过使用运算符“.”实现对自己方法的调用。

例如,在例 3-14 中定义了 Box 类。在 Box 类中定义了三个 double 类型的成员变量和一个 volume()方法,将来每个具体对象的内存空间中都保存有自己的三个变量和一个方法的引用,并由它的 volume()方法来操纵自己的变量,这就是面向对象的封装特性的体现。要访问或调用一个对象的变量或方法需要首先创建这个对象,然后用算符""调用该对象的某个变量或方法。

【例 3-14】应用创建类的实例对象计算长方体的体积,其代码如下:

```
/* 构造长方体 */
package com.ex02_14
import android.app.Activity
import android.os.Bundle;
import android.widget.TextView
public class Ex02_14Activity extends Activity
{
 public void onCreate (Bundle savedInstanceState)
 {
     super.onCreate (savedInstanceState)
     Box box = nev Box ();
     Textview txt - new TextView (this);
     double v;
     v = vbox.volume();
     txt.setText {"长方体体积为:  " + v};
     txt.setTextsize (25);
     setContentviev(txt);
 }

/* 创建一个内部类 */
class Box
{
  double width, height, depth;
```

```
24      Box( );
25      {
26       width  =  10;
27       height  =  10;
28       depth  =  10;
29      }
30      double volume( );
31       {
32          return width height  *  depth;
33         }
34      }
35    }
```

(1)本例中第 11 行创建了一个内部类 Box,其中第 21 ~ 29 行为构造方法,构造方法的特点是方法名与类名相同,且在类的前面无返回类型,第 30 ~ 33 行为普通方法;

(2)程序的第 11 行应用“对象名 = new 构造方法( )”创建 Box 类的实例对象 box;

(3)程序的第 12 行创建 TextView 类的实例对象 txt;

(4)程序的第 14 行为调用对象的普通方法。

程序的运行结果如图 3 – 10 所示。

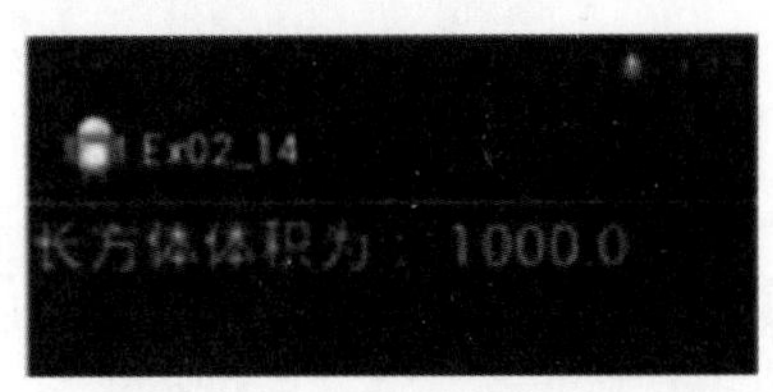

图 3 – 10　程序的运行结果

3.4.3　接口

接口是类的一种(抽象类),只包含常批和方法的定义,没有变量和具体方法的实现,且其方法都是抽象方法。接口的用处体现在以下几个方面:

(1)通过接口实现不相关类的相同行为,而无须考虑这些类之间的关系;

(2)通过接口指明多个类需要实现的方法;

(3)通过接口了解对象的交互界面,而无须了解对象所对应的类。

1. 接口定义的一般格式

接口的定义包括接口声明和接口体。

接口定义的一般格式如下:

```
[public] interface 接口名[extends 父接口名]
{
    …  //接口体
}
```

extends 子句与类声明的 extends 子句基本相同,不同的是一个接口可有多个父接口,用逗号隔开,而一个类只能有一个父类。

2. 接口的实现

在类的声明中用 implementa 子句来表示一个类使用某个接口,在类体中可以使用接口中定义的常量,而且必须实现接口中定义的所有方法。一个类可以实现多个提口,在 implementa 子句中用逗号隔开。

3.4.4 包

在 Java 语言中,每个类都会生成一个字节码文件,该字节码文件名与类名相同,这样可能会发生同名类的冲突。为解决这个问题,Java 采用包来管理类名空间。包不仅提供了一种类名管理机制,还提供了一种面向对象方法的封装机制。包将类和接口封装在一起,方便了类和技口的管理与调用,如 Java 的基础类都对装在 java. lang 包中、所有与网络相关的类都封装在 java. net 包中等。程序设计人员也可以将自己编写的类和接口根据需要封装到一个包中。

1. 包的定义

把一个源程序归入列某个包的方法用 package 来实现。

package 语句的一般格式如下:

```
package 包名;
```

例如,要编写一个 MyTestjava 源文件,并且文件存放在当前运行目录的子目录 abc\test 下,语句如下:

```
package abe. test
public class yTest
{
…
}
```

在源文件中,package 是源程序的第一条语句。包名一定是当前运行目录的子目录,一个包内的 Java 代码可以访问该包的所有类及类中的所有变量和方法。

2. 包的引用

如果要使用包中的类，必须用关键字 import 导入这些类所在的包。

import 语句的一般格式如下：

```
import 包名.类名;
```

当要引用包中所有的类成接口时，类名可以用通配符“ * ”代替。

3.5　XML 语法简介

扩展标记语言（Extensible Markup Langunge，XML）是一套定义语义标记的规则，这些标记将文档分成许多部件并对这些部件加以标识。XML 的语法规则既简单又严格，熟悉 HTML 的读者会发现它的语法与 HTML 很相似，非常容易学习和使用。

1. XML 文档结构

下面看一个 XML 文档实例，其代码如下：

```
1  <?xml version = "1.0" encoding = "utf-8"?>
2  <bookstore>
3     <book catogory = "计算机">
4      <title lang = "中文">Java 语言程序设计</title>
5      <author>张思民</author>
6      <year>2012</year>
7      <price 39.00</price>
8     </book>
9  </bookstore>
```

（1）第 1 行是 XML 声明，描述文档定义的版本是 XML 1.0 版和所使用的编码方式是 utf-8 编码；

（2）第 2 行定义文档的根元素为 <booksore>，根元素类似 HTML 文档中 <HTML> 开头标记；

（3）第 3 行定义根的子元素 <book>，并定义了 book 的属性 categorya"计算机"；

（4）第 4 ~7 行分别定义元素 <book> 的 4 个子元素（title、author、year、price）；

（5）第 8 行定义元素的结尾 <book>；

（6）第 9 行定义根元素的结尾 <bookstore>。

从上述实例中可以看出，XML 文档由文档声明、元素、属性、文本、安体、注释等内容组成。

XML 文档是一种树结构，必须包含一个根元素，从“根部”开始，然后扩展到“枝叶”部分。描述一本书的 XML 文档结构如图 3－11 所示。

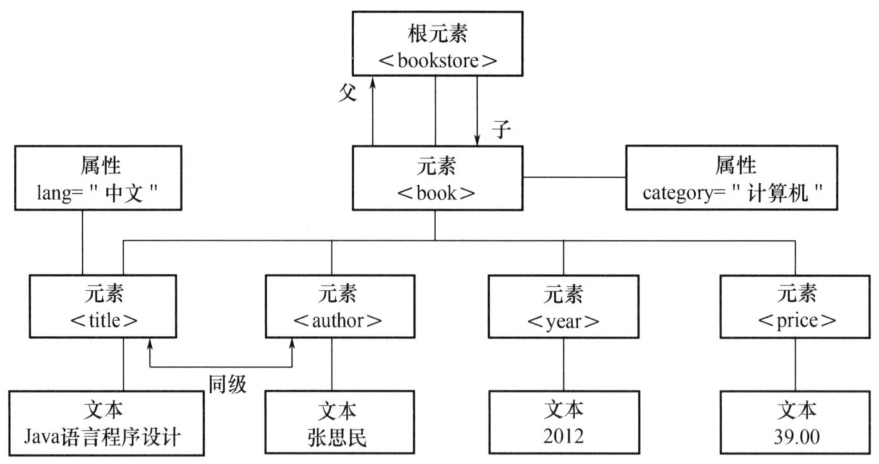

图 3－11　描述一本书的 XML 文档结构

2. 元素

元素是 XML 文档的基本组成部分，其实就是标记内容。XML 文档中共有四类元素：空元素、仅含文本的元素、包含其他元素的元素、混合元素。

（1）空元素。如果一个元素中没有任何文本内容，那么它就是一个空元素，如 < book > < /book >。

（2）仅含文本的元素。有些元素中仅含文本内容，如 < author > 张思民 < /author >。

（3）包含其他元素的元素。一个元素中可以包含其他元素，该元素称为父元素，被包含的元素称为子元素。例如：

```
< book category = "计算机" >
< title lang - "中文" > Java 语言程序设计 < /title >
< author > 张思民 < /author >
< year > 2012 < /year >
< price 39.00 < /price >
< /book >
```

（4）混合元素。混合元素既包含文本内容又包含子元素。

3. 属性

XML 元素可以拥有属性。属性是对标识进行进一步的描述和说明，一个标识可以有多个属性。在 XML 中，属性值必须用单引号或双引号括起来，其基本格式如下：

<元素名 属性名 - "属性值" >

例如：

<title lang = "中文" >

4. 注释

注释以“<！ - -”开始，以“ - - >”结束，注释内的任何标记都会被忽略。注释可以出现在 XML 文档的任何位置，其基本格式如下：

<！ - -注释内容 - - >

【例 3 - 15】系统自动生成的应用程序界面布局文件 main. xml 语句分析。

main. xml 布局文件的代码如下：

```
1   <? xml version = "1.0"  encoding "utf - 8"?  >
2   <LinearLayout xmlns:android = "http://schemas. android. com/apk/res/android"
3     android:orientation - "vertical
4     android:layout_width = "fill parent
5     android:layout_height = "fill parent"
6      >
7   <TextView
8     android:layout width = "fill_parent
9     android:layout height - "wrap_content
10   android:text = "estring/hello"
11   />
12    </LinearLayout >
```

（1）第 1 行定义 XML 文档声明，说明该文档的版本是 XML 1.0 版，所使用的编码方式是 utf - 8 编码。

（2）第 2 行定义根元素 <LinearLayout >，到第 12 行结束，该元素说明界面布局的排列方式，其中 xmlns 为根元素的属性，其属性是一个名为 android 的命名空间，其值是固定的，如下：

xmlns:android = http://schemas. android. com/apk/res/android

该网址中有该文件所使用的全部元素的定义，在编写该文件时，如果不注明命名空间，编译器并不会报错，但在程序运行时可能会发生错误。

（3）第 3 ~ 5 行均为根元素的属性值。第 3 行说明布局按从下到下的垂直方式排列组件，第 4 行定义布局宽度，第 5 行定义布局高度。

（4）第 7 ~ 11 行定义元素 <TextView >，该元素是一个文本组件，第 8 ~ 10 行均为说明该元素的属性，第 8、9 行定义文本组件的宽和高，第 10 行定义文本组件的文本内容。

【**例 3－16**】系统自动生成的应用程序配置文件 AndroidManifest. xml 语句分析。

AndroidManifest. xml 文件的代码如下：

```
1   <? xml version - "1.0" encoding = "utf - 8"?  >
2   <manifest xmlns:android = "http://schemas. android. com/apk/res/android"
3     package = "com. HelloAndroid"
4     android:versionCode = "1"
5     android:versionName - "1.0" >
6      <uses - sdk android:minSdkVersion - "15"/>
7      <application
8        android:icon = "@ drawable/ic launcher"
9        android:label - "@ string/app_name" >
10       <activity
11         android:label - "estring/app name
12         android:name - . HelloAndroidActivity" >
13          <intent - filter >
14          <action android:name - "android. intent. action. MAIN"/>
15      <category android:name - "android. intent. category. LAUNCHER"/>
16      </intent - filter
17      </activity
18      </application >
19    </manifest >
```

（1）第 1 行定义 XML 文档声明，说明该文档的版本是 XML 1.0 版，所使用的编码方式是 utf－8 编码。

（2）第 2 行定义根元素 <manifest>，到第 19 行结束，xmins:android 为命名空间属性，

（3）第 3～5 行定义根元素属性。第 3 行指定应用程序唯一的包名 package，该应用程序的包名为“com. HelloAndroid"。第 12 行 activity 元素指定应用程序名称为 HelloAndoidActivity，这只是简化名称，完整的应用程序名称应该加上包名，即 com. HelloAndroid. HelloAndroidActivity。

（4）第 6 行指定运行的最低版本号。

（5）第 7～18 行定义子元素 <application>。

（6）第 10～17 行定义 plication 的子元素 <activity>。

（7）第 13～16 行定义 activity 的子元素 <intent － fitter>，该元素为指定应用程序的启动条件和运行程序的入口。

第 4 章　Android 用户界面设计

4.1　用户界面组件包 widget 和 View 类

4.1.1　用户界面组件包 widget

Android 系统为开发人员提供了丰富多彩的用户界面组件，使用这些组件可以设计出炫丽的界面。大多数用户界面组件放置在 android. widget 包中。widget 包中的常用组件见表 4－1。

表 4－1　widget 包中的常用组件

感应检测	说明
Button	按钮
CalenderView	日历视图
CheckBox	复选框
EditText	文本编辑框
ImageView	显示图像或图标，并提供缩放、着色等各种图像处理方法
ListView	列表框视图
MapView	地图视图
RadioGroup	单选按钮组
Spinner	下拉列表
TextView	文本标签
WebView	网页浏览器视图
Toast	消息提示

4.1.2　View 类

View 是用户界面组件的共同父类，几乎所有的用户界面组件都是继承 View 类实现的，如 TextView、Button、EditText 等。

对于 View 类及其子类的属性,可以在界面布局文件中设置,也可以通过成员方法在Java代码文件中动态设置。View 类的常用属性和方法见表 4 -2。

表 4 -2　View 类的常用属性和方法

感应检测	对应方法	说明
android:background	setBackgroundColor (int color)	设置背景颜色
android:id	setld(int)	为组件设置可通过 findViewByld 方法获取的标识符
android:alpha	setAlpha(float)	设置透明度,取值范围为 0 ~ 1
	findViewByld(int id)	与 id 所对应的组件建立关联
android:visibility	setVisibility(int)	设置组件的可见性
Android:clickable	setClickable(boolean)	设置组件是否响应单击事件

4.2　文本标签与按钮

4.2.1　文本标签

文本标签(TextView)用于显示文本内容,是最常用的组件之一。文本标签常用方法见表4 -3。

表 4 -3　文本标签常用方法

方法	功能
getText0);	获取文本标签的文本内容
setTexu(CharSequence text);	设置文本标签的文本内容
setTextSize(float);	设置文本标签的文本大小
setTextColor(int color);	设置文本标签的文本颜色

文本标签常用的 XML 文件元素属性见表 4 -4。

表 4 -4　文本标签常用的 XML 文件元素属性

元素属性	功能
android:id	文本标签标识
android:layout width	文本标签(TextView)的高度,通常取值“fill_parent”(屏幕宽度)或以像素 px 为单位的固定值

续表 4 – 4

元素属性	功能
android:layout height	文本标签(TextView) 的高度,通常取值"wrap_ content" (文本的高)或以像素 px 为单位的固定值
android:text	文本标签(TextView) 的文本内容
android:textSize	文本标签(TextView) 的文本大小

【例 4 – 1】创建名为 Ex03_ 01 的新项目,包名为 com. ex03_ 01。打开系统自动生成的项目框架,需要设计的文件如下:

①界面布局文件 activity_ main. xml;

②控制文件 MainActityjava;

③资源文件 strings. xml。

(1)设计界面布局文件 activity_ main. xml。在界面布局文件 activity. main. xml 中加入文本标签 TextView,设置文本标签组件的 id 属性,在界面布局中设置文本标签如图 4 – 1 所示。

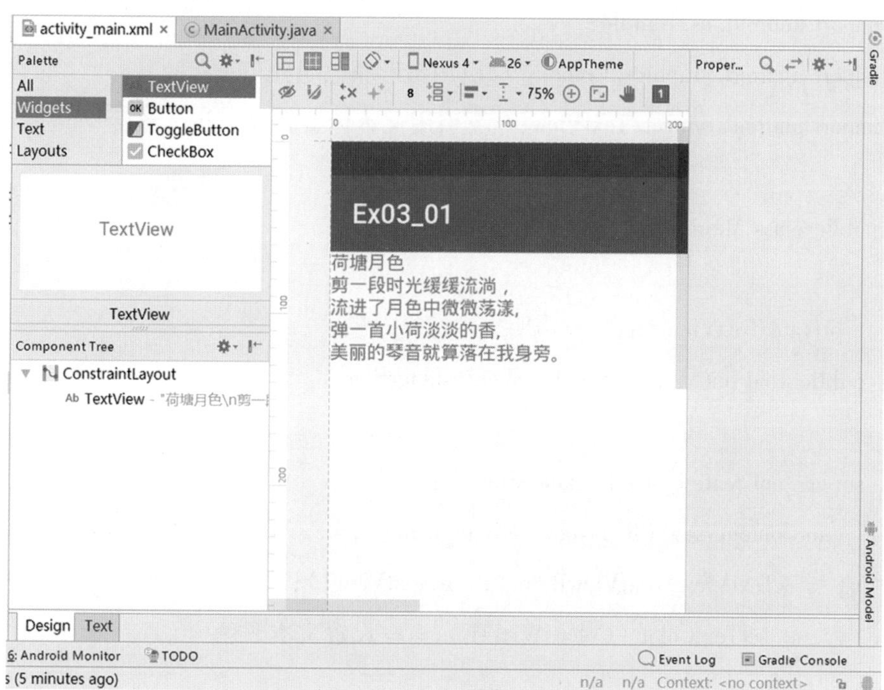

图 4 – 1　在界面布局中设置文本标签

activity_main. xml 代码如下:

```
<? xml version = "1.0" encoding = "utf-8"? >
```

```
<LinearLayout xmlns:android = "http://schemas.android.com/apk/res/android"
android:layout_width = "fill_parent"
android:layout_height = "fill_parent "
android:orientation = "vertical"  >
<TextView
   android:id = "@ +id/textView1"    //设置文本标签的 id 属性值
android:layout_width = "fill_parent"
   android:layout_height = "wrap_content "
android:text = "@string/hello"  />
 </LinearLayout >
```

(2)设计控制文件 MainActivity.java。在控制文件 MainActivity.java 中添加文本标签组件,并将界面布局文件中所定义的文本标签元素属性值赋给文本标签,与界面布局文件中的文本标签建立关联,程序代码如下:

```
package com.ex03_01;
import android.app.Activity;
import android.os.Bundle;
import android.graphics.Color;   //引用图形颜色组件
import android.widget.TextView;   //引用文本标签组件

public class MainActivity extends Activity
{
   private TextView txt;   //声明文本标签对象
public void onCreate (Bundle savedInstanceState)
{
super . onCreate (savedInstanceState) ;
    setContentView (R.layout.activity_main) ;   //与界面中的文本关联
txt  = (TextView)findViewById (R.id.textView1);
       txt.setTextColor (Color.WHITE);   //设置文本颜色
}
}
```

(3)设计资源文件 strings.xml。修改资源文件 strings.xml 中属性为"hello"的元素项的文本内容,程序代码如下:

```
<? xml version = "1.0"  encoding = "utf-8"?  >
```

```
<resources>
<string name "hello" > \n    荷塘月色
                             \n 剪一段时光缓缓流淌,
                             \n 流进了月色中微微荡漾,
                             \n 弹一首小荷淡淡的香,
                             \n 美丽的琴音就落在我身旁。
</string>
<string name = "app name" >Ex03_01 </string>
</resources>
```

保存项目,配置应用程序的运行参数。程序的运行结果如图 4 – 2 所示。

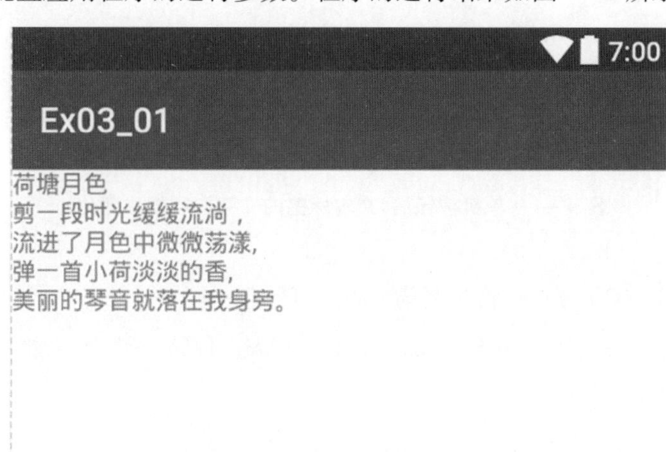

图 4 – 2　程序的运行结果

4.2.2　按钮

按钮(Button) 用于处理人机交互事件,在一般应用程序中经常会用到。

由于按钮(Button) 是文本标签 TextView 的子类,因此按钮与文本标签的继承关系如图 4 – 3 所示。按钮继承了文本标签 TextView 的所有方法和属性。

```
java. lang. Object
  └ android. view. View
       └ android. widget. TextView
            └ android. widget. Button
```

图 4 – 3　按钮与文本标签的继承关系

按钮在程序设计中最常用的方式是实

现 OnClickListener 监听接口，当单击按钮时，通过 OnClickListener 监听接口触发 onClick0 事件，实现用户需要的功能。OnClickListener 接口有一个 onClick( ) 方法，在按钮 Button 实现 OnClickListener 接口时，一定要重写这个方法。

按钮调用 OnClickListener 接口对象的方法如下：

按钮对象. setOnClickListener(OnClickListener 对象）；

【例 4 －2】编写程序，实现单击按钮页面标题及文本标签的文字内容发生变化的功能（图 4 －4）。

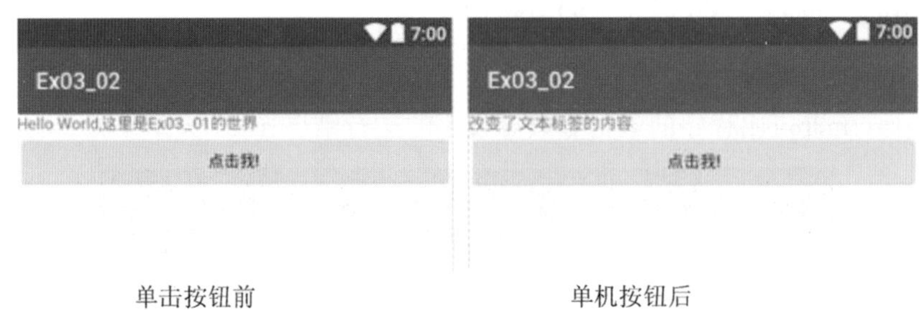

单击按钮前　　　　单机按钮后

图 4 －4　单机按钮后，文本标签的文字内容发生变化

创建名为 Ex03_02 的新项目，包名为 com. ex03_ 02。

（1）设计界面布局文件 activity_ main. xml。在界面布局文件中添加一个按钮，将其 id 设置为 button1，程序代码如下：

```
<? xml version = "1.0"  encoding = "utf - 8"?  >
<LinearLayout xmlns:android =  "http://schemas.android.com/apk/res/android"
android:layout_width = "fill_parent"
android:layout_height = "fill_parent"
android:orientation = "vertical"  >
<TextView
android:id = "@ +id/textview1"           //设置文本标签的 id 属性值
android:layout_width = "fill_ parent"
android:layout_height =  "wrap_ content"
android:text = "@ string/hello"  / >
<Button
    android:id = "@ +id/buttonl"       //设置按钮的 id 属性值
    android:layout_width = "fill_ parent"
    android:layout_height = "wrap_content"
    android:text = " @ string/button"  / >
```

```
</LinearLayout>
```

(2)设计控制文件 MainActivity. java。在控制文件 MainActivity. java 中设计一个实现按钮监听接口的内部类 mClick，当单击按钮时，触发 onClick( )事件，程序代码如下：

```
package com. ex03_02;
import android. app. Activity;
import android. os. Bundle;
import android. view. View;
import android. view. View. OnClickListener;
import android. widget. TextView;
import android. widget. Button;

public class MainActivity extends Activity
 {
  private TextView txt;
  private Button btn;
  public void onCreate (Bundle savedInstanceState)
  {
  super. onCreate(savedInstancestate) ;
  setContentView (R. layout. activity. main) ;
  txt = (TextView) findViewById(R. id. textViewl);
  btn = (Button) findViewById (R. id. button1);
  btn. setonClicklistener (new mClick());        //注册监听接口
  }
  class mClick implements onClickListener //定义实现监听接口的内部类
  {
  public void onClick(View v)
  {
  Ex03_ 02Activity. this. setTitle("改变标题");
  txt. setText (R. string. newStr);
  }
  }
}
```

(3)设计资源文件 strings. xml,程序代码如下:

```
<? xml version = "1.0"  encoding = "utf - 8"?  >
<resources >
<string name = "hello" >Hello World,这是 Ex03_ 02 的界面!  </string >
<string name = "app_ name" >Ex03_ 02 </string >
<string name =  "button" >点击我!  </string >
<string name = "newStr" >改变了文本标签的内容 </string >
</resources >
```

【例 4 -3】编写程序,实现单击按钮改变文本标签的文字及背景颜色的功能(图 4 -5)。

点击按钮前　　　　点击按钮后

图 4 -5　点击按钮后,文本标签的文字及背景颜色发生变化

本例涉及颜色定义,Android 系统在 android. graphics. Color 中定义了 12 种常见的颜色常数,颜色常数见表 4 -5。

表 4 -5　颜色常数

感应检测	对应方法	说明
Color. BLACK	0xff000000	黑色
Color. BLUE	0xff00ff00	蓝色
Color. CYAN	0xff00ffff	青绿色
Color. DKGRAY	0xff444444	灰黑色
Color. GRAY	0xff888888	灰色
Color. GREEN	0xff0000ff	绿色
Color. LTGRAY	0xffcccccc	浅灰色
Color. MAGENTA	0xffff00ff	红紫色
Color. RED	0xffff0000	红色
Color. TRANSPAEENT	0x00ffffff	透明
Color. WHITE	0xffffffff	白色
Color. YELLOW	0xffffff00	黄色

创建名为 Ex03_ 03 的新项目,包名为 com. ex03_03。

(1)设计界面布局文件 activity_main. xml。

在 XML 文件中表示颜色的方法有多种。

①#RGB。用 3 位十六进制数分别表示红、绿、蓝颜色。

②#ARGB。用 4 位十六进制数分别表示透明度以及红、绿、蓝颜色。

③#RRGGBB。用 6 位十六进制数分别表示红、绿、蓝颜色。

④#AARRGGBB。用 8 位十六进制数分别表示透明度以及红、绿、蓝颜色。

用 8 位十六进制数表示透明度以及红、绿、蓝颜色的程序代码如下:

```
<? xml version = "1.0" encoding = "utf - 8"? >
<LinearLayout xm1ns:android = "http://schemas. android. com/apk/res/android"
android: layout_width = "fill_ parent"
android: layout_height = "fill_ parent"
android: background = "4ff7f7c"
android: orientation = "vertical" >
    <TextView
    android:id = "@ + id/textView1 "
    android:layout_width = "fill_ parent "
          android:layout_height = "wrap_content"
     android:textColor = "ff0000"         //采用 8 位十六进制数表示颜色
          android: text = "@ string/hello"/ >
     < Button
        android:id = "@ + id/button1"
        android:layout_width = "wrap_content"
        android:layout_height = "wrap_content"
        android:text = "@ string/button"/ >
</LinearLayout >
```

(2)设计控制文件 MainActivity. java,程序代码如下:

```
package com. ex03_03;
import android. app. Activity;
import android. graphics. Color;
import android. os. Bundle;
import android. view. View;
import android. view. View. OnClickListener;
```

```
import android. widget. Button;
import android. widget. TextView;

public class MainActivity extends Activity
{
/* * Caliled when the activity is first created.  */
private TextView txt;
private Button btn;
@ Override
   public void onCreate (Bundle savedInstanceState)
     {
super. onCreate( savedInstanceState) ;
   setContentView(R. layout. activity. main) ;
   btn = (Button)findViewById(R. id. button1) ;
   txt = (TextView)findViewById(R. id. textView1) ;
   btn. setOnc1ickListener( new click( ) ) ;  //注册监听接口
}
class click implements OnClicklistener4  //定义实现监听接口的内部类
     {
public void onClick (view v)
{
   int BLACK  = 0xffcccccc;
   txt. setText("改变了文字及背景颜色") ;
   txt. setrextColor( Color. YELLOW) ;  //采用颜色常数设置文字颜色
   txt. setBackgroundColor( BLACK) ;  //设置文本标签的背景颜色
    }
     }
}
```

(3)设计资源文件 strings. xml,程序代码如下:

```
<? xm1 version = "1.0" encoding = "utf-8"?  >
<resources>
<string name = "hello" >Hello World, MainActivity!  </string>
<string name = "app_name" >Ex03_03 </string>
```

```
<string name = "button" >点击我,改变文字背景颜色</string>
</resources>
```

4.3 文本编辑框

文本编辑框 EditText 用于接收用户输入的文本信息内容。文本编辑框 EditText 继承于文本标签 TextView。文本编辑框 EditText 的继承关系如图 4-6 所示。

```
android.view.View
  └─ android.widget. TextView
       └─ android.widget. Edit Text
```

图 4-6 文本编辑框 EditText 的继承关系

文本编辑框 EditText 主要继承文本标签 TextView 的方法,文本编辑框 EditText 的常用方法见表 4-6。

表 4-6 文本编辑框 EditText 的常用方法

方法	功能
EditText(Context context)	构造方法,创建文本编辑框对象
getText()	获取文本编辑框的文本内容
setText(CharSequence text)	设置文本编辑框的文本内容

文本编辑框 EditText 的常用 XML 文件元素属性见表 4-7。

表 4-7 文本编辑框 EditText 的常用 XML 文件元素属性

元素属性	说明
android:editable	设置是否可编辑,其值为 true 或 false
android:numeric	设置 TextView 只能输入数字,其参数默认值为 false
android:password	设置密码输入,字符显示为圆点,其值为 true 或 false
android:phoneNumber	设置只能输入电话号码,其值为 true 或 false

定义框 EditText 元素的 android:numeric 属性,其取值只能是下列常量(可由"|"连接多个常量)。

(1)integer。可以输入数值。

(2)signed。可以输入带符号的数值。

(3)decimal。可以输入带小数点的数值。

【例 4－4】设计一个密码验证程序,密码验证程序运行界面如图 4－7 所示。

图 4－7　密码验证程序运行界面

创建名为 Ex03_04 的新项目,包名为 com. ex03_04。

(1)设计界面布局文件 activity main. xml。在界面布局中设置一个编辑框,用于输入密码,再设置一个按钮,判断密码是否正确,设置两个文本标签,其中一个显示提示信息“请输入密码”,另一个显示密码正确与否,程序代码如下:

```
<? xml version = "1.0" encoding = "utf -8"? >
<LinearLayout xmlns:android = "http://schemas. android. com/apk/res/android"
android:layout_width = "fill_parent"
android:layout_height = "fill_parent"
android:orientation = "vertical" >
 <! - -建立一个 TextView - - >
<TextView
android:id = "@ +id/myTextView01"
android:layout_width = "fill_parent"
android:layout_height = "41px"
android:layout_x = "33px"
android:layout_y = "106px"
android:text = "请输入密码:"
android:textsize = "24sp"
/>
 <! - -建立一个 EditText - - >
<EditText
     android:id = "@ +id/myEditText"
```

```
        android:layout_width = "180px"
    android:layout_height = "wrap content"
    android:1ayout_x = "29px"
    android:layout_y = "33px"
    android:inputType = "text"
    android:textSize = *24sp"/>
  <! --建立一个 Button -->
    <Button
    android:id = "@+id/myButton"
    android:layout_width = "100px"
    android:layout_height = "wrap_content"
    android:text = "确定"
    android:textSize = "24sp"
    />
  <! --建立一个 TextView -->
    <TextView
    android:id = "@+id/myTextView02"
    android:layout_width = "180px"
    android:layout_height = "41px"
    android:layout_x = "33px"
    android:layout_y = "106px"
    android:textSize = "24sp"
    />
    </LinearLayout>
```

(2)设计控制文件 MainActivity. java。在控制文件 MainActivity. java 中主要是设计按钮的监听事件,当单击按钮后,从文本编辑框中获取输入的文本内容,与密码“abc123”进行比较,程序代码如下:

```
package com. ex03_04;
import android. app. Activity;
import android. os. Bundle;
import android. view. View;
import android. view. View. OnClickListener;
import android. widget. EditText;
```

```
import android. widget. TextView;
import android. widget. Button;
public class MainActivity extends Activity
{
private EditText edit;
private TextView txt1,txt2;
private Button mButton01;
@ Override
public void onCreate( Bundle savedInstanceState)
{
super. onCreate( savedInstanceState) ;
setContentView( R. layout. activity_main) ;
txt1 = (TextView) findViewById( R. id. myTextView01);
txt2 = (TextView) findViewById (R. id. myTextView02);
edit = (EditText) findViewById (R. id. myEditText);
mButton01 = (Button) findViewById (R. id. myButton);
mButton01. setOnClickListener( new mClick( ) ) ;
}
class mClick imp1ements OnClickListener    //实现监听接口的内部类
{
public void onClick( View v)
{
String passwd;
passwd = edit. getText( ). toString( ); //获取文本编辑框中的文本
if( passwd. equals( "abc123" ) )   //比较两个字符串是否相等
txt2. setText( "欢迎进入快乐大本营!" ) ;
else
txt2. setText( "非法用户,请立刻离开!" );
}
}
}
```

4.4 布 局 管 理

Android 系统按照 MVC（Model - View - Cotoller）设计模式，将应用程序的界面设计与功能控制设计分离，从而可以单独地修改用户界面，而不需要去修改程序代码。应用程序的用户界面通过 XML 定义组件布局来实现。

Android 系统的布局管理是指在 XML 布局文件中设置组件的大小、间距、排列及对齐方式等。Android 系统中常见的布局方式有五种，分别是 LinearLayout、FrameLayout、TableLayout、RelativeLayout、AbsoluteLayout。

4.4.1 布局文件的规范与重要属性

1. 布局文件的规范

Android 系统应用程序的 XML 布局文件有以下规范。

（1）布局文件作为应用项目的资源存放在 res\layout 目录下，其扩展名为. xml。

（2）布局文件的根结点通常是一个布局方式，在根结点内可以添加组件作为结点。

（3）布局文件的根结点必须包含一个命名空间，如下：

xmlns: android = "http: //schemas . android. com/apk/ res/android"

（4）如果要在实现控制功能的 Java 程序中控制界面中的组件，则必须为界面布局文件中的组件定义一个 ID，其定义格式如下：

android:id = "@ + id/ <组件 ID >"

2. 布局文件的重要属性

在一个界面布局中会有很多元素，这些元素的大小和位置由其属性决定。下面简述布局文件中的几个重要属性。

（1）设置组件大小的属性。

①wrap_ content。根据组件内容的大小决定组件的大小。

②fill_parent（或 match parent）。使组件填充父组件容器的所有空间。

（2）设置组件大小的单位。

①px（pixels）。像素，即屏幕上的发光点。

②dp（或 dip，即 device independent pixels）。设备独立像素，一种支持多分辨率设备的抽象单位，与硬件相关。

③sp（scaled pixels）。比例像素，设置字体大小。

（3）设置组件的对齐方式。

在布局文件中，由 android:gravity 属性控制组件的对齐方式，其属性值有上（top）、下

(bottom)、左(letf)、右(right)、水平方向居中(center, horizontal)、垂直方向居中(center,vertical) 等。

4.4.2 常见的布局方式

1. 线性布局

线性布局 LinearLayout 是 Android 系统中常用的布局方式之一,它将组件按照水平或垂直方向排列。在 XML 布局文件中,由根元素 LinearLayout 来标识线性布局。

在布局文件中,由 android:orientation 属性来控制排列方向,其属性值有水平(horizontal)和垂直(vertical) 两种。

(1)设置线性布局为水平方向:

android: orientation = "horizontal"

(2)设置线性布局为垂直方向:

android: orientation = "vertical"

【例 4 –5】线性布局应用示例。

创建名为 Ex03_05 的新项目,包名为 com. ex03_ 05。生成项目框架后,修改界面布局文件 activity_ main. xml 的代码如下:

```
<? xml version = "1.0" encoding = "utf – 8"? >
<LinearLayout xmlns:android = "http://schemas. andrcid. com/apk/res/android"
android:layout_width = "fill_parent"
android:layout_height = "fill_parent"
android:orientation = "vertical" >
<! – – android:orientation = "horizontal" – – >
<Button
    android:id = "@ +id/mButton1"
    android:layout_width = "60px"
    android:layout_height = "wrap_content "
    android:text = "按钮 1" />
<Button
    android:id = "@ +id/mButton2"
    android:layout_width = "60px "
    android:layout_height = "wrap_content"
    android:text = "按钮 2" />
<Button
```

```
18      android:id = "@ + id/mButton3"
19      android:layout_width = "60px"
20      android:layout_height = "wrap_content"
21      android:text = "按钮 3" / >
22    < Button
23      android:id = "@ + id/mButton4"
24      android:layout_width = "60px"
25      android:layout_height = "wrap_content"
26      android:text = "按钮 4" / >
27    < /LinearLayout >
```

程序运行的结果如图 4 - 8(a)所示。如果将代码中的第 5 行 android:orientation = "vertical"(垂方向的线性布局)更改为 android:orientation = "horizontal" (水平方向的线性布局),则运行结果如图 4 - 8(b)所示。

(a)垂直方向的线性布局　　　　(b)水平方向的线性布局

图 4 - 8　线性布局的程序运行结果

2. 帧布局

帧布局 FrameLayout 是将组件放置到左上角位置,当添加多个组件时,后面的组件将遮盖之前的组件。在 XML 布局文件中,由根元素 FrameLayout 来标识帧布局。

【例 4 - 6】帧布局应用示例。

创建名为 Ex03_06 的新项目,包名为 com. ex03_06。生成项目框架后,将事先准备的图像文件 img. png 复制到 res\drawable - hdpi 目录下。

(1)设计界面布局文件 activity_main. xml,程序代码如下:

```
< ? xml version = "1.0" encoding = "utf - 8" ? >
< FrameLayout
xmlns:android = "http://schemas. android. com/apk/res/android"
```

```
android:layout_width = "fill_parent"
android:layout_height = "fill_parent" >
 <ImageView
android:id = "@ +id/mImageView"
andrcid:layout_width = "60px"
android:layout_height = "wrap_content"
 />
 <TextView
android:layout_width = "wrap_content"
android:layout_height = "wrap_content"
android:text = "快乐大本营"
android:textSize = "18sp"
 />
 </FrameLayout>
```

(2)设计控制文件 MainActivity. java,程序代码如下:

```
package com. ex03_ 06;
import android. app. Activity;
import android. os. Bundle;
import android. widget. ImageView;
public class MainActivity extends Activity
{
ImageView imageview;
@ Override
public void onCreate(Bundle savedInstanceState)
{
    super. onCreate(savedInstanceState) ;
setContentView(R. layout. activity_main) ;
imageview = (ImageView)this. findViewById(R. id. mImageView);
imageview. setImageResource(R. drawable. img);
}
}
```

帧布局的程序运行结果如图 4 -9 所示,可见在界面布局文件中添加的文本框组件遮挡了之前的图像组件。

3. 表格布局

表格布局 TableLayout 是将页面划分成由行、列构成的单元格。在 XML 布局文件中，由根元素 TableLayout 来标识表格布局。

表格的列数由 android:shrinkColumns 定义。例如，android:shrinkColumns = "0, 1, 2" 表示表格为 3 列，其列编号为第 1、2、3。

表格的行由 <TableRow> </TableRow> 定义。组件放置到哪一列，由 android:layout_column 指定列编号。

【例 4 – 7】表格布局应用示例。设计一个 3 行 4 列的表格布局，如图 4 – 10 所示。

图 4 – 9　帧布局的程序运行结果

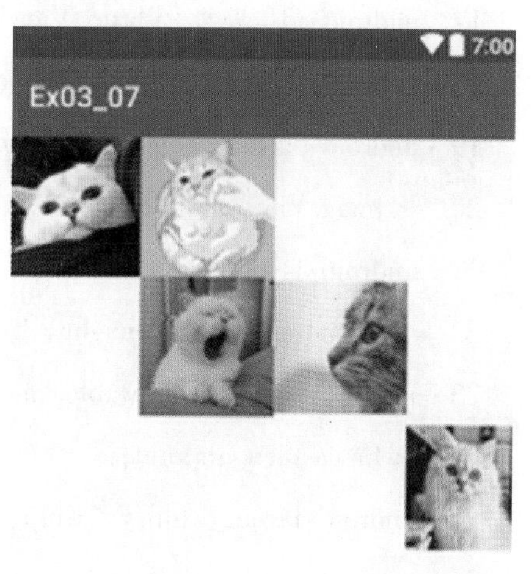

图 4 – 10　3 行 4 列的表格布局

创建名为 Ex03_07 的新项目，包名为 com. ex03_ 07。生成项目框架后，将准备好的图像文件 img1. png、img2. png、img3. png、img4. png、img5. png 复制到 res\drawable – hdpi 目录下。

(1)设计表格的界面布局文件 activity main. xml。在如图 4 – 10 所示的界面布局中，由于有显示图片的空白单元格，因此可以使用文本标签组件将其文字内容设置为空，这样显示出来的就是空白的单元格了，该文件的代码如下：

```
<? xml version = "1.0"encoding = "utf - 8"? >
<TableLayout xmlns:android = "http://schemas. android. com/apk/res/android"
android:layout_width = "fill_parent"
android:layout_height = "fill_parent" >
<TableRow >    <! - - 第 1 行 - - >
<ImageView android:id = "@ + id/mImageView1"        //第 1 列
android:layout_width = "wrap_content"
android:layout_height = "wrap_content
```

```
android:src = "@ drawable/img1" / > ”
 < ImageView android:id = "@ +id/mImageView2"      //第 2 列
   android:layout_width = "wrap_content”
android:layout_height = "wrap content"
android:src = "@ drawable/img2" / >
 </TableRow >
 < TableRow >   <! - -第 2 行 - - >
 < TextView
android:id = "@ +id/textView1"           //第 1 列空白单元格
android:layout_width = "wrap_content "
android: layout_height =  "wrap_content" / >
 < ImageView android:id = "@ +id/mImageView3"      //第 2 列
android:layout_width = "wrap_content”
     android:layout_height = "wrap_content”
android:src = "@ drawable/img3" / >
 < ImageView android:id = "@ +id/mImageView4"      //第 3 列
android:layout_width = "wrap_content”
android:layout_height = "wrap_ content”
android:src = "@ drawable/img4" / >
 </TableRow >
 < TableRow >   <! - -第 3 行 - - >
 < TextView
   android:id = "@ +id/textView2"          //第 1 列空白单元格
   android:layout_width = "wrap_content"
   android:layout_height = "wrap_content"/ >
 < TextView
   android:id = "@ +id/textView3"          //第 2 列空白单元格
   android:layout_width = "wrap_content"
   android:layout_height = "wrap_content"/ >
 < TextView
   android:id = "@ +id/textView4”          //第 3 列空白单元格
   android:layout_width = "wrap_content"
   android:layout_height = "wrap_content"/ >
```

```
<ImageView android:id = "@ +id/mImageView5"        //第 4 列
  android:layout_width = "wrap_content"
  android:layout_height = "wrap_content"
  android:src = "@ drawable/img5"/>
</TableRow>
</TableLayout>
```

(2)设计控制文件 MainActivity. java,程序代码如下:

```
package com. ex03_07;
import android. app. Activity;
import android. os. Bundle;
import android. widget. ImageView;
public class MainActivity extends Activity
{
ImageView img1,img2,img3,img4,img5;
@ Override
public void onCreate(Bundle savedInstanceState)
 {
 super. onCreate(savedInstanceState) ;
 setContentView(R. layout. activity_main) ;
 img1 = (ImageView) this. findViewById(R. id. mImageView1);
 img2 = (ImageView) this. findViewById(R. id. mImageView2);
 img3 = (ImageView) this. findViewById(R. id. mImageView3);
 img4 = (ImageView) this. findViewById(R. id. mImageView4);
 img5 = (ImageView) this. findViewById(R. id. mImageView5);
 img1. setImageResource(R. drawable. img1);
 img2. setImageResource(R. drawable. img2);
 img3. setImageResource(R. drawable. img3);
 img4. setImageResource(R. drawable. img4);
 img5. setImageResource(R. drawable. img5);
 }
}
```

4. 相对布局

RelativeLayout 是一个视图容器,其显示所有子视图在相对的位置。每个子视图的位置可以通过其相对于其他兄弟视图(如在另一个子视图的左边或者下面)的位置或者其在 RelativeLayout 区域(与底部对齐或者中间偏左)的相对位置进行说明。RelativeLayout 示意图如图 4-11 所示。

图 4-11 RelativeLayout **示意图**

RelativeLayout 能够消除嵌套视图容器并保持你的布局层次扁平化,这可以改善应用的性能,因此它是一个设计用户界面的强有力工具。如果发现自己正在使用一些嵌套的 LinearLayout,则可以用一个单独的 RelativeLayout 来替换它们。在 XML 布局文件中,由根元素 RelativeLayout 来标识相对布局。由于相对布局属性较多,因此下面简单介绍几种常用属性。

设置该控件与父元素右对齐:

android:layout_alignParentRight = "true"

设置该控件在 id 为 re_edit0 控件的下方:

android:layout_below = "@ id/re_edit_0”

设置该控件在 id 为 re image 0 控件的左边:

android:layout_toLeftOf = "@ id/re_image_0"

设置当前控件与 id 为 name 控件的上方对齐:

android:layout_alignTop = "@ id/name"

设置偏移的像素值:

android:layout_marginRight = "30dip"

该布局方式的属性较多,下面简单归纳一下。

(1)属性值为 true 或 false。

android:layout. centerHrizontal

android:layout. centerVertical

android:layout. centerInparent

android:layout. alignParentBottom

android:layout. alignParentLeft

android:layout. alignParentRight

android:layout. alignParentTop

android:layout. alignWithParentIfMissing

(2)属性值必须为 id 的引用名“@ id/id - name”。

android:layout_below

android:layout_above

android:layout_toLeftOf

android:layout_toRightOf

android:layout_alignTop

(3)属性值为具体的像素值,如 30dp。

android:layout. marginBottom

android:layout. marginLeft

android:layout. marginRight

android:layout. marginTop

【例 4 -8】创建名为 Ex03_08 的新项目,包名为 com. ex03_08。生成项目框架后,修改界面布局文件 activity. main. xml 的代码如下:

```
<? xml version = "1.0" encoding = "utf - 8"? >
<RelativeLayout xmlns:android = "http://schemas. android. com/apk/ res/android"
android:layout. width = "fill_parent"
android:layout. height = "fill_parent" >
<TextView
android:id = "@ + id/label"
android:layout. width = "fill_parent"
android:layout. height = "wrap_content"
android:textSize = "24sp"
android:text = "相对布局"/ >
<EditText
android:id = "@ + id/edit"
android:layout_width = "fill_parent"
android:layout_height = "wrap_content"
android:background = "@ android:drawable/editbox_background"
android:layout_below = "@ id/label"/ >    //在文本标签的下方
```

```
<Button
android:id="@+id/ok"
android:layout_width="wrap_content"
android:layout_height="wrap_content"
android:layout_below="@id/edit"    //在文本编辑框的下方
android:layout_alignParentRight="true"    //与父容器右对齐
android:layout_marginLeft="10dip"
android:text="OK"/>
<Button
android:layout_width="wrap_content"
android:layout_height="wrap_content"
android:layout_toleftOf="@id/ok"    //在 OK 按钮的左方
android:layout_alignTop="@id/ok"    //与 OK 按钮顶部对齐
android:text="Cancel"/>
</RelativeLayout>
```

相对布局的程序运行结果如图 4－12 所示。

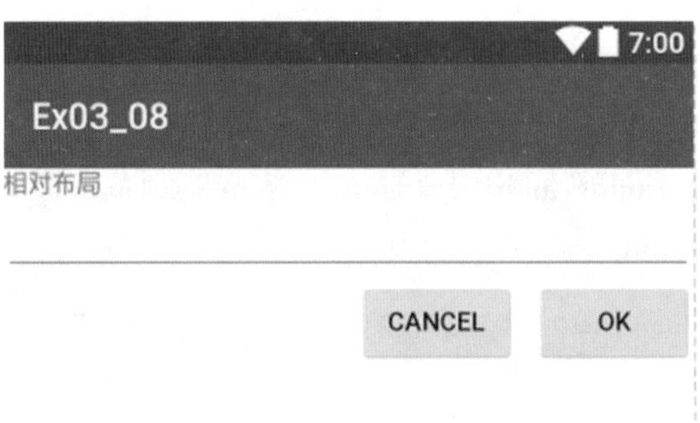

图 4－12　相对布局的程序运行结果

5. 绝对布局

绝对布局 AbsoluteLayout 是在界面布局文件中指定组件在屏幕上的坐标位置。在 XML 布局文件中，由根元素 AbsoluteLayout 来标识绝对布局。

如果要正确应用绝对布局安排组件的位置，需要了解 Android 图形界面的坐标系统。

在一个二维的 Android 图形界面坐标系中，该坐标的原点在组件的左上角，坐标的单位是像素。*X* 轴在水平方向上从左至右，*Y* 轴在垂直方向上从上向下，坐标系统如图 4－13 所示。

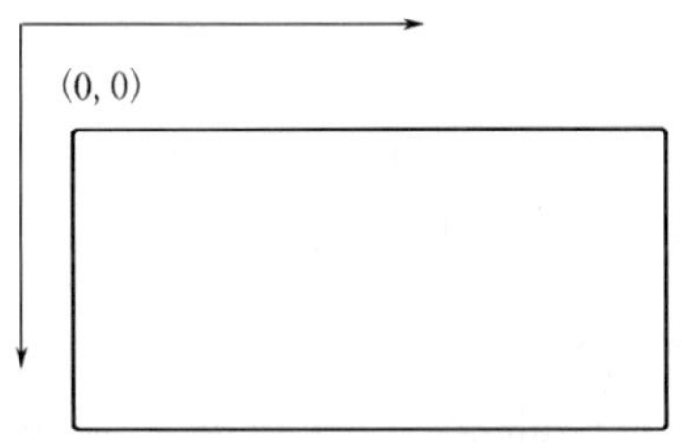

图 4－13　坐标系统

【例 4 –9】绝对布局示例。

设一个文本标签在屏幕上的坐标为(40, 150)，修改界面布局文件 activity_ main. xml 的代码如下：

```
<? xml version = "1.0" encoding = "utf -8"? >
<AbsoluteLayout xmlns:android = "http://schemas. android. com/apk/xes/android"
android:layout_width = "fill_parent"
android:layout_height = "fill_parent"
android:orientation = "vertical" >
<TextView
android:layout_width = "fill_parent"
android:layout_height = "wrap_content"
android:layout_x = "40dp"
android:layout_y = "150dp"
android:text = "欢迎进入 Android 世界!" />
</Absolutelayout >
```

绝对布局的程序运行结果如图 4 –14 所示。

6. 网格布局

GridView 是一个视图容器，GridView 示意图如图 4 –15 所示，其用来在两个维度的可滚动网格中显示子项。采用 ListAdapter 可以自动将网格项插入进布局。

图 4 –14　绝对布局的程序运行结果

图 4 –15　GridView 示意图

【例 4 –10】创建一个缩略图的网格。当选择一个子项时，一个浮动框消息将显示所选择图片的位置。

(1)创建一个 HelloGridView 的新工程。

(2)搜集一些你自己喜欢的图片，将这些图片文件保存到项目的 res/drawable/目录中。

(3)打开 res/layout/main. xml 文件，插入如下代码：

```
<?xml version="1.0" encoding="utf-8"?>
<GridView xmlns:android="http://schemas.android.com/apk/res/android"
    android:id="@+id/gridview"
    android:layout_width="fill_parent"
    android:layout_height="fill_parent"
    android:columnWidth="90dp"
    android:numColumns="auto_fit"
    android:verticalSpacing="10dp"
    android:horizontalSpacing="10dp"
    android:stretchMode="columnWidth"
    android:gravity="center"
/>
```

这个 GridView 将填充整个屏幕。属性的含义是相当明显的。

打开 HelloGridView. java,并且将如下代码插入 onCreate( )方法中:

```
public void onCreate(Bundle savedInstanceState) {
    super.onCreate(savedInstanceState);
    setContentView(R.layout.main);
    GridView gridview = (GridView) findViewById(R.id.gridview);
    gridview.setAdapter(new ImageAdapter(this));
    gridview.setOnItemClickListener(new OnItemClickListener() {
            public void onItemClick(AdapterView<?> parent,
View v, int position, long id) {
Toast.makeText(HelloGridView.this, "" + position,
 Toast.LENGTH_SHORT).show();
            }
    });
}
```

在将 main. xml 布局设置成内容视图之后,采用 findViewById(int)从布局中获得 GridView,然后 setAdapter( )方法将自定义的适配器 ImageAdapter 作为网格中所有子项显示的数据源,ImageAdapter 在下一步被创建。

当网格中的子项被点击时,为了做一些事情,一个新的接口 AdapterView. OnItemClickListener 调用了 setOnItemClickListener( )方法。这个匿名实例定义了 onItemClick( )回调方法去显示 toast 对象,这个对象显示了被选中子项(在一个真实的场景中,位置能够被用来获得全尺寸的图片进而去执行其他任务)的索引位置(从 0 开始)。

7. 创建一个继承 BaseAdapter 的新类 ImageAdapter

```
public class ImageAdapter extends BaseAdapter {
    private Context mContext;
    public ImageAdapter(Context c) {
        mContext = c;
    }
    public int getCount() {
        return mThumbIds.length;
    }
    public Object getItem(int position) {
        return null;
    }
    public long getItemId(int position) {
        return 0;
    }
    // create a new ImageView for each item referenced by the Adapter
    public View getView(int position, View convertView, ViewGroup parent) {
        ImageView imageView;
        if (convertView == null) {
            // if it's not recycled, initialize some attributes
            imageView = new ImageView(mContext);
            imageView.setLayoutParams(new GridView.LayoutParams(85, 85));
            imageView.setScaleType(ImageView.ScaleType.CENTER_CROP);
            imageView.setPadding(8, 8, 8, 8);
        } else {
            imageView = (ImageView) convertView;
        }
        imageView.setImageResource(mThumbIds[position]);
        return imageView;
    }
    // references to our images
    private Integer[] mThumbIds = {
```

```
            R. drawable. sample_2, R. drawable. sample_3,
            R. drawable. sample_4, R. drawable. sample_5,
            R. drawable. sample_6, R. drawable. sample_7,
            R. drawable. sample_0, R. drawable. sample_1,
            R. drawable. sample_2, R. drawable. sample_3,
            R. drawable. sample_4, R. drawable. sample_5,
            R. drawable. sample_6, R. drawable. sample_7,
            R. drawable. sample_0, R. drawable. sample_1,
            R. drawable. sample_2, R. drawable. sample_3,
            R. drawable. sample_4, R. drawable. sample_5,
            R. drawable. sample_6, R. drawable. sample_7
    };
}
```

首先,实现一些必要的继承自 BaseAdapter 的方法。构造函数和 getCount( )是显而易见的。正常来说,getItem(int)应该返回在适配器中被指定的实际对象,但是在这个例子中它被忽略了。同样,getItemId(int)应该返回子项的行 Id,但这里不需要。

getView 是第一个必须的方法,这个方法为每个加入到 ImageAdapter 的图片创建了一个新的 View。当这个方法被调用之后,一个 View 被返回,其通常来说是一个可回收的对象(在这之后至少可以被调用一次),这样如果这个对象是空的,就可以去查看。如果它是空的,一个 ImageView 就可以通过期望的图片显示属性实例化和配置。

setLayoutParams(ViewGroup. LayoutParams) 设置视图的宽度和高度,这保证了无论画布的大小,都可以根据情况去调整和裁剪每张图片去适应这些尺寸。

setScaleType(ImageView. ScaleType)表明(如果必要)图片应该按照中心去裁剪。

setPadding(int, int, int, int)定义四周的内边距(如果图片具有不同的宽高比,当它不能匹配给定的尺寸去适应 ImageView 时,那么较小的内边距将引起较大的图片裁剪)。

如果传入 getView 的 View 不为空,那么利用可回收 View 对象实例化本地 ImageView。

在 getView 方法的最后,将 position 整数传入 qetview 方法中,将其用于从 mThumbIds 数组中选择图片,mThumbIds 数组是 ImageView 的图片数据源集合。

剩下的内容是用来定义画布资源的 mThumbIds 数组。

8. 执行应用程序

通过调整 GridView 和 ImageView 元素的属性来测试它们的行为。例如,不采用 setLayoutParams(ViewGroup. LayoutParams),而试着使用 setAdjustViewBounds(boolean)。

4.5 进度条和选项按钮

4.5.1 进度条

进度条 ProgressBar 能以形象的图示方式直观地显示某个过程的进度。进度条 ProgressBar 的常用属性和方法见表 4－8。

表 4－8 进度条 ProgressBar 的常用属性和方法

属性	方法	功能
android:max	setMax(int max)	设置进度条的变化范围为 0 ~ max
android:progress	setProgress(int progress)	设置进度条的当前值(初始值)
android:progressby	incrementProgressBy(int diff)	设置进度条的变化步长值

【例 4－11】进度条应用示例。

在界面设计中安排一个进度条组件，并设置两个按钮，用于控制进度条的进度变化，进度条进度控制如图 4－16 所示。

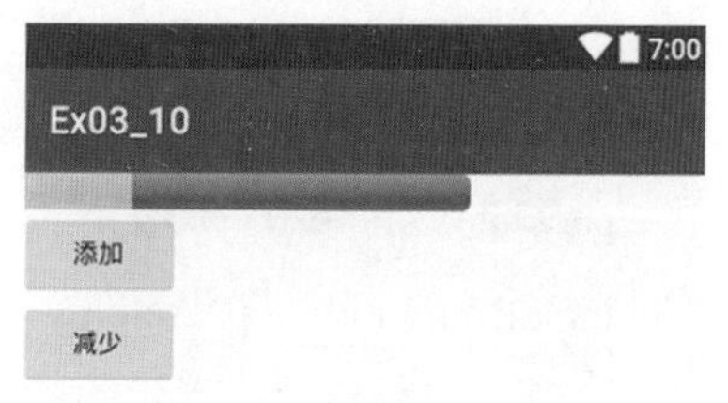

图 4－16 进度条进度控制

程序设计步骤如下：

(1)在界面布局文件中声明 ProgressBar；

(2)在 Activity 中获得 ProgressBar 实例；

(3)调用 ProgressBar 的 incrementProgressBy() 方法增加或减少进度。

程序代码如下。

(1)设计界面布局文件 activity. main. xml 的代码如下：

```
<? xml version = "1.0" encoding = "utf-8"? >
<LinearLayout xmlns:android = "http//schemas. android. com/apk/res/ android"
android:layout_width = "fill_parent"
android:layout_height = "fill_parent"
android:orientation = "vertical" >
<ProgressBar
android:id = "@ + id/ProgressBar01"
style = "@ android:style/Widget. ProgressBar. Horizontal"
```

```
android:layout_width = "250dp"
android:layout_height = " wrap_content"
android:max = "200"
android:progress = "50"  >
</ProgressBar >
< Button
android:id = "@  + id/button1"
android:layout_width = " wrap_content"
android:layout_height = " wrap_content"
android:text = "@ string/btn1"  / >
< Button
android:id = "@  + id/button2"
android:layout_width = " wrap_content"
android:layout_height = " wrap_content"
android:text = "@ string/btn2"/ >
</LinearLayout >
```

(2)控制文件 MainActivity. java 的代码如下:

```
package com. ex03_10;
import android. app. Activity;
import android. os. Bundle;
import android. view. View;
import android. view. View. OnClicklistener;
import android. widget. Button;
import android. widget. ProgressBar;
public class MainActivity extends Activity
{
ProgressBar progressBar;
Button btn1,btn2;
@ Override
public void onCreate( Bundle savedInstanceState)
{
super. onCreate( savedInstanceState) ;
setContentView( R. layout. activity. main) ;
```

```
progressBar = (ProgressBar)findViewById(R.id.ProgressBar01);
btn1 = (Button)findviewById(R.id.button1);
btn2 = (Button) findViewById(R.id.button2);
btn1.setOnClickListener(new click1());
btn2.setOnClickListener(new click2());
}
class click1 implements OnClickListener
{
  public void onClick(View v)
{ progressBax.incrementProgressBy(5); }   //增加进度
}
class click2 implements OnClickListener
{
public void onClick(View v)
{ progressBar.incrementProgressBy( -5); }   //减少进度
}
}
```

4.5.2 选项按钮

1. 复选框

复选框 CheckBox 用于多项选择,用户可以一次性选择多个选项。复选框 CheckBox 是按钮 Button 的子类,其属性和方法继承于按钮 Button。复选框 CheckBox 的常用方法见表 4 -9。

表 4 -9 复选框 CheckBox 的常用方法

方法	功能
isChecked()	判断选项是否被选中
getText()	获取复选框的文本内容

【例 4 -12】复选框应用示例。

在界面设计中,安排三个复选框和一个普通按钮,选择选项后,单击按钮,在文本标签中显示所选中的选项文本内容,复选框应用示例如图 4 -17 所示。

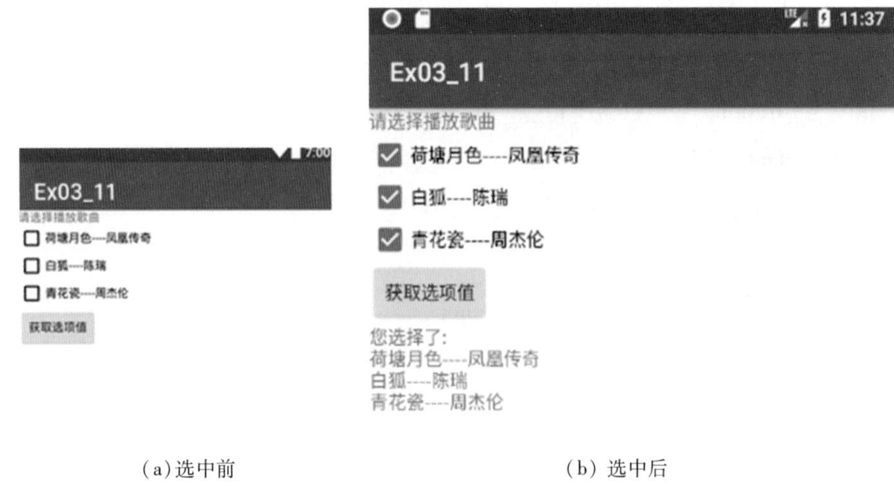

(a)选中前　　　　(b) 选中后

图 4－17　复选框应用示例

程序设计步骤如下:

(1)在界面布局文件中声明复选框 CheckBox;

(2)在 Activity 中获得复选框 CheckBox 实例;

(3)调用 CheckBox 的 isChecked0 方法判断该选项是否被选中,如果被选中,则调用 gef-Text( )方法获取选项的文本内容。

程序代码如下。

(1)设计界面布局文件 activity_main. xml 的代码如下:

```
<? xml version = "1.0"  encoding = "utf - 8"?  >
<LinearLayout xmlns:android = "http://schemas. android. com/apk/res/android"
android:layout_width = "fill_parent"
android:layout. height = "fill_parent"
android:orientation = "vertical"  >
<TextView
android:layout. width = "fill_parent"
android:layout_height = "wrap_content"
android:text = "@ string/hello"
android: textSize = "20sp"/ >
<CheckBox
android:id = "@ id/check1”
android:layout_width = "fill_parent"
android:layout_height = "wrap_content"
```

```
android:textSize = "20sp"
android:text = "@ string/one"  / >
 < CheckBox
android:id = "@  + id/check2"
android:layout_width = "fill_parent"
android:layout_height = " wrap_content"
android:textSize = "20sp"
android:text = "@ string/two"/ >
 < CheckBox
android:id = "@  + id/check3"
android:layout_width = "fill_parent"
android:layout_height = " wra_content"
android:textSize = "20sp"
android:text = "@ string/three"  / >
 < Button
android:id = "@  + id/button"
android:layout_width = " wrap_content"
android:layout_height = " wrap_content"
androld:textSize = "20sp"
android:text = "@ string/btn"  / >
 < TextView
andro1d:id = "@  + id/textView2  "
android:layout_width = "fill_parent"
androld:layout_height = " wrap_content"
android:text = " "
 android:textSize = "20sp"/ >
 < /LinearLayout >
```

(2)在 strings. xml 文件中设置要使用的字符串,代码如下:

```
< ? xml version = "1.0"  encoding = "utf - 8"?  >
< resources >
< string name = "hello" > 请选择播放歌曲: < /string >
< string name = "app name" > Ex03_11 < /string >
< string name = "one" > 荷塘月色 - - - - 凤凰传奇 < /string >
```

```
<string name = "two" >白狐 - - - - 陈瑞 </string >
<string name =  "three" >青花瓷 - - - - 周杰伦 </string >
<string name =  "btn" >获取选项值 </string >
</resources >
```

(3)设计控制文件 MainActivity. java。在控制文件 MainActivity. java 中建立组件与界面布局文件中相关组件的关联,按钮的事件处理代码如下:

```
package com. ex03_11;
import android. app. Activity;
import android. os. Bundle;
import android. view. View;
import android. view. View. OnClickListener;
import android. widget. Button;
import android. widget. CheckBox;
import android. widget. TextView;
public class MainActivity extends Activity
{
CheckBox ch1 ,ch2,ch3;
Button okBtn;
TextView txt;
@ Override
public void onCreate( Bundle savedInstanceState)
{
super. onCreate( savedInstanceState) ;
setContentView( R. layout. activity. main) ;
ch1 = ( CheckBox) findViewById( R. id. check1) ;
ch2 = ( CheckBox) findViewById( R. id. check2) ;
ch3 = ( CheckBox) findViewById( R. id. check3) ;
okBtn = ( Button) findViewById( R. id. button) ;
txt = ( TextView) findViewById( R. id. textView2) ;
okBtn. setOnClickListener( new click( ) ) ;
}
   class click implements OnClickListener
{
```

```
public void onClick( View v)
{
String str = " " ;
if( ch1. isChecked( ) ) str = str + " \n" + ch1. getText( ) ;
if( ch2. isChecked( ) ) str = str + " \n" + ch2. getText( ) ;
if( ch3. isChecked( ) ) str = str + " \n" + ch3. getText( ) ;
txt. setText( "您选择了" + str) ;
}
}
}
```

2. 单选组件与单选按钮

单选组件 RadioGroup 用于多项选择中只允许任选其中一项的情形。单选组件 RadioGroup 由一组单选按钮 RadioButton 组成。单选按钮 RadioButton 是按钮 Button 的子类。单选按钮 RadioButton 的常用方法见表 4 – 10。

表 4 – 10　单选按钮 RadioButton 的常用方法

方法	功能
isChecked( ) ;	判断选项是否被选中
getText( ) ;	获取单选按钮的文本内容

【例 4 – 13】单选按钮应用示例。

在界面设计中,安排两个单选按钮、一个文本编辑框和一个普通按钮,选择选项后,单击按钮,在文本标签中显示文本编辑框及选中选项的文本内容,单选按钮示例如图 4 – 18 所示。

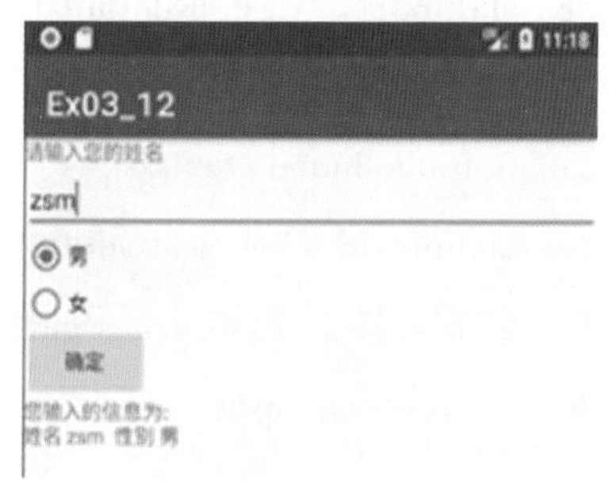

图 4 – 18　单选按钮示例

程序设计步骤如下:

(1)在界面布局文件中声明单选组件 RadioGroup 和单选按钮 RadioButton;

(2)在 Activity 中获得单选按钮 RadioButton 实例;

(3)调用 RadioButton 的 isChecked( )方法判断该选项是否被选中,如果被选中,则调用 getText( )方法获取选项的文本内容。

程序代码如下。

(1)设计界面布局文件 activity main. xml 的代码如下:

```
<? xml version = "1.0" encoding = "utf - 8"? >
<LinearLayout xmlns:android = "http://schemas. android. com/apk/res/android"
android:layout_width = "fill_parent"
android:layout_height = "fill_parent"
android:orientation = "vertical" >
<TextView
android:layout_width = "fill_parent"
layout_height = "wrap_content"
android:textSize = "20sp"
android:text = "@ string/hello" / >
<EditText
android:id = "@ +id/edit1"
android:layout_width = "fill. parent"
android:layout_height = "wrap_content"
android:inputType = "text"
android:textSize = "20sp" / >
<RadioGroup
android:layout_width = "fill_parent"
android:layout_height = "wrap_content" >
<RadioButton
android:id = "@ +id/boy01"
android:text = "@ string/boy"/ >
<RadioButton
android:id = "@ +id/girl01"
android:text = "@ string/girl" / >
</RadioGroup >
<Button
android:id = "@ +id/myButton"
android:layout_width = "wrap_content"
android:layout_height = "wrap_content"
android:text = "确定"
android:textSize = "20sp"
```

```
/ >
 < TextView
android:id = "@ + id/text02"
android:layout_width = "fill_parent"
android:layout_height = "wrap_content"
android:textSize = "20sp"
/ >
 < /LinearLayout >
```

（2）在 strings. xml 文件中设置要使用的字符串，代码如下：

```
< ? xml version = "1.0" encoding = "utf - 8"? >
< resources >
< string name = "hello" > 请输入您的姓名: < /string >
< string name = "app_name" > Ex03_12 < /string >
< string name = "boy" > 男 < /string >
< string name = "girl" > 女 < /string >
< /resources >
```

（3）设计控制文件 MainActivity. java。在控制文件 MainActivity. java 中建立组件与界面布局文件中相关组件的关联，按钮的事件处理代码如下：

```
package com. ex03_12;
import android. app. Activity;
import android. os. Bundle;
import android. view. view;
import android. view. View. OnClickListener;
import android. widget. Button;
import android. widget. EditText;
import android. widget. RadioButton;
import android. widget. TextView;
public class MainActivity extends Activity
{
Button okBtn;
EditText edit;
TextView txt;
RadioButton r1,r2;
```

```
@ Override
public void onCreate( Bundle savedInstanceState)
{
super. onCreate( savedInstanceState) ;
setContentView( R. layout. activity_main) ;
edit = ( EditText) findViewById ( R. id. edit1) ;
okBtn = ( Button) findViewById ( R. id. myButton) ;
txt  = ( TextView) findViewById( R. id. text02) ;
r1 = ( RadioButton) findViewById( R. id. boy01) ;
r2 = ( RadioButton) findViewById( R. id. girl01) ;
 okBtn = setOnClickListener( new mClick( ) ) ;
}
class mClick implements OnClickListener
{
public void onClick ( View v)
{
CharSequence str  =  "" , name =  "";
name = edit. getText ( ) ;
if ( r1. isChecked( ) )  //第 1 个单选按钮被选中
str = r1. getText( ) ;
if ( r2. isChecked( ) )  //第 2 个单选按钮被选中
str = r2. getText( ) ;
txt. setText( "您输入的信息为:\n 姓名 " + name + "\t 性别" + str);
}
}
}
```

4.6 图像显示与画廊组件

4.6.1 图像显示 ImageView 类

ImageView 类用于显示图片或图标等图像资源,并提供图像缩放及着色(渲染)等图像处理功能。

ImageView 类的常用属性和对应方法见表 4－11。

表 4－11　ImageView 类的常用属性和对应方法

元素属性	对应方法	说明
android:maxHeight	setMaxHeight(int)	为显示图像提供最大高度的可选参数
android:maxWidth	setMaxWidth(int)	为显示图像提供最大宽度的可选参数
android:ScaleType	setScaleType(lmageView. ScaleType)	控制图像适合 ImageView 大小的显示方式
android:src	setlmagcResource(int)	获取图像文件的路径

ImageView 类的 ScaleType 属性值见表 4－12。

表 4－12　ImageView 类的 ScaleType 属性值

ScaleType 属性值常量值	值	说明
matrix	0	用矩阵来绘图
fitXY	1	拉伸图片(不按宽高比例)以填充 View 的宽高
fitStart	2	按比例拉伸图片,拉伸后图片的高度为 View 的高度,且显示在 View 的左边
fitCenter	3	按比例拉伸图片,拉伸后图片的高度为 View 的高度,且显示在 View 的中间
fitEnd	4	按比例拉伸图片,拉伸后图片的高度为 View 的高度,且显示在 View 的右边
center	5	按原图大小显示图片,但图片的宽高大于 View 的宽高时,截取图片中间部分显示
centerCrop	6	按比例放大原图,直至等于某边 View 的宽高显示
centerInside	7	当原图宽高等于 View 的宽高时,按原图大小居中显示,否则将原图缩放至 View 的宽高居中显示

【例 4－14】显示图片示例。

程序设计步骤如下:

(1)将事先准备好的图片序列 img1. jpg、img2. jpg、…、img6. jpg 复制到 res\drawable－hdpi 目录下;

(2)在布局文件中声明图像显示组件 ImageView;

(3)在 Activity 中获得相关组件实例;

(4)通过触发按钮事件,调用 OnClickListener 接口的 onClick( )方法显示图像。

程序代码如下。

(1)设计界面布局文件 activity. main. xml。在界面设计中,安排两个按钮和一个图像显示组件 ImageView,单击按钮,可以翻阅浏览图片。布局设计如图 4 - 19 所示。

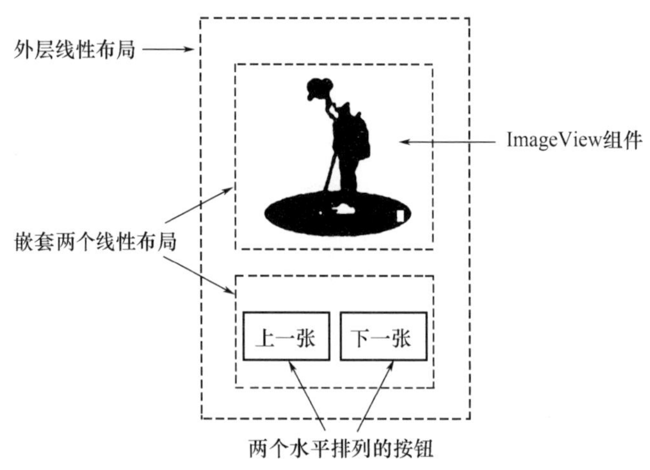

图 4 - 19　布局设计

程序代码如下:

```
<?xml version="1.0" encoding="utf-8"?>
<LinearLayout xmlns:android="http://schemas.android.com/apk/res/android"
android:layout_width="fill_parent"
android:layout_height="fill_parent"
android:gravity="center|fill"
android:orientation="vertical" >
<LinearLayout
android:layout_width="fill_parent"
android:layout_height="wrap_content"
android:gravity="center" >
<ImageView
android:id="@+id/img"
android:layout_width="240dp"
android:layout_height="240dp"
android:layout_centerVertical="true"
android:src="@drawable/img1" />
```

```
</LinearLayout>
<LinearLayout
android:layout_width = "fill_parent"
android:layout_height = "wrap_content"  >
<Button
android:id = "@ +id/btn_last"
android:layout_width = "150dp"
android:layout_height = "wrap_content"
android:text = "L - Z"  />
<Button android:id = "@ +id/btn_next"
android:layout_width = "150dp"
android:layout_height = "wrap_content "
android:text = "T - K"  />
</LinearLayout>
</LinearLayout>
```

(2)设计控制文件 MainActivity. java。在控制文件 MainActivity. java 中建立组件与界面布局文件中相关组件的关联,按钮的事件处理代码如下:

```
package com. ex03_13;
import android. app. Activity;
import android. os. Bundle;
import android. view. View;
import android. view. View. OnClickListener;
import android. widget. Button;
import android. widget. ImageView;
public class MainActivity extends Activity {
ImageView img;
Button btn,last,btn_next;
//存放图片 id 的 int 数组
private int[ ] imgs = {    //数组元素为资源目录中的图片序列
R. drawable. img1,
R. drawable. img2,
R. drawable. img3,
R. drawable. img4,
```

```
R. drawable. img5,
R. drawable. img6 };
int index = 1;
@ Override
public void onCreate(Bundle savedInstanceState) {
super. onCreate(savedInstanceState);
setContentView(R. layout. activity. main);
img. (ImageView)findViewById(R. id. img);
btn_last. (Button)findViewById(R. id. btn_last);
btn_next. (Button)findViewById(R. id. btn_ next);
btn_last. setOnClickListener (new mClick());      //注册监听接口
btn_next. setOnClickListener (new mClick());      //注册监听接口
}
class mClick implements OnClickListener  //定义一个类实现监听接口
{
public void onClick(View v)
{
  if (v = btn_last)                           //下一张按钮事件
{
if(index >0 && index < imgs. length)
{
index - -;
img. setImageResource(imgs[index]);
} else (index = imgs. length + 1; }
}
if(v = = btn_next)                            //上一张按钮事件
{
if(index >0&&index < imgs. length - 1)
{
index + +;
img. setImageResource(imgs[index]);
} else {index = imgs. length - 1;  }
}
```

```
    }
}
```

图像显示示例的程序运行结果如图 4－20 所示。

图 4－20　图像显示示例的程序运行结果

4.6.2　画廊组件 Gallery 与图片切换器 ImageSwitcher

Gallery 是 Android 中控制图片展示的组件，它可以横向显示一列图像。Gallery 的常用属性及方法见表 4－13。

表 4－13　Gallery 的常用属性及方法

元素属性	对应方法	说明
android:spacing	setSpacing(int)	设置图片之间的间距，以像素为单位
android: unsclectcdAlpha	setUnselecedAlpha(float)	设置未选中图片的透明度(Alpha)
android:animationDuration	setAnimationDuration(int)	设置布局变化时动画转换所需的时间(毫秒级)，仅在动画开始时计时
	onTouchEvent(MotionEventenent)	触摸屏幕时触发 MotionEvent 事件
	onDown(MotionEvent e)	按下屏幕时触发 MotionEvent 事件

Gallery 经常与图片切换器 ImageSwitcher 配合使用，用图片切换器 ImageSwitcher 展示图片效果。使用 ImageSwitcher 时必须用 ViewFactory 接口的 makeView()方法创建视图。ImageSwitcher 的常用方法见表 4 – 14。

表 4 – 14　ImageSwitcher 的常用方法

方法	说明
setInAnimation(Animation inAnimation)	设置动画对象进入屏幕的方式
setOutAnimation(Animation outAnimation)	设置动画对象退出屏幕的方式
setImageResource(int resid)	设置显示的初始图片
showNext( )	显示下一个视图
showPrevious( )	显示前一个视图

【例 4 – 15】画廊展示图片示例。

在界面设计中，安排一个画廊组件 Gallery 和一个图片切换器 ImageSwitcher，单击画廊中的小图片，可以在图片切换器中显示放大的图片，展示图片示例如图 4 – 21 所示。

图 4 – 21　展示图片示例

程序设计步骤如下：

(1)在界面布局文件中声明画廊组件 Gallery 和图片切换器 ImageSwitcher，采用表格布局；

(2)把事先准备好的图片文件 img1. jpg、img2. jpg、…、img8. jpg 复制到项目的资源目录 res\drawable – hdpi 中，在 Activity 中创建一个图像文件数组 imgs[ ]，其数组元素为图片文件；

(3)在 Activity 中创建画廊组件 Gallery 和图片切换器 ImageSwitcher 组件的实例对象；

(4)在 Activity 中创建一个实现 ViewFactory 接口的内部类，重写 makeVie( )方法建立 imageView 图像视图，图片切换器 ImageSwitcher 通过该图像视图显示放大的图片；

(5)在 Activity 中创建一个 BaseAdapter 适配器，用于安排放在画廊 Gallery 中的图片文件及显示方式。

程序代码如下：

(1)设计界面布局文件 activity_ main. xml 的代码如下：

```
<? xml version = "1.0" encoding: = "utf-8"? >
<TableLayout android:id = "@ +id/TableLayout01"
android:layout_width = "wrap_content"
android:layout_height = "wrap_content"
xmlns:android = "http://schemas. android. com/apk/res/android"
android:layout_gravity = "center" >
<TextView
android:layout_width = "fill_parent"
android:layout_height = "wrap_content"
android:textSize = "20sp"
android:text = "estring/hello" / >
<Gallery android:id = "@ +id/Gallery01"
android:layout_width = "wrap_content"
android:layout_height = "wrap_content"
android:spacing = "10dp"/ >
<ImageSwitcher android:id = "@ +id/ImageSwitcher01"
android:layout_width = "wrap_content"
android:layout_height = "wrap_content" >
</ ImageSwitcher >
/TableLayout >
```

(2)设计控制文件 MainActivity. java。在控制文件 MainActivity. java 中创建图像文件序列数组，并编写按钮的事件处理代码。通过 ViewFactory 接口建立 imageView 图像视图，并实现 OnItemSclectedListener 接口来选择图片，程序代码如下：

```
package com. ex03_14;
import android. app. Activity;
import android. os. Bundle;
```

```
import android.view.View;
import android.view.ViewGroup;
import android.view.animation.AnimationUtils;
import android.widget.AdapterView;
import android.widget.BaseAdapter;
import android.widget.Gallery;
import android.widget.ImageSwitcher;
import android.widget.ImageView;
import android.widget.AdapterView.OnItemSelectedListener;
import android.widget.ViewSwitcher.ViewFactory;

public class MainActivity extends Activity
{
private ImageSwitcher imageSwitcher;
Gallery gallery;
private int[] imgs = {
R.drawable.img1,
R.drawable.img2,
R.drawable.img3,
R.drawable.img4,
R.drawable.img5,
R.drawable.img6,
R.drawable.img7,
R.drawable.img8,
};

@Override
public void onCreate(Bundle savedInstanceState)
{
super.onCreate(savedInstanceState);
setContentView(R.layout.activity_main);
imageswitcher = (ImageSwitcher)findViewById(R.id.ImageSwitcher01);
imageSwitcher.setEactory(new viewFactory());
```

```
imageSwitcher. setInAnimation( AnimationUtils    //设置淡入方式
                . loadAnimation (this,android. R. anim. fade_in)) ;
imageSwitcher. setOutAnimation( AnimationUtils   //设置淡出方式
                    . loadAnimation( this,android. R. anim. fade_out)) ;
imageswitcher. setImageResource( R. drawable. img1) ; //设置初始图片
gallery = (Gallery)findViewById( R. id. Gallery01)  ;
gallery. setOnItemSelectedListener(
new onItemSelectedlistener( )) ;   //设置监听,获取选择的图片
gallery. setspacing( 10) ;//设定画廊图片之间的间隔
gallery. setAdapter( new baseAdapter( )) ;   //设置显示方式的适配器
}
//通过 viewFactory 接口建立一个 imageView 图像视图
class viewFactory implements ViewFactory
{
@ Override
public View makeView ( )
{
ImageView imageView. new ImageView( MainActivity. this) ;
imageView. setScaleType( ImageView,ScaleType. FIT_CENTER) ;
return imageView;
}
}
//实现选项监听接口,获取选择的图片
class onItemSelectedListener implements OnItemSelectedListener
{
@ override
public void onItemSelected( AdapterView < ? > parent, //监听选项
View view,int position,long id)
{
imageSwitcher. setImageResource(
  lery. getItemIdAtPosition( position)) ;
}
@ Override
```

```
public void onNothingSelected(AdapterView<?> arg0) {    }
}
//设置一个适配器,安排放在画廊 gallery 中的图片文件及显示方式
class baseAdapter extends BaseAdapter
{
//取得 gallery 内的图片数量
public int getCount()
{return imgs.length;}
public Object getItem(int position)
{    return null;        }
//取得 gallery 内选择的某一张图片文件
public long getItemId(int position)
{        return imgs[position];    }imagevien
//将选择的图片放在 imageview,且设定显示方式为居中,大小为 60 px×60 px
public View getView (int position,View convertView,ViewGroup parent)
{
    ImageView imageView = new ImageView (MainActivity.this);
    imageView.setImageResource(imgs[position]);
    imageView.setScaleType(ImageView.ScaleType.FIT_CENTER) ;
imageView.setLayoutParams(new Gallery.LayoutParams(60, 60));
return imageView;
}
}
}
```

4.7 消 息 提 示

在 Android 系统中,可以用 Toast 来显示帮助或提示消息。该提示消息以浮于应用程序之上的形式显示在屏幕上。因为它并不获得焦点,所以不会影响用户的其他操作。使用消息提示组件 Toast 的目的就是尽可能不中断用户操作,并使用户看到提供的信息内容。Toast 类的常用方法见表 4-15。

表 4－15　Toast 类的常用方法

方法	说明
Toast(Context contex!)	Toast 的构造方法,构造一个空的 Toast 对象
makeText(Context context, CharSequence text, int duration)	以特定时长显示文本内容,参数 text 为显示的文本,参数 duration 为显示时间,较长时间取值 LENGTH_LONG,较短时间取值 LENGTH_SHORT
getView( )	返回视图
setDuration(int duration)	设置存续时间
setView(View view)	设置要显示的视图
setGravity(int gravity, int xOffset, int yOffset)	设置提示信息在屏幕上的显示位置
setText(int resId)	更新 makeText( )方法所设置的文本内容
show( )	显示提示信息
LENGTH_LONG	提示信息显示较长时间的常量
LENGTH_SHORT	提示信息显示较短时间的常量

【例 4－16】消息提示 Toast 分别按默认方式、自定义方式和带图标方式显示的示例。将事先准备好的图标文件 icon. jpg 复制到 res\drawable－hdpi 目录下,以作提示消息的图标之用。

(1)设计界面布局文件 activity_ main. xml。在界面设计中设置一个文本标签和三个按钮,分别对应消息提示 Toast 的三种显示方式,程序代码如下:

```
<? xml version = "1.0" encoding = "utf-8"? >
<LinearLayout xmlns:android = "http://schemas.android.com/apk/res/android"
android:layout_width = "fill_parent"
android:layout_height = "fill_parent"
android:orientation = "vertical" >
<TextView
android:layout_width = "fill_parent"
android:layout_height = "wrap_content"
android:gravity = "center"   //居中显示文本
android:text = "消息提示 Tost"
   android:textSize = "24sp" />
<Button
```

```
android:id="@+id/btn1"
android:layout_height="wrap_content"
android:layout_width="fill_parent"
android:text="默认方式"
android:textSize="20sp"/>
<Button
android:id="@+id/btn2"
android:layout_height="wrap_content"
android:layout_width="fill_parent"
android:text="自定义方式"
android:textSize="20sp"/>
<Button
android:id="@+id/btn3"
android:layout_height="wrap_content"
android:layout_width="fill_parent"
android:text="带图标方式"
android:textSize="20sp"/>
</LinearLayout>
```

(2)设计控制文件 MainActivity.java 的代码如下:

```
package com.ex03_15;
import android.app.Activity;
import android.os.Bundle;
import android.view.Gravity;
import android.view.View;
import android.view.View.OnClickListener;
import android.widget.Button;
import android.widget.ImageView;
import android.widget.LinearLayout;
import android.widget.ListView;
import android.widget.Toast;

public class MainActivity extends Activity
{
```

```
ListView list; ;
 Button btn1,btn2,btn3;
 @ Override
 public void onCreate( Bundle savedInstanceState)
 {
 super. onCreate( savedInstanceState) ;
 setContentView( R. layout. activity main) ;
 btn1 = ( Button) findViewById( R. id. btn1) ;
 btn2 = ( Button) findViewById( R. id. btn2) ;
 btn3 = ( Button) findViewById( R. id. btn3) ;
 btn1. setonClickListener( new mClick( ) ) ;  //为按钮注册事件监听器
 btn2. setOnClickListener( new mClick( ) ) ;
 btn3. setonClickListener( new mClick( ) ;
 }

 class mClick implements OnClickListener
 {
 Toast toast;
 LinearLayout toastView;
 ImageView imageCodeProject;
 @ Override
 public void onClick( View v)
 {
 if (v = =btn1)  //居中显示文本
 {
 Toast. makeText( getApp1icationContext( ),//设置提示消息内容
 "默认 Toast 方式",
   Toast. LENGTH_SHORT). show( ) ;
 }
 else if( v = =btn2)
 {
 toast =Toast. makeText( getApplication( ),
 "自定义 Toast 的位置",
```

```
toast. setGravity( Gravity. GENTER,0,0);

toast. show( );
}
else if (v = = btn3)
{
toast = Toast. makeText( getApplicationContext( ),
"带图标的 Toast",
Toast. LENGTH_SHORT);
toast. setGravity( Gravity. CENTER,0,80) ;
toastView = (LinearLayout)toast. getview( );   //定义视图
imageCodeProject = new ImageView( MainActivity. this);
imageCodeProject. setImageResource( R. drawable. icon);
toastView. addView( inageCodeProject,0); //在视图中添加图标
toast. show( ):
}
}
}
}
```

消息提示 Toast 三种方式的程序运行结果如图 4－22 所示。

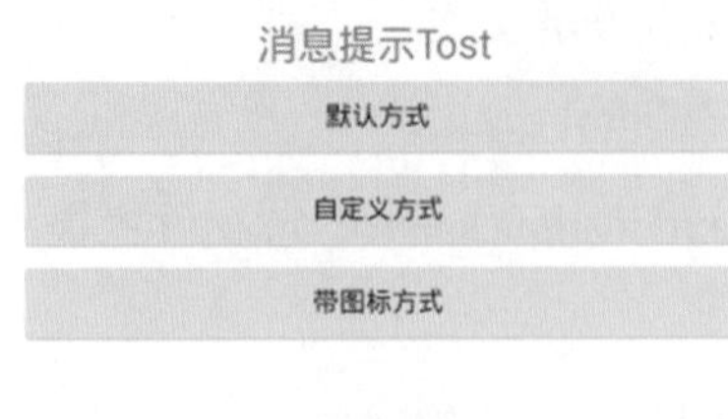

图 4－22　消息提示 Toast 三种方式的程序运行结果

4.8　列 表 组 件

4.8.1　列表组件 Spinners 类

SpinnerView 一次显示列表中的一项,并可以使用户在其中进行选择。

【例 4 - 17】通过实例介绍 SpinnerView 的使用方法。

(1)编辑布局文件 fragment_main. xml。

```
<RelativeLayout xmlns:android = "http://schemas. android. com/apk/res/android"
  xmlns:tools = "http://schemas. android. com/tools"
  android:layout_width = "match_parent"
  android:layout_height = "match_parent"
  android:paddingBottom = "@ dimen/activity_vertical_margin"
  android:paddingLeft = "@ dimen/activity_horizontal_margin"
  android:paddingRight = "@ dimen/activity_horizontal_margin"
  android:paddingTop = "@ dimen/activity_vertical_margin"
  tools:context = "com. example. spinner. MainActivityMYMPlaceholderFragment" >

    <TextView
        android:id = "@ + id/textView1"
        android:layout_width = "wrap_content"
        android:layout_height = "wrap_content"
        android:layout_alignLeft = "@ + id/planets_spinner"
        android:layout_alignParentTop = "true"
        android:text = "黑龙江"
        android:textSize = "18dp" />

    <Spinner
      android:id = "@ + id/planets_spinner"
      android:layout_width = "fill_parent"
      android:layout_height = "wrap_content"
      android:layout_below = "@ + id/textView1"
      android:layout_centerHorizontal = "true" />      </RelativeLayout>
```

(2)编辑资源文件 strings. xml。

```
        < resources >
        < string - array name = "planets_array" >
        < item >哈尔滨 </item >
        < item >齐齐哈尔 </item >
        < item >牡丹江 </item >
        < item >佳木斯 </item >
        < item >大庆 </item >
        < item >伊春 </item >
        < item >双鸭山 </item >
        < item >鹤岗 </item >
    </string - array >
    </resources >
```

(3)创建适配器。

```
public static class PlaceholderFragment extends Fragment {
@ Override
public View onCreateView( LayoutInflater inflater, ViewGroup container,
                    Bundle savedInstanceState) {
View rootView  =  inflater. inflate( R. layout. fragment_main, container, false) ;
Spinner spinner  =  (Spinner) rootView. findViewById( R. id. planets_spinner) ;
      ArrayAdapter < CharSequence > adapter  =  ArrayAdapter
          . createFromResource( getActivity( ) ,R. array. planets_array, android
          . R. layout. simple_spinner_item) ;
      adapter. setDropDownViewResource( android. R
                  . layout. simple_spinner_dropdown_item) ;
  spinner. setAdapter( adapter) ;
  return rootView;
  }
}
```

(4)设置监听器。

```
AdapterView. OnItemSelectedListener listener =  new
AdapterView. OnItemSelectedListener( ) {
  public void onItemSelected( AdapterView < ?  > parent, View view,
```

```
                        int pos, long id) {
            Toast.makeText(getActivity(), String
    .valueOf(pos), Toast.LENGTH_SHORT).show();
            }
            public void onNothingSelected(AdapterView<?> parent) {
            Toast.makeText(getActivity(), "Nothing",
        Toast.LENGTH_SHORT).show();
              }
          };
          spinner.setOnItemSelectedListener(listener);
```

Spinner 控件应用示例的程序运行结果如图 4－23 所示。

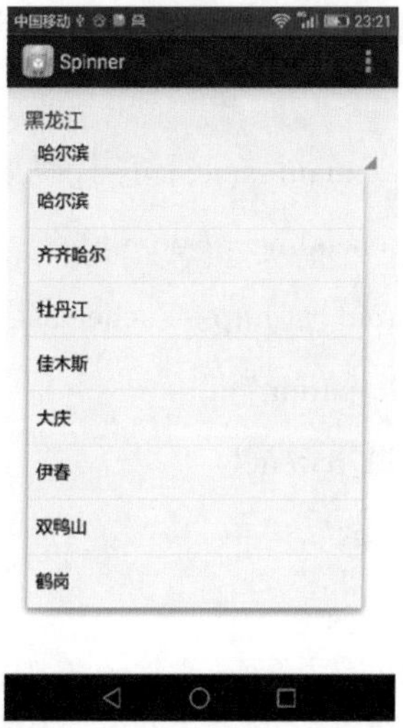

图 4－23　Spinner 控件应用示例的程序运行结果

4.8.2　列表组件 ListView 类

ListView 类是 Android 程序开发中经常用到的组件，该组件必须与适配器配合使用，由适配器提供显示样式和显示数据。

ListView 类的常用方法见表 4－16。

表 4－16　ListView 类的常用方法

常用方法	说明
ListView(Context context)	构造方法
setAdapter(ListAdapter adapter)	设置提供数组选项的适配器

续表 4－16

常用方法	说明
addHeaderView(View v)	设置列表项目的头部
addHeaderView(View v)	设置列表项目的底部
setOnItemClickListener(AdapterwntmClicklistener)	注册单击选项时执行的方法,该方法继承于父类 android. widget. AdapterView

【例 4－18】列表组件示例。

在界面设计中设置一个文本标签和一个列表组件 ListView,程序设计步骤如下:

(1)在界面布局文件中声明列表组件 ListView;

(2)在 Activity 中获得相关组件实例;

(3)通过触发列表的选项事件,调用 mClick 类的 onClick0 方法显示相应提示内容。

程序代码如下。

(1)设计界面布局文件 activity. main. xml,程序代码如下:

```
<? xml version ="1. 0" encoding ="utf－8"? >
<LinearLayout xmlns:android =" http://schemas. android. com/apk/res/android"
android:layout_width =" fill_parent"
android:layout_height =" fill_parent"
android:orientation =" vertical" >
<TextView
android:layout_width =" fill_parent"
android:layout_height =" wrap_content"
android:text =" 凤凰传奇"
android:textSize ="24sp" / >
<ListView
android:id =" @ + id/ListView01"
android:layout_height =" wrap_content"
android:layout_width =" fill_parent" / >
</LinearLayout >
```

(2)设计控制文件 MainActivity. java,程序代码如下:

```
package com. ex03_16;
import android. app. Activity;
import android. os. Bundle;
```

```
import android. view. View;
import android. widget. AdapterView;
import android. widget. AdapterView. OnItemClickListener;
import android. widget. ArrayAdapter;
import android. widget. Listview;
import android. widget. TextView;
import android. widget. Toast;
public class MainActivity extends Activity
{
ListView list;
@ Override
public void onCreate( Bundle savedInstanceState)
{
super. onCreate( savedInstanceState) ;
setContentView( R. layout. activity main) ;
list = ( ListView) findViewById( R. id. ListView01) ;
//定义数组
String[ ] data  = {
"(1)荷塘月色",
"(2)最炫民族风",
"(3)天蓝蓝",
"(4)最美天下",
"(5)自由飞翔",
};
//为 ListView 设置数组适配器 ArrayAdapter
list. setAdapter( new ArrayAdapter < String > ( this,
android. R. layout. simple_list_item_1 ,data) ) ;
//为 ListView 设置列表选项监听器
list. setonItemClickListener( new mItemClick( ) ) ;
}
//定义列表选项监听器的事件
class mItemClick implements OnItemClickListener
{
```

```
@ Override
    public void onItemClick( AdapterView < ?  > arg0, View arg1 , int arg2, long arg3)
{
Toast. makeText( MainActivity, "您选择的项目是:"        //提示信息
 + (  (TextView)  arg1). getText( ) ,Toast. LENGTH SHORT). show( ) ;
}
}
}
```

语句说明:

(1) android. R. layoutsimple_list_item_1 是一个 Android 系统内置的 ListView 布局方式。

①android. R. layout. simple_list_item_1:一行 text。

②android. R. layout. simple_ list_item_2:一行 title,一行 text。

③android. R. layout. simple_list_item_single_choice:单选按钮。

④android. R. layout. simple_list_item_multiple_choice:多选按钮。

(2) OnItemClickListener 是一个接口,用于监听列表组件选项的触发事件。

(3) Toast. makeText( ). show( ) 显示提示消息框。

列表组件示例的程序运行结果如图 4 – 24 所示。

图 4 – 24　列表组件示例的程序运行结果

4.8.3　列表组件 ListActivity 类

当整个 Activity 中只有一个 ListView 组件时,可以使用 ListActivity。其实,ListActivity 和一个只包含一个 ListView 组件的普通 Activity 没有太大区别,只是实现了一些封装而已。ListActivity 类继承于 Activity 类,默认绑定了一个 ListView 组件,并提供了一些与 ListView 处理相关的操作。

ListActivity 类常用的方法为 getListView( ),该方法返回绑定的 ListView 组件。

【例 4 –19】ListActivity 应用示例。

(1)设计界面布局文件 activity_main. xml,其代码如下:

```
<? xml version = "1.0" encoding = "utf - 8"? >
<LinearLayout xmlns:android = "http://schemas. android. com/apk/res/android"
android:layout_width = "fill_parent"
android:layout_height = "fill_parent"
android:orientation = "vertical" >
<ListView
android:id = "@ + id/android:list"
android:layout_height = "wrap_content"
android:layout_width = "fill_parent" />
</LinearLayout >
```

说明:ListActivity 布局文件中的 ListView 组件的 id 应设为"@ + id/android:list"。

(2)设计控制文件 MainActivityjava,其代码如下:

```
package com. ex03_17;
import android. app. ListActivity;
import android. os. Bundle;
import android. view. View;
import android. widget. AdapterView;
import android. widget. ArrayAdapter;
import android. widget. ListView;
import android. widget. TextView;
import android. widget. Toast;
import android. widget. AdapterView. OnItemClickListener;
public class MainActivity extends ListActivity
{
```

```
@ Override
public void onCreate ( Bundle savedInstancestate)
{
super. onCreate( savedInstanceState)
setContentView( R. layout. activity. main) ;
//定义数组
String[ ] data = {
"(1)荷塘月色",
"(2)最炫民族风",
"(3)天蓝蓝",
"(4)最美天下",
"(5)自由飞翔",
};
//获取列表项
ListView list = getListview( ) ;
//设置列表项的头部
TextView header = new TextView( this) ;
header. setText( "凤凰传奇经典歌曲") ;
header. setTextSize(24) ;
list. addHeaderView( header) ;
//设置列表项的底部
extView foot = new TextView( this) ;
foot. setText( "请选择") ;
foot. setTextsize(24) ;
list. addFooterView( foot) ;
setListAdapter( new ArrayAdapter < String > ( this,
android. R. layout. simple_list_item_1 ,data) ) ;
list. setOnItemClickListener( new mItemClick( ) ) ;
}
//定义列表选项监听器
class mItemClick implements OnItemClickListener
{
@ Override
```

```
public void onItemClick(AdapterView<?> arg0, View arg1, int arg2, long arg3)
{
Toast.makeText(getApplicationContext(),
"您选择的项目是:" + ((TextView)arg1).getText(),
Toast.LENGTH_SHORT).show();
}
}
}
```

ListActivity 应用示例的程序运行结果如图 4 - 25 所示。

图 4 - 25　ListActivity **应用示例的程序运行结果**

4.9　滑动抽屉组件

在日常生活中,当杂乱的物品很多时,可以把这些物品分类整理好放在不同的抽屉中,这样在使用物品时,打开抽屉,里面的东西便一目了然了。在 Android 系统中,也可以把多个程序放到一个应用程序的抽屉里。如图 4 - 26(a)所示,单击“向上”按钮(称为手柄),将打开抽屉;如图 4 - 26(b)所示,单击“向下”按钮,将关闭抽屉。

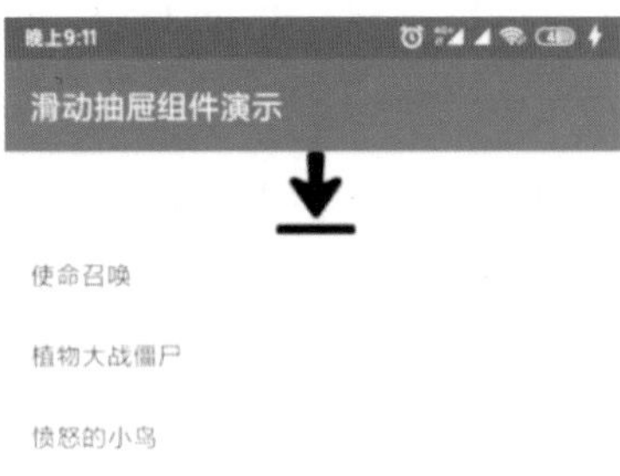

(a)单击“向上”按钮,将打开抽屉　　(b)单机“向下”按钮,将关闭抽屉

图 4－26　滑动抽屉示例

使用 Android 系统提供的 SlidingDraw 组件可以实现滑动抽屉的功能。先看一下 SlidingDraw 类的重要方法和属性,SlidingDraw 类重要的 XML 属性见表 4－17。

表 4－17　SlidingDraw 类重要的 XML 属性

属性	说明
android:allowSingleTap	设置通过手柄打开或关闭滑动抽屉
android:animateOnClick	单击手柄时,是否加入动画,默认为 true
android:handle	指定抽展的手柄 handle
android:content	隐藏在抽屉里的内容
android:orientation	滑动抽展内的对齐方式

SlidingDraw 类的重要方法见表 4－18。

表 4 – 18　SlidingDraw 类的重要方法

方法	说明
animateOpen( )	关闭抽实现动画
animateOpen( )	打开时实现动画
getContent( )	获取内容
getHandle( )	获取手柄
setOnDrawerOpenListener (SlidingDrawer. onDrawerOpenListener onDrawerOpenListener)	打开抽屉的监听器
setOnDrawerCloseListener (SlidingDrawer. onDrawerCloseListener onDrawerCloseListener)	关闭抽屉的监听器
setOnDrawerScrollListener (SlidingDrawer. onDrawerScrollListener onDrawerScrollListener)	打开/关闭切换时的监听器

【例 4 – 20】实现如图 4 – 26 所示滑动抽屉 SlidingDraw 组件的应用示例。

事先准备好两个图标文件，分别命名为 up. jpg 和 down. jpg，将它们复制到 resldrawabl – hdpi 目录下，以作滑动抽屉的手柄之用。

(1)设计界面布局文件 Activity_main. xml。

在 XML 文件中设置一个 SlidingDraw 组件，然后设置一个图标按钮 ImageButton 作为抽屉手柄。activity main. xml 的代码如下：

```
<Linearlayout xmlns:android = "http://schemas. android. com/apk/res/ android"
xmlns:tools = "http://schemas. android. com/tools"
android:id = "@ + id/Linearlayout1"
android:layout_width = "match_parent"
android:layout_height = "match_parent"
android:orientation = "vertical" >
<! – –设置 handle 和 content 的 id – – >
<SlidingDrawer
android:layout_width = "fill_parent"
android:layout_height = "fill_parent"
android:handle = "@ + id/handle"
android:content = "@ + id/content"
android:orientation = "vertical"
android:id = "@ + id/slidingdrawez" >
```

```
<! --设置 handle,就是用一个图标按钮来处理滑动抽屉事件-->
<ImageButton
android:id="@id/handle"
android:layout_width="50dip"
android:layout_height="44dip"
android:src="@drawable/up" />
<! --设置抽屉内容,当拖动抽屉时候就会看到-->
<LinearLayout
android:id="@id/content"
android:layout_width="fill_parent"
android:layout_height="fill_parent"
android:background="#66cccc"
android:focusable="true" >
</LinearLayout>
</SlidingDrawer>
</LinearLayout>
```

(2)设计控制文件 MainActivity.java。在控制程序 MainActivity.java 中主要是实现滑动抽屉的几个监听事件,程序代码如下:

```
package com.example.ex03_18;
import android.os.Bundle;
import android.app.Activity;
import android.widget.ArrayAdapter;
import android.widget.ImageButton;
import android.widget.LinearLayout;
import android.widget.ListView;
import android.widget.SlidingDrawer;
import android.widget.Toast;

public class MainActivity extends Activity
{
SlidingDrawer mDrawer;
ImageButton imgBtn;
ListView listView;
```

```
LinearLayout layout ;
String data[ ] = new string[ ]("E#B"" ,"HitHP" ,"t##IMYM" );
@ Override
public void onCreate( Bundle savedInstanceState)
{
super. onCreate( savedInstanceState) ;
setContentView( R. layout. activity,main) ;
layout = (LinearLayout) findViewById( R. id. content) ;
listView. new ListView( MainActivity. this) ;
listView. setAdapter( new ArrayAdapter < String > (
MainActivity. this,
android. R. layout. simple_expandable_list_item_1 ,
data) ) ;
layout. addView( listView) ;
imgBtn = (ImageButton) findViewById( R. id. handle) ;
mDrawer = (SlidingDrawer) findViewById( R. id. slidingdrawer) ;
mDrawer. setOnDrawerOpenListener( new mOpenListener( ) ) ;
mDrawer. setOnDrawerCloseListener( new mCloselistener( ) ) ;
mDrawer. setOnDrawerScrollListener( new mScrollListener( ) ) ;
}

class mOpenListener implements slidingDrawer. OnDrawerOpenListener
{
@ override
public void onDrawerOpened ( )                    //打开抽屉时处罚
{
imgBtn. setImageResource( R. drawable. down) ;
}
}

class mCloseListener implements SlidingDrawer. OnDrawerCloseListener
{
@ override
```

```
public void onDrawerClosed( )                    //关闭抽屉时处罚
{
imgBtn. setImageResource( R. drawable. up) ;
}
}

class mScrollListener implements SlidingDrawer. OnDrawerScrollListener
{
@ Override
public void onScrollEnded( )
{
Toast. makeText( MainActivity. this,"结束拖动", //关闭切换时触发
Toast. LENGTH SHORT). show( ) ;
}
@ Override
public void onScrollstarted( )
{                                                //打开切换时处罚
Toast. makeText( MainActivity. this,"窗口拖动开始",
Toast. LENGTH SHORT). show( ) ;
}
}
}
```

第 5 章　活动、意图与广播

5.1　活动与碎片

一个碎片（Fragment）表示活动（Activity）中用户界面的一个行为或是一部分。可以组合多个碎片到一个活动中，从而实现多区域 UI，也可以在多个活动中重用一个碎片。可以将碎片想象成活动的一个模块，它有自己的生命周期，接收自己的输入事件，当活动运行时，可以增加或删除它（有点像“子活动”，可以在多个不同的活动中重用它）。

碎片必须总是被嵌入到一个活动中，碎片的生命周期直接受到其宿主活动生命周期的影响。例如，当活动暂停时，活动中所有的碎片也处于暂停状态；而当活动被销毁时，所有的碎片也随之销毁。但是，当一个活动正在运行（即活动处于 resumed 生命周期状态）时，可以单独控制每个碎片，如增加或删除它们。当执行一个碎片事务处理时，其也可以将它添加到回退栈（back stack），回退栈由活动管理，在活动中的每一个回退栈条目均是一次碎片事务处理发生的记录。回退栈允许用户通过点击 Back 按钮来恢复一次碎片事务处理（即向后导航）。

当增加一个碎片作为其活动布局的一部分时，它就处于活动视图层次体系内部的视图组（ViewGroup）中，碎片定义了它自己的视图布局。可以通过在活动的布局文件中声明碎片来向活动中插入一个碎片，采用一个 <fragment> 标签元素来完成碎片声明，或者是在应用程序的代码中将碎片加入到一个当前的视图组（ViewGroup）。但是一个碎片不是一定要作为活动布局的一部分，它也可以为活动隐身工作。

5.1.1　设计原理

Android 在 Android 3.0（API level 11）中引入了碎片概念，主要为了在大屏幕（如平板电脑）上支持更加动态和灵活的 UI 设计。由于一个平台电脑的屏幕比手机屏幕大得多，因此有更大的空间去组合与互换 UI 组件。有了碎片之后，可以不必去管理视图层次体系的复杂变化。通过将活动的布局划分成碎片，可以改变活动在执行时的外观，并且在由活动管理的回退栈中保存这些变化。

例如，一个新闻应用程序能够使用一个碎片在左边显示一列文章，使用另一个碎片在右边显示一篇文章，两个碎片并排出现在一个活动中，每一个碎片都有自己的生命周期回调方法集合,并且控制自己的用户输入事件。这样,就不需要再用一个活动去选择文章,用另一个活动去阅读文章,用户能够在同一个活动内部选择并阅读文章。采用碎片在一个活动中设计两个 UI 模块如图 5－1 所示。

读者应该将每一个碎片设计成模块化、可重用的活动组件,这是因为每一个碎片利用自己的生命周期回调函数定义了它们自己的布局和行为,可以在多个活动中包含同一个碎片,因此应该考虑支持重用的设计,避免直接从一个碎片操作另一个碎片。这是非常重要的,因为一个模块化的碎片允许针对不同屏幕大小改变其自己的碎片组合。当设计应用去同时支持平板和手机时,可以基于可获得的屏幕空间,通过不同的布局配置重用其碎片,以此来优化用户体验。例如,在手机上可能必须将碎片组合分离以提供一个单独区域 UI,因为一个以上的窗口不适合在用一个活动中。

例如,继续以新闻应用程序为例,当它运行在平台电脑大小的设备上时,应用程序可以嵌入两个碎片在活动 A 中。但是,在一个手机大小的屏幕上,由于没有足够的空间容纳两个碎片,因此活动 A 只能包含显示文章列表的碎片,当用户选择一篇文章时,它启动活动 B,其包含用来阅读文章的碎片。这样,通过在不同的组合方案中重用碎片应用能够同时支持平板电脑和手机,如图 5－1 所示。

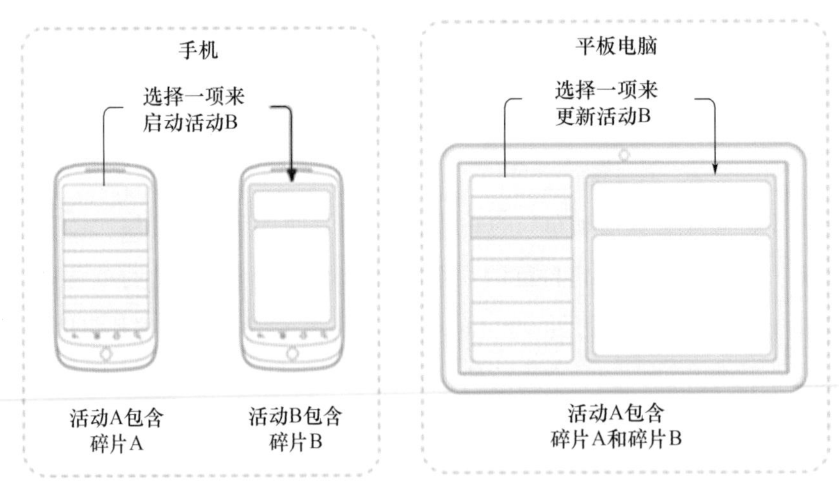

图 5－1 采用碎片在一个活动中设计两个 UI 模块

5.1.2 创建并添加碎片

要创建一个碎片,必须创建一个 Fragment(或是继承自它的子类)的子类。Fragment 类的代码看起来很像 Activity。它与活动一样都有回调函数,如 onCreate()、onStart()、onPause()和 onStop()。事实上,如果正在利用碎片来转化一个现成的 Android 应用,其可以简单地将活动回调函数中的代码分别移植到碎片的回调函数中。

一般来说，至少需要实现以下几个生命周期方法。

(1) onCreate( )。

在创建碎片时系统会调用此方法。在实现代码中，应该初始化其想要在碎片中保持的那些必要组件，当碎片处于暂停或者停止状态之后可重新启用它们。

(2) onCreateView( )。

在为碎片初次绘制用户界面时系统会调用此方法。为碎片绘制用户界面，这个函数必须要返回 View 对象，它是碎片布局的根。如果碎片不提供 UI，也可以让这个函数返回 null。

(3) onPause( )。

在用户正在离开碎片时，系统首先会调用此方法（当然这并不总是意味着碎片正在被销毁）。应该在这个方法中提交一些在当前用户会话之外应该被持久化保存的变化（因为用户可能不再返回）。

大部分应用都应该为每个碎片至少实现这三个方法，此外还有一些其他的回调方法，这些方法也应该实现，它们可以帮助控制碎片生命周期的各个阶段。碎片的生命周期如图 5－2 所示。

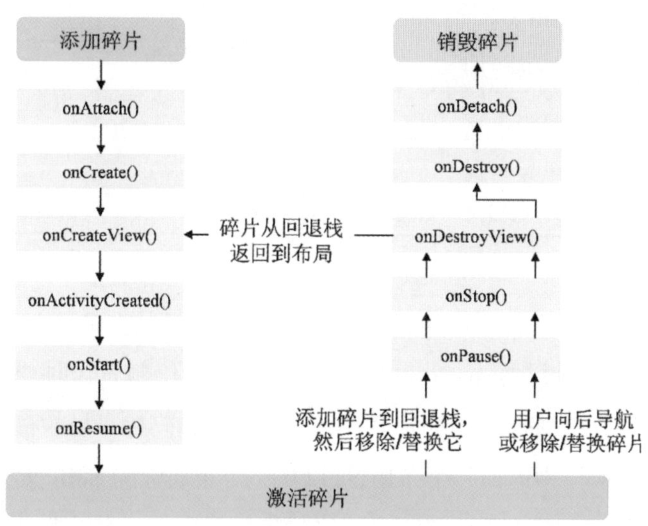

图 5－2　碎片的生命周期

与活动一样，碎片也有以下三种状态。

(1) Resumed。碎片在运行中的活动可见。

(2) Paused。另一个活动处于前台且获得焦点，但这个碎片的宿主活动仍然可见。

(3) Stopped。碎片不可见（可能是宿主活动已经停止，也可能是碎片已经从活动上移除，但已经被添加到回退栈）。

如图 5－2 所示，除 onCreate( )、onStart( ) 和 onPause( ) 等与活动类似的回调方法外，碎

片还有以下一些额外的生命周期回调方法。

①onAttach( )。当碎片被绑定到活动时被调用。

②onActivityCreated( )。当活动的 onCreate( )函数返回时被调用。

③onDestroyView( )。当与碎片关联的视图层次体系正被移除时被调用。

④onDetach( )。当碎片正与活动解除关联时被调用

碎片的生命周期实际上受其宿主活动影响。宿主活动生命周期对碎片生命周期的影响见表 5－1,可以看出活动的状态是如何决定碎片可能接收到哪个回调方法的。例如,当 activity 接收到 onPause( )时,这个 activity 之中的每个 fragment 都会接收到 onPause( );当活动处于 resumed 状态时,可以在活动中自由的添加或者移除碎片。因此,只有当活动处于 resumed 状态时,碎片的生命周期才可以独立变化。

表 5－1　宿主活动生命周期对碎片生命周期的影响

活动状态	碎片回调方法
Created	onAttach( )、onCreate( )、onCreateView、onActivityCreated( )
Started	onStart( )
Resumed	onResume( )
Paused	onPause( )
Stopped	onStop( )
Destroyed	onDestroyView( )、onDestroy( )、onDetach( )

碎片经常被用来作为活动用户界面的一部分,其将自己的布局添加到活动中。为向碎片提供布局,必须实现回调方法 onCreateView( ),当为碎片绘制布局时,Android 系统就调用这个方法。这个方法的实现必须返回一个 View 对象,这是碎片布局的根。

为从 onCreateView( )中返回一个布局,可以从一个布局资源 XML 文件中加载它,onCreateView( )提供了一个 LayoutInflater 对象。例如,有一个 Fragment 子类,从 one_fragment. xml 文件中载入一个布局:

```
public static class OneFragment extends Fragment {
public View onCreateView(LayoutInflater inflater, ViewGroup container,
                                    Bundle savedInstanceState) {
        // Inflate the layout for this fragment
        return inflater. inflate(R. layout. one_fragment, container, false);
    }
}
```

在上面这个例子中,参数 container 是父视图组(来自于活动布局),碎片布局就被添加到其中。参数 savedInstanceState 是 Bundle 类型的对象,如果碎片当前处于 resumed 状态,那么它提供先前碎片实例的数据。

函数 inflate()的三个参数含义如下。

(1)参数 1 表示想要 inflate 的布局的资源 ID。此处,R. layout. one_fragment 是一个布局资源的引用,这个布局资源以名字 one_fragment. xml 存放在应用程序资源中。

(2)参数 2 表示被 inflate 的布局的父视图组 ViewGroup。此处,传入 container 很重要,这是为了让系统将布局参数应用到被 inflate 的布局的根视图中,由其将要嵌入的父视图指定。

(3)参数 3 表示在 inflate 期间被 inflate 的布局是否应该附上 ViewGroup(第二个参数),是一个布尔值。此处,传入的是 false,因为系统已经将被 inflate 的布局插入到容器中(container),传入 true 会在最终的布局里创建一个多余的 ViewGroup。

一般来说,碎片构建了其宿主活动的部分界面,它被作为活动整体视图体系的一部分而嵌入进去。在活动布局中添加碎片有两种方法:静态文件添加和动态代码添加。

1. 在活动的布局文件里声明碎片

在这种情况下,可以像为视图一样为碎片指定布局属性。例如,下面是在一个活动中包含两个碎片的布局文件:

```
<?xml version="1.0" encoding="utf-8"?>
<LinearLayout xmlns:android="http://schemas.android.com/apk/res/android"
android:orientation="horizontal"
android:layout_width="match_parent"
android:layout_height="match_parent">
<fragment android:name="edu.hrbust.news.ArticleListFragment"
android:id="@+id/list"
android:layout_weight="1"
android:layout_width="0dp"
android:layout_height="match_parent" />
<fragment android:name=" edu.hrbust.news.ArticleReaderFragment"
android:id="@+id/viewer"
android:layout_weight="2"
android:layout_width="0dp"
android:layout_height="match_parent" />
</LinearLayout>
```

标签元素 <fragment> 中的 android:name 属性指定了布局中实例化的 Fragment 类。当系统创建活动布局时,它实例化了布局文件中指定的每一个碎片,并为它们调用 onCreateView( ) 函数,以获取每一个碎片的布局。系统直接在 <fragment> 元素的位置插入碎片返回的 View 对象。

注意:每个碎片都需要一个唯一的标识,如果重启活动,系统可用其来恢复碎片(并且可用来捕捉碎片的事务处理,如移除碎片)。为碎片提供 ID 有以下三种方法:

(1)用 android:id 属性提供一个唯一的标识;

(2)用 android:tag 属性提供一个唯一的字符串;

(3)如果上述两个属性都没有,系统会使用其容器视图的 ID。

2. 通过编码将碎片添加到已存在的 ViewGroup 中

在活动运行的任何时候,都可以将碎片添加到活动布局中,只需要简单指定用来放置碎片的 ViewGroup。

必须使用 FragmentTransaction 的 API 来对活动中的碎片进行事务处理(如添加、移除或者替换碎片)。可以像下面这样从 Activity 中取得 FragmentTransaction 的实例:

```
FragmentManager fragmentManager = getFragmentManager( );
FragmentTransaction fragmentTransaction = fragmentManager. beginTransaction( );
```

FragmentManager 可以帮助对活动中的碎片进行管理,使用 FragmentManager 还可以做以下事情:

(1)使用 findFragmentById( )或者 findFragmentByTag( )获取活动中存在的碎片;

(2)使用 popBackStack( )从后退栈中弹出碎片;

(3)使用 addOnBackStackChangedListener( )注册一个监听后退栈变化的监听器。

可以用 add( )函数添加碎片,并指定要添加的碎片以及要将其插入到哪个视图之中:

```
OneFragment fragment = new OneFragment ( );
fragmentTransaction. add(R. id. fragment_container, fragment);
fragmentTransaction. commit( );
```

在 add( )函数中,参数 1 是碎片被放置的 ViewGroup,它由资源 ID 指定;参数 2 是要添加的碎片。只要通过 FragmentTransaction 做了更改,就应当调用 commit( )方法使变化生效。

此外,还可以添加无界面的碎片。上面的例子是如何将碎片添加到活动中,目的是提供一个用户界面。然而,也可以使用碎片为活动提供后台动作,却不呈现多余的用户界面。

想要添加没有界面的碎片,可以使用 add(Fragment, String)函数(为碎片提供一个唯一的字符串"tag",而不是视图 ID),这样就添加了碎片。但是,因为它没有关联到活动布局中的视图,所以就接收不到对 onCreateView( )的调用,因此就不需要实现这个方法。

为无界面碎片提供字符串标签并不是专门针对无界面碎片的,也可以为有界面碎片提供字符串标签,但是对于无界面碎片,字符串标签是识别它的唯一方法。如果之后想从活动中取到碎片,则需要使用函数 findFragmentByTag( )。

5.1.3　处理碎片事物

在活动中使用碎片的一大特点是具有添加、删除、替换以及利用它们执行其他动作,以响应用户交互的能力。提交给活动的每一系列变化称为事务,并且可以用 FragmentTransaction 中的 APIs 处理。也可以将每一个事务保存在由活动管理的后退栈中,并且允许用户导航回退碎片变更(类似于活动的导航回退)。

每项事务是在同一时间内要执行的一系列的变更。可以为一个给定的事务用相关方法设置想要执行的所有变化,如 add( )、remove( )和 replace( ),然后用 commit( )将事务提交给活动。

然而,在调用 commit( )之前,为将事务添加到碎片事务后退栈中,可能会想调用 addToBackStatck( )。这个后退栈由活动管理,并且允许用户通过按 BACK 键回退到前一个碎片状态。

例如,下面的代码是如何使用另一个碎片代替一个碎片,并且将之前的状态保留在后退栈中:

```
//创建一个新的碎片和事务
Fragment newFragment = new OneFragment ( );
FragmentTransaction transaction = getFragmentManager( ). beginTransaction( );
//用新的碎片替换另一个碎片,将事务添加到回退栈
transaction. replace( R. id. fragment_container, newFragment);
transaction. addToBackStack( null);
//提交事务
transaction. commit( );
```

在这个例子中,newFragment 替换了当前在布局容器中用 R. id. fragment_container 标识的所有的碎片(如果有的话),替代的事务被保存在后退栈中,因此用户可以回退该事务,可通过按 BACK 键还原之前的碎片。

如果添加多个变更事务(如另一个 add( )或者 remove( ))并调用 addToBackStack( ),那么在调用 commit( )之前的所有应用的变更作为一个单独的事务添加到后台栈中,并且 BACK 键可以将它们一起回退。

将变更添加到 FragmentTransaction 中的顺序注意以下两点:

(1)必须要在最后调用 commit( );

(2)如果正将多个碎片添加到同一个容器中,那么添加顺序决定了它们在视图体系里

显示的顺序。

在执行删除碎片事务时,如果没有调用 addToBackStack( ),那么事务一提交,碎片就会被销毁,而且用户也无法回退它。然而,当移除一个碎片时,如果调用了 addToBackStack( ),那么之后碎片会被停止;如果用户回退,它将被恢复过来。

5.2 使用 Intent 链接 Activity

Intent 是一个动作的完整描述,包含了动作的产生组件、接收组件和传递的数据信息。Intent 也可称为一个在不同组件之间传递的消息,这个消息在到达接收组件后,接收组件会执行相关的动作。Intent 为 Activity、Service 和 BroadcastReceiver 等组件提供了交互能力。Intent 的用途包括:启动 Activity 和 Service;在 Android 系统上发布广播消息。其中,广播消息可以是接收到特定数据或消息,也可以是手机的信号变化或电池的电量过低等信息。

5.2.1 启动 Activity

在 Android 系统中,应用程序一般都有多个 Activity,Intent 可以实现不同 Activity 之间的切换和数据传递。启动 Activity 方式有以下两种。

(1)显式启动。必须在 Intent 中指明启动的 Activity 所在的类,或者指明启动的 Activity 的意图筛选器的名称。前者用于在同一个应用程序内部实现活动调用,后者用于在不同的应用程序之间实现活动调用。

(2)隐式启动。Android 系统根据 Intent 的动作和数据来决定启动哪一个 Activity。也就是说,在隐式启动时,Intent 中只包含需要执行的动作和所包含的数据,而无须指明具体启动哪一个 Activity,选择权由 Android 系统和最终用户来决定。

1. 显示启动 Activity

在同一个应用程序内部,使用 Intent 显式启动 Activity 的步骤如下:

(1)创建一个 Intent;

(2)指定当前的应用程序上下文以及要启动的 Activity;

(3)把创建好的这个 Intent 作为参数传递给 startActivity( )方法。

具体实现代码如下:

```
Intent intent = new Intent(this, ActivityToStart.class);
startActivity(intent);
```

下面用 IntentDemo 示例说明如何使用 Intent 启动新的 Activity。IntentDemo 示例包含两个 Activity,分别是 IntentDemoActivity 和 NewActivity。程序默认启动的 Activity 是 IntentDemoActivity,在用户点击“启动本程序内部的 Activity”按钮后,程序启动的 Activity 是 NewActivity。

在 IntentDemo 示例中使用了两个 Activity，因此需要在 AndroidManifest. xml 文件中注册这两个 Activity。注册 Activity 应使用 < activity > 标签，嵌套在 < application > 标签内部。

AndroidManifest. xml 文件代码如下：

```
<? xml version = "1.0" encoding = "utf - 8"? >
<manifest xmlns:android = "http://schemas. android. com/apk/res/android"
package = "edu. hrbust. IntentDemo"
android:versionCode = "1"
android:versionName = "1.0" >
<application android:icon = "@drawable/icon" android:label = "@string/app_name" >
<activity android:name = ". IntentDemo"
android:label = "@string/app_name" >
<intent - filter >
<action android:name = "android. intent. action. MAIN" / >
<category android:name = "android. intent. category. LAUNCHER" / >
</intent - filter >
</activity >
<activity android:name = ". NewActivity"
android:label = " NewActivity" >
<intent - filter >
<action android: name = "edu. hrbust.  IntentDemo. NewActivity" / >
<category android:name = "android. intent. category. DEFAULT" / >
</intent - filter >
</activity >
</application >
<uses - sdk android:minSdkVersion = "14" / >
</manifest >
```

Android 应用程序中，用户使用的每个组件都必须在 AndroidManifest. xml 文件中的 < application > 节点内定义。在上面的代码中，< application > 节点下共有两个 < activity > 节点，分别代表应用程序中所使用的两个 Activity，即 IntentDemoActivity 和 NewActivity。

在上面 IntentDemo 应用程序中的 AndroidManifest. xml 文件代码中对 NewActivity 进行如下定义：

(1)". NewActivity"是 NewActivity 的类名；

(2)" NewActivity"是 NewActivity 的标签名称;

(3)"edu. hrbust. IntentDemo. NewActivity"是 NewActivity 的意图筛选器的名称;

(4)"android. intent. category. DEFAULT"是 NewActivity 的意图筛选器的类别。

IntentDemoActivity. java 文件中包含了使用 Intent 启动 Activity 的核心代码:

```
Button button = (Button)findViewById(R. id. btn);
button. setOnClickListener(new OnClickListener(){
public void onClick(View view){
Intent intent = new Intent(IntentDemoActivity. this, NewActivity. class);
startActivity(intent);
}
});
```

在点击事件的处理函数中,Intent 构造函数的第 1 个参数是应用程序上下文,在这里就是 IntentDemoActivity;第 2 个参数是接收 Intent 的目标组件,这里使用的是显式启动方式,直接指明了需要启动的 Activity。

在不同的应用程序之间,使用 Intent 显式启动 Activity 的步骤如下:

(1)创建一个 Intent;

(2)将要启动的 Activity 的意图筛选器的名称作为创建 Intent 时的参数;

(3)把创建好的这个 Intent 作为参数传递给 startActivity()方法。

具体实现代码如下:

```
Intent intent = new Intent("ActivityToStart 的意图筛选器名称");
startActivity(intent);
```

下面用 IntentModifyDemo 示例说明如何在 IntentModifyDemo 应用程序中启动 Intent-Demo 应用程序中的 NewActivity,实现跨应用程序的 Activity 调用。

程序默认启动的 Activity 是 IntentModifyDemo 应用程序中的 IntentModifyDemoActivity ,在用户点击"启动外部应用程序中的 Activity"按钮后,程序启动的 Activity 是 IntentDemo 应用程序中的 NewActivity。

IntentModifyDemoActivity. java 文件中包含了使用 Intent 启动 Activity 的核心代码:

```
Button button = (Button)findViewById(R. id. btn);
button. setOnClickListener(new OnClickListener(){
public void onClick(View view){
Intent intent = new Intent("edu. hrbust. IntentDemo. NewActivity");
startActivity(intent);
}
```

```
7   });
```

在点击事件的处理函数中，字符串"edu. hrbust. IntentDemo. NewActivity"是要启动的活动的意图筛选器的名称，IntentModifyDemoActivity 将通过这个名称来调用 NewActivity。在实际开发过程中，应该使用公司的反向域名作为意图筛选器的名称，以减少与另外一个应用程序具有相同意图筛选器名称的可能性。

2. 隐式启动 Activity

隐式启动的好处在于不需要指明需要启动哪一个 Activity，而由 Android 系统来决定，这样有利于降低组件之间的耦合度。选择隐式启动 Activity，Android 系统会在程序运行时解析 Intent，并根据一定的规则对 Intent 和 Activity 进行匹配，使 Intent 上的动作、数据与 Activity 完全吻合。

匹配的组件可以是程序本身的 Activity，也可以是 Android 系统内置的 Activity，还可以是第三方应用程序提供的 Activity。因此，这种方式强调了 Android 组件的可复用性。

如果程序开发人员希望启动一个浏览器，查看指定的网页内容，却不能确定具体应该启动哪一个 Activity，则可以使用 Intent 的隐式启动方式，由 Android 系统在程序运行时决定具体启动哪一个应用程序的 Activity 来接收这个 Intent。程序开发人员可以将浏览动作和 Web 地址作为参数传递给 Intent，Android 系统则通过匹配动作和数据格式找到最适合于此动作和数据格式的组件。

具体实现代码如下：

```
Intent intent = new Intent(Intent.ACTION_VIEW, Uri.parse(urlString));
startActivity(intent);
```

Intent 构造函数的第 1 个参数是 Intent 需要执行的动作；第 2 个参数是 URI，表示需要传递的数据。在上面的代码中，Intent 的动作是 Intent. ACTION_VIEW，数据是 Web 地址，使用 Uri. parse(urlString)方法，可以简单地把一个字符串解释成 Uri 对象。

Android 系统在匹配 Intent 时，首先根据动作 Intent. ACTION_VIEW 得知需要启动具备浏览功能的 Activity，但具体是浏览电话号码还是浏览网页，还需要根据 URI 的数据类型来做最后判断。如果利用"http://www. hrbust. edu. cn"为 urlString 赋值，那么因为数据提供的是 Web 地址，所以最终可以判定 Intent 需要启动具有网页浏览功能的 Activity。在缺省情况下，Android 系统会调用内置的 Web 浏览器。

WebViewIntentDemo 示例说明了如何隐式启动 Activity，WebViewIntentDemo 示例执行效果如图 5 - 3 所示。当用户在文本框中输入 Web 地址后，通过点击"浏览此 URL"按钮，程序根据用户输入的 Web 地址生成一个 Intent，并以隐式启动的方式调用 Android 内置的 Web 浏览器，并打开指定的 Web 页面。本例输入的 Web 地址为 http://www. hrbust. edu. cn。

图 5－3　WebViewIntentDemo 示例执行效果

5.2.2　在 Activity 之间传递数据

在上一节的 IntentDemo 示例中，通过 startActivity(Intent) 方法启动 Activity，启动后的两个 Activity 之间相互独立，没有任何的关联。但在很多情况下，后启动的 Activity 是为了让用户对特定信息进行选择，在后启动的 Activity 关闭时，这些信息需要返回给先前启动的 Activity。后启动的 Activity 称为为"子 Activity"，先启动的 Activity 称为"父 Activity"。如果需要将子 Activity 的信息返回到父 Activity，则可以使用 Sub－Activity 的方式去启动子Activity。

获取子 Activity 的返回值，一般可以分为以下三个步骤：

①以 Sub－Activity 的方式启动子 Activity；

②设置子 Activity 的返回值；

③在父 Activity 中获取返回值。

下面详细介绍每一个步骤的过程和代码实现。

以 Sub－Activity 方式启动子 Activity，需要调用 startActivityForResult(Intent, requestCode) 函数，参数 Intent 用于决定启动哪个 Activity，参数 requestCode 是请求码。因为所有子 Activity 返回时，父 Activity 都调用相同的处理函数，所以父 Activity 使用 requestCode 来确定数据是哪一个子 Activity 返回的。

(1) 以 Sub－Activity 的方式启动子 Activity。

①显式启动子 Activity 的代码如下：

```
int SUBACTIVITY1 = 1;
Intent intent = new Intent(this, SubActivity1.class);
// Intent intent = new Intent(SubActivity1 的意图筛选器的名称);
startActivityForResult(intent, SUBACTIVITY1);
```

②隐式启动子 Activity 的代码如下:

```
int SUBACTIVITY2 = 2;
Uri uri = Uri.parse("content://contacts/people");
Intent intent = new Intent(Intent.ACTION_PICK, uri);
startActivityForResult(intent, SUBACTIVITY2);
```

(2)设置子 Activity 的返回值。

在子 Activity 调用 finish()函数关闭前,调用 setResult()函数设定需要返回给父 Activity 的数据。setResult()函数有两个参数:一个是结果码,一个是返回值。结果码表明了子 Activity 的返回状态,通常为 Activity.RESULT_OK(正常返回数据)或者 Activity.RESULT_CANCELED(取消返回数据),也可以是自定义的结果码,结果码均为整数类型。

返回值封装在 Intent 中,也就是说子 Activity 通过 Intent 将需要返回的数据传递给父 Activity。数据主要以 Uri 形式返回给父 Activity,此外还可以附加一些额外信息,这些额外信息用 Extra 的集合表示。以下代码说明了如何在子 Activity 中设置返回值:

```
Uri data = Uri.parse("tel:" + tel_number);
Intent result = new Intent(null, data);
setResult(RESULT_OK, result);
finish();
```

(3)在父 Activity 中获取返回值。

当子 Activity 关闭后,父 Activity 会调用 onActivityResult()函数,用于获取子 Activity 的返回值。如果需要在父 Activity 中处理子 Activity 的返回值,则重载此函数即可。onActivityResult()函数的语法如下:

```
public void onActivityResult(int requestCode, int resultCode, Intent data);
```

其中,第 1 个参数 requestCode 是请求码,用来判断第 3 个参数是哪一个子 Activity 的返回值;resultCode 用于表示子 Activity 的数据返回状态;data 是子 Activity 的返回数据,返回数据类型是 Intent。返回数据的用途不同,Uri 数据的协议也不同,也可以使用 Extra 方法返回一些原始类型的数据。

以下代码说明如何在父 Activity 中处理子 Activity 的返回值:

```
private static final int SUBACTIVITY1 = 1;
private static final int SUBACTIVITY2 = 2;
```

```
3  @Override
4  public void onActivityResult(int requestCode, int resultCode, Intent data){
5  Super.onActivityResult(requestCode, resultCode, data);
6  switch(requestCode){
7  case SUBACTIVITY1:
8  if (resultCode == Activity.RESULT_OK){
9  Uri uriData = data.getData();
10  }else if (resultCode == Activity.RESULT_CANCEL){
11  }
12  break;
13  case SUBACTIVITY2:
14  if (resultCode == Activity.RESULT_OK){
15  Uri uriData = data.getData();
16  }
17  break;
18  }
19  }
```

在上面代码中,第 1 行和第 2 行是两个子 Activity 的请求码,在第 7 行对请求码进行匹配。第 8 行和第 10 行对结果码进行判断,如果返回的结果码是 Activity. RESULT_OK,则在代码的第 9 行使用 getData( )函数获取 Intent 中的 Uri 数据;如果返回的结果码是 Activity. RESULT_CANCELED,则放弃所有操作。

ActivityCommunication 示例说明了如何以 Sub - Activity 方式启动子 Activity,以及如何使用 Intent 进行组件间通信。当用户点击“启动 Activity1”和“启动 Activity2”按钮时,程序将分别启动子 SubActivity1 和 SubActivity2。

SubActivity1 提供了一个输入框,以及“接受”和“撤销”两个按钮。如果在输入框中输入信息后点击“接受”按钮,程序会把输入框中的信息传递给其父 Activity,并在父 Activity 的界面上显示;如果用户点击“撤销”按钮,则程序不会向父 Activity 传递任何信息。

SubActivity2 主要目的是说明如何在父 Activity 中处理多个子 Activity,因此仅提供了用于关闭 SubActivity2 的“关闭”按钮。

ActivityCommunication 示例。父 Activity 的代码在 ActivityCommunication. java 文件中,界面布局在 main. xml 中;两个子 Activity 的代码分别在 SubActivity1. java 和 SubActivity2. java 文件中,界面布局分别在 subactivity1. xml 和 subactivity2. xml 中。

ActivityCommunicationActivity. java 文件的核心代码如下:

```
public class ActivityCommunicationActivity extends Activity {
private static final int SUBACTIVITY1 = 1;
private static final int SUBACTIVITY2 = 2;
TextView textView;
@ Override
public void onCreate(Bundle savedInstanceState) {
super. onCreate(savedInstanceState);
setContentView(R. layout. main);
textView = (TextView)findViewById(R. id. textShow);
final Button btn1 = (Button)findViewById(R. id. btn1);
final Button btn2 = (Button)findViewById(R. id. btn2);
btn1. setOnClickListener(new OnClickListener() {
public void onClick(View view) {
Intent intent = new Intent(ActivityCommunication. this, SubActivity1. class);
startActivityForResult(intent, SUBACTIVITY1);
}
});
btn2. setOnClickListener(new OnClickListener() {
public void onClick(View view) {
Intent intent = new Intent(ActivityCommunication. this, SubActivity2. class);
startActivityForResult(intent, SUBACTIVITY2);
}
});
}
@ Override
protected void onActivityResult(int requestCode, int resultCode, Intent data) {
super. onActivityResult(requestCode, resultCode, data);
switch(requestCode) {
case SUBACTIVITY1:
```

```
34 if (resultCode == RESULT_OK){
35 Uri uriData = data.getData();
36 textView.setText(uriData.toString());
37 }
38 break;
39 case SUBACTIVITY2:
40 break;
41 }
42 }
43 }
```

在代码的第 2 行和第 3 行分别定义了两个子 Activity 的请求码。在代码的第 16 行和第 23 行以 Sub－Activity 的方式分别启动两个子 Activity。代码第 29 行是子 Activity 关闭后的返回值处理函数,其中 requestCode 是子 Activity 返回的请求码,与第 2 行和第 3 行定义的两个请求码相匹配;resultCode 是结果码,在代码第 32 行对结果码进行判断,如果等于 RESULT_OK,则在第 35 行代码获取子 Activity 返回值中的数据;data 是返回值,子 Activity 需要返回的数据就保存在 data 中。

SubActivity1. java 的核心代码如下:

```
public class SubActivity1 extends Activity {
@Override
public void onCreate(Bundle savedInstanceState) {
super.onCreate(savedInstanceState);
setContentView(R.layout.subactivity1);
final EditText editText = (EditText)findViewById(R.id.edit);
Button btnOK = (Button)findViewById(R.id.btn_ok);
Button btnCancel = (Button)findViewById(R.id.btn_cancel);

btnOK.setOnClickListener(new OnClickListener(){
public void onClick(View view){
String uriString = editText.getText().toString();
Uri data = Uri.parse(uriString);
Intent result = new Intent(null, data);
setResult(RESULT_OK, result);
finish();
```

```
17 }
18 });
19
20 btnCancel.setOnClickListener(new OnClickListener() {
21 public void onClick(View view) {
22 setResult(RESULT_CANCELED, null);
23 finish();
24 }
25 });
26 }
27 }
```

代码第13行将EditText控件的内容作为数据保存在Uri中,并在第14行代码中构造Intent。在第15行代码中,RESUIT_OK作为结果码,通过调用setResult()函数,将result设定为返回值。最后,在代码第16行调用finish()函数关闭当前的子Activity。

SubActivity2.java的核心代码如下:

```
1 public class SubActivity2 extends Activity {
2 @Override
3 public void onCreate(Bundle savedInstanceState) {
4 super.onCreate(savedInstanceState);
5 setContentView(R.layout.subactivity2);
6
7 Button btnReturn = (Button)findViewById(R.id.btn_return);
8 btnReturn.setOnClickListener(new OnClickListener() {
9 public void onClick(View view) {
10 setResult(RESULT_CANCELED, null);
11 finish();
12 }
13 });
14 }
15 }
```

在SubActivity2的代码中,第10行的setResult()函数仅设置了结果码,第2个参数为null,表示没有数据需要传递给父Activity。

5.3 Intent 过滤器

隐式启动 Activity 时，并没有在 Intent 中指明 Activity 所在的类，因此 Android 系统一定存在某种匹配机制，使 Android 系统能够根据 Intent 中的数据信息找到需要启动的 Activity。这种匹配机制是依靠 Android 系统中的 Intent 过滤器（Intent Filter）来实现的。

Intent 过滤器是一种根据 Intent 中的动作（Action）、类别（Category）和数据（Data）等内容对适合接收该 Intent 的组件进行匹配和筛选的机制。

Intent 过滤器可以匹配数据类型、路径和协议，还可以确定多个匹配项顺序的优先级（Priority）。

应用程序的 Activity、Service 和 BroadcastReceiver 组件都可以注册 Intent 过滤器。这样，这些组件在特定的数据格式上可以产生相应的动作。

为使组件能够注册 Intent 过滤器，通常在 AndroidManifest. xml 文件的各个组件下定义 < intent – filter > 节点，然后在 < intent – filter > 节点中声明该组件所支持的动作、执行的环境和数据格式等信息。当然，也可以在程序代码中动态地为组件设置 Intent 过滤器。

< intent – filter > 节点支持 < action > 标签、< category > 标签和 < data > 标签，分别用来定义 Intent 过滤器的“动作”“类别”和“数据”。< intent – filter > 节点支持的标签、属性和说明见表 5 – 2。

表 5 – 2　< intent – filter > 节点支持的标签、属性和说明

标签	属性	说明
< action >	android:name	指定组件所能响应的动作，用字符串表示，通常由 Java 类名和包的完全限定名构成
< category >	android:category	指定以何种方式去服务 Intent 请求的动作
< data >	android:host	指定一个有效的主机名
	android:mimetype	指定组件能处理的数据类型
	android:path	有效的 URI 路径名
	android:port	主机的有效端口号
	android:scheme	所需要的特定协议

< category > 标签用来指定 Intent 过滤器的服务方式，每个 Intent 过滤器可以定义多个 < category > 标签，程序开发人员可以使用自定义的类别，或使用 Android 系统提供的类别。

Android 系统提供的类别见表 5 – 3。

表 5 – 3　Android 系统提供的类别

值	说明
ALTERNATIVE	Intent 数据默认动作的一个可替换的执行方法
SELECTED_ALTERNATIVE	与 ALTERNATIVE 类似,但替换的执行方法不是指定的,而是被解析出来的
BROWSABLE	声明 Activity 可以由浏览器启动
DEFAULT	为 Intent 过滤器中定义的数据提供默认动作
HOME	设备启动后显示的第一个 Activity
LAUNCHER	在应用程序启动时首先被显示

Intent 到 Intent 过滤器的映射过程称为"Intent 解析"。Intent 解析可以在所有的组件中找到一个可以与请求的 Intent 达成最佳匹配的 Intent 过滤器。Android 系统中 Intent 解析的匹配规则如下。

(1) Android 系统把所有应用程序包中的 Intent 过滤器集合在一起,形成一个完整的 Intent 过滤器列表。

(2) 在 Intent 与 Intent 过滤器进行匹配时,Android 系统会将列表中所有 Intent 过滤器的"动作"和"类别"与 Intent 进行匹配,任何不匹配的 Intent 过滤器都将被过滤掉。没有指定"动作"的 Intent 过滤器可以匹配任何的 Intent,但是没有指定"类别"的 Intent 过滤器只能匹配没有"类别"的 Intent。

(3) 把 Intent 数据 Uri 的每个子部与 Intent 过滤器 <data> 标签中的属性进行匹配,如果 <data> 标签指定了协议、主机名、路径名或 MIME 类型,那么这些属性都要与 Intent 的 Uri 数据部分进行匹配,任何不匹配的 Intent 过滤器均被过滤掉。

(4) 如果 Intent 过滤器的匹配结果多于一个,则可以根据在 <intent - filter> 标签中定义的优先级标签对 Intent 过滤器进行排序,优先级最高的 Intent 过滤器将被选择。

IntentResolutionDemo 示例说明了如何在 AndroidManifest. xml 文件中注册 Intent 过滤器,以及如何设置 <intent - filter> 节点属性来捕获指定的 Intent。

AndroidManifest. xml 的完整代码如下:

```
<?xml version="1.0" encoding="utf-8"?>
<manifest xmlns:android="http://schemas.android.com/apk/res/android"
package="edu.brbust.IntentResolutionDemo"
```

```
4   android:versionCode = "1"
5   android:versionName = "1.0" >
6    <application android:icon = "@drawable/icon" android:label = "@string/app_name" >
7    <activity android:name = ".IntentResolutionDemo"
8   android:label = "@string/app_name" >
9    <intent-filter>
10   <action android:name = "android.intent.action.MAIN" />
11   <category android:name = "android.intent.category.LAUNCHER" />
12   </intent-filter>
13   </activity>
14   <activity android:name = ".ActivityToStart"
15  android:label = "@string/app_name" >
16   <intent-filter>
17   <action android:name = "android.intent.action.VIEW" />
18   <category android:name = "android.intent.category.DEFAULT" />
19   <data android:scheme = "schemodemo" android:host = "edu.brbust" />
20   </intent-filter>
21   </activity>
22   </application>
23   <uses-sdk android:minSdkVersion = "14" />
24   </manifest>
```

在代码的第 7 行和第 14 行分别定义了两个 Activity。第 9 ~ 12 行是第 1 个 Activity 的 Intent 过滤器,动作是 android. intent. action. MAIN,类别是 android. intent. category. LAUNCHER。由此可知,这个 Activity 是应用程序启动后显示的缺省用户界面。第 16 ~ 20 行是第 2 个 Activity 的 Intent 过滤器,过滤器的动作是 android. intent. action. VIEW,表示根据 Uri 协议,以浏览的方式启动相应的 Activity,类别是 android. intent. category. DEFAULT,表示数据的默认动作,数据的协议部分是 android:scheme = "schemodemo",数据的主机名称部分是 android:host = "edu. brbust"。

IntentResolutionDemo. java 文件中定义了一个 Intent 用来启动另一个 Activity,这个 Intent 与 Activity 设置的 Intent 过滤器是完全匹配的。IntentResolutionDemo. java 文件中 Intent 实例化和启动 Activity 的代码如下:

```
Intent intent = new Intent(Intent.ACTION_VIEW, Uri.parse("schemodemo://edu.hrbust/path"));
```

```
2  startActivity(intent);
```

代码第 1 行定义的 Intent 动作为 Intent. ACTION_VIEW，与 Intent 过滤器的动作 android. intent. action. VIEW 匹配，Uri 是“schemodemo://edu. brbust/path”，其中的协议部分为“schemodemo”，主机名部分为“edu. hrbust”，也与 Intent 过滤器定义的数据要求完全匹配。因此，代码第 1 行定义的 Intent 在 Android 系统与 Intent 过滤器列表进行匹配时会与 AndroidManifest. xml 文件中 ActivityToStart 定义的 Intent 过滤器完全匹配。

AndroidManifest. xml 文件中每个组件的 <intent - filter> 都被解析成一个 Intent 过滤器对象。当应用程序安装到 Android 系统时，所有的组件和 Intent 过滤器都会注册到 Android 系统中。这样，Android 系统便可以将任何一个 Intent 请求通过 Intent 过滤器映射到相应的组件上。

5.4　广播与 BroadcastReceiver

Intent 的另一种用途是发送广播消息，应用程序和 Android 系统都可以使用 Intent 发送广播消息，广播消息的内容可以是与应用程序密切相关的数据信息，也可以是 Android 的系统信息，如网络连接变化、电池电量变化、接收到短信或系统设置变化等。

使用 Intent 发送广播消息非常简单，只需创建一个 Intent 并调用 sendBroadcast( ) 函数就可以把 Intent 携带的信息广播出去。但需要注意的是，在构造 Intent 时必须定义一个全局唯一的字符串，用来标识其要执行的动作，通常使用应用程序包的名称。如果要在 Intent 传递额外数据，可以用 Intent 的 putExtra( ) 方法。

下面的代码构造用于广播消息的 Intent，并添加了额外的数据，然后调用 sendBroadcast( ) 发送广播消息：

```
String UNIQUE_STRING = "edu. hrbust. BroadcastReceiverDemo";
Intent intent = new Intent(UNIQUE_STRING);
intent. putExtra("key1", "value1");
intent. putExtra("key2", "value2");
sendBroadcast(intent);
```

如果应用程序注册了 BroadcastReceiver，则可以接收到指定的广播消息。BroadcastReceiver 是广播接收者，它本质上是一种全局的监听器，用于监听系统全局的广播消息。由于 BroadcastReceiver 是一种全局的监听器，因此它可以非常方便地实现系统不同组件之间的通信。下面给出创建 BroadcastReceiver 的步骤。

1. 创建 BroadcastReceiver 的子类

由于 BroadcastReceiver 本质上是一种监听器，因此创建 BroadcastReceiver 的方法也非常简单，只需要创建一个 BroadcastReceiver 的子类然后重写 onReceive（Context context，Intent intent）方法即可。

示例代码如下：

```
public class MyBroadcastReceiver extends BroadcastReceiver {
@Override
public void onReceive(Context context, Intent intent) {
//TODO: React to the Intent received.
}
}
```

2. 注册 BroadcastReceiver

在实现了 BroadcastReceiver 之后，接下就应该指定该 BroadcastReceiver 能匹配的 Intent，即注册 BroadcastReceiver。注册 BroadcastReceiver 的方式有两种：静态文件注册和动态代码注册。前者是在 AndroidManifest. xml 配置文件中注册，通过这种方式注册的广播为常驻型广播，也就是说即使应用程序关闭了，一旦有相应事件触发，程序还是会被系统自动调用运行；后者是通过代码在. Java 文件中进行注册。通过这种方式注册的广播为非常驻型广播，即它会跟随 Activity 的生命周期，所以在 Activity 结束前需要调用 unregisterReceiver(receiver)方法移除它。程序代码如下：

```
//通过代码的方式动态注册 MyBroadcastReceiver
MyBroadcastReceiver receiver = new MyBroadcastReceiver();
IntentFilter filter = new IntentFilter();
filter.addAction("android.intent.action.MyBroadcastReceiver");
//注册 receiver
registerReceiver(receiver, filter);
```

当这个 Activity 销毁时，要主动撤销注册，否则会出现异常，其方法如下：

```
@Override
protected void onDestroy() {
        // TODO Auto-generated method stub
        super.onDestroy();
        //当 Activity 销毁时取消注册 BroadcastReceiver
        unregisterReceiver(receiver);
}
```

当 Android 系统接收到与注册 BroadcastReceiver 匹配的广播消息时，Android 系统会自动调用这个 BroadcastReceiver 接收广播消息。在 BroadcastReceiver 接收到与之匹配的广播消息后,onReceive( )方法会被调用,但 onReceive( )方法必须要在 5 s 内执行完毕,否则 Android 系统会认为该组件失去响应,并提示用户强行关闭该组件。

BroadcastReceiverDemo 示例说明了如何在应用程序中静态注册 BroadcastReceiver 组件,并指定接收广播消息的类型。在此示例中,通过点击"发送广播消息"按钮后,EditText 控件中内容将以广播消息的形式发送出去,示例内部的 BroadcastReceiver 将接收这个广播消息,并显示在用户界面的下方。

BroadcastReceiverDemo. java 文件中包含发送广播消息的代码,其关键代码如下:

```
public class BroadcastReceiverDemoActivity extends Activity {
private EditText entryText ;
private Button button;
    @ Override
    public void onCreate( Bundle savedInstanceState) {
        super. onCreate( savedInstanceState) ;
        setContentView( R. layout. main) ;
        entryText  =  (EditText)findViewById( R. id. entry) ;
        button  =  (Button)findViewById( R. id. btn) ;
         button. setOnClickListener( new OnClickListener( ) {
            public void onClick( View view) {
            Intent intent  =  new Intent( "edu. hrbust. BroadcastReceiverDemo" ) ;
            intent. putExtra( "message" , entryText. getText( ). toString( ) ) ;
            sendBroadcast( intent) ;
            }
         }) ;
    }
}
```

在配置文件 AndroidManifest. xml 中静态注册 BroadcastReceiver 组件的代码如下:

```
 <? xml version = "1.0"  encoding = "utf-8"?  >
 < manifest xmlns:android = "http://schemas. android. com/apk/res/android"
package = "edu. hrbust. BroadcastReceiverDemo"
android:versionCode = "1"
android:versionName = "1.0" >
  < application  android: icon  = " @ drawable/icon"  android: label  = " @ string/app_name" >
```

```
7 <activity android:name=".BroadcastReceiverDemo"
8 android:label="@string/app_name">
9 <intent-filter>
10 <action android:name="android.intent.action.MAIN" />
11 <category android:name="android.intent.category.LAUNCHER" />
12 </intent-filter>
13 </activity>
14 <receiver android:name=".MyBroadcastReceiver">
15 <intent-filter>
16 <action android:name="edu.hrbust.BroadcastReceiverDemo" />
17 </intent-filter>
18 </receiver>
19 </application>
20 <uses-sdk android:minSdkVersion="14" />
21 </manifest>
```

在代码的第 14 行中创建了一个 <receiver> 节点，在第 16 行中声明了 Intent 过滤器的动作为"edu. hrbust. BroadcastReceiverDemo"，这与 BroadcastReceiverDemo. java 文件中 Intent 的动作相一致，表明这个 BroadcastReceiver 可以接收动作为"edu. hrbust. BroadcastReceiver-Demo"的广播消息。

MyBroadcastReceiver. java 文件创建了一个自定义的 BroadcastReceiver，其核心代码如下：

```
1 public class MyBroadcastReceiver extends BroadcastReceiver {
2 @Override
3 public void onReceive(Context context, Intent intent) {
4 String msg = intent.getStringExtra("message");
5 Toast.makeText(context, msg, Toast.LENGTH_SHORT).show();
6 }
7 }
```

代码第 1 行首先继承了 BroadcastReceiver 类，并在第 3 行重载了 onReveive( ) 函数。当接收到 AndroidManifest. xml 文件定义的广播消息后，程序将自动调用 onReveive( ) 函数进行消息处理。代码第 4 行通过调用 getStringExtra( ) 函数，从 Intent 中获取标识为 message 的字符串数据，并在第 5 行使用 Toast( ) 函数将信息显示在界面中。

第 6 章　Android 后台服务

Service 是 Android 后台服务组件,适用于开发长时间运行的应用功能。学习本章,可以了解后台服务的基本原理,掌握进程内服务与跨进程服务的使用方法,有助于深入理解 Android 系统进程间通信机制。

6.1　Service

6.1.1　Service 的特征

1. Service(服务)的特点

(1)Service 是一个在后台运行的 Activity,它不是一个单独的进程。

(2)若要实现和用户的交互,需要通过通知栏或者是通过发送广播,由 UI 去接收显示。

(3)它的应用十分广泛,尤其是在框架层,应用更多的是对系统服务的调用。

2. 服务的作用

(1)用于处理一些不干扰用户使用的后台操作,如下载、网络获取、播放音乐,可以通过 INTENT 来开启,同时也可以绑定到宿主对象(调用者如 ACTIVITY 上)来使用。

(2)如果说 Activity 是显示前台页面的信息,那么 Service 就是在后台进行操作的。如果 Service 需要和前台 UI 进行交互,则可以用发送广播或者通知栏的方式。

3. 生命周期

(1)service 整体的生命时间从 onCreate()被调用开始,到 onDestroy()方法返回为止。与 activity 一样,service 在 onCreate()中进行它的初始化工作,在 onDestroy()中释放残留的资源。

(2)startService()的方式是 onCreate()→onStartCommand()→onStart()→onDestroy()。

(3)BindService()的方式是 onCreate()→onBinder()→onUnbind()→onDestroy(),onUnbind()方法返回后就结束了。

6.1.2　Service 的启动方法

Service 自己不能运行,需要通过某一个 Activity 或其他 Context 对象来调用。

(1)通过 startService。

调用 Context. startService( ),Service 会经历 onCreate→onStart,服务结束时调用 onDestroy 方法。

如果是调用者自己直接退出而没有调用 stopService,Service 会一直在后台运行。

①通过调用 Context. startService( )启动 Service,通过调用 Context. stopService( )或 Service. stopSelf( )停止 Service。因此,Service 一定是由其他组件启动的,但停止过程可以通过其他组件或自身完成。

②在启动方式中,启动 Service 的组件不能获取到 Service 的对象实例,因此无法调用 Service 中的任何函数,也不能获取到 Service 中的任何状态和数据信息。

③如果仅以启动方式使用 Service,则这个 Service 需要具备自管理的能力,且不需要通过函数调用向外部组件提供数据或功能。

(2)通过 bindService。

调用 Context. bindService( ),Service 只会运行 onCreate, 这时服务的调用者和服务绑定在一起,调用者退出了,Service 就会调用 onUnbind→onDestroyed。bindService 方法将调用者和服务绑定在一起,使调用方调用服务上的其他方法。

①Service 的使用是通过服务链接(Connection)实现的,服务链接能够获取 Service 的对象实例,因此绑定 Service 的组件可以调用 Service 中实现的函数,或直接获取 Service 中的状态和数据信息。

②使用 Service 的组件通过 Context. bindService( )建立服务链接,通过 Context. unbindService( )停止服务链接。

③如果在绑定过程中 Service 没有启动,Context. bindService( )会自动启动 Service,而且同一个 Service 可以绑定多个服务链接,这样可以同时为多个不同的组件提供服务。

startService 和 bindService 采用不同的方法,service 的生命周期也不同。

startService 启动,其生命周期不会因启动它的组件 Destroy 而消亡,而是依赖于 mainThread(即应用主线程),一旦主线程退出,即代表整个应用退出,活动也会跟着销毁。

bindService 启动,其生命周期依赖启动它的组件,组件销毁时,服务也随之销毁。

这两种使用方法并不是完全独立的,在某些情况下可以混合使用。以 MP3 播放器为例,在后台工作的 Service 通过 Context. startService( )启动某个音乐播放,但在播放过程中如果用户需要暂停音乐播放,则需要通过 Context. bindService( )获取服务链接和 Service 对象实例,进而通过调用 Service 对象实例中的函数暂停音乐播放过程,并保存相关信息。在这种情况下,如果调用 Context. stopService( )并不能够停止 Service,则需要在所有的服务链接关闭后,Service 才能够真正停止。

注意:平常使用较多的是 startService 方法,可以把一些耗时的任务放到后台去处理,当处理完成后，可以通过广播或者通知栏来通知前台。

通过以下代码可以深入理解启动、绑定服务，启动、绑定服务示例代码生成界面如图 6－1 所示。

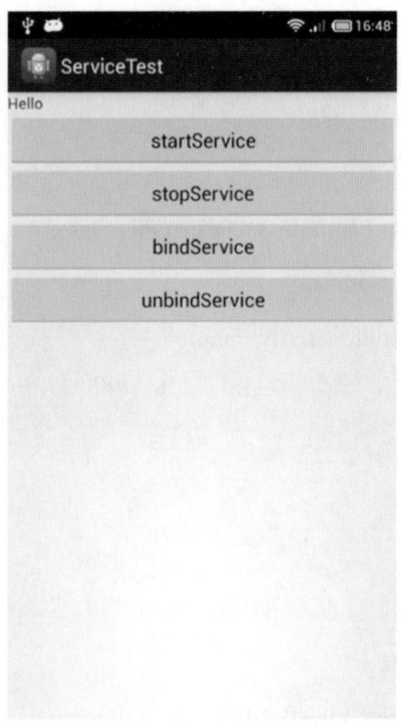

图 6－1　启动、绑定服务示例代码生成界面

（1）MainActivity. java 类的代码如下：

```
Package com. example. servicetest;
import com. example. servicetest. service. MyService;
import android. app. Activity;
import android. content. ComponentName;
import android. content. Intent;
import android. content. ServiceConnection;
import android. os. Bundle;
import android. os. IBinder;
import android. util. Log;
import android. view. View;
import android. view. View. OnClickListener;
import android. widget. Button;
public class MainActivity extends Activity implements OnClickListener{
```

```
/* *标志位*/
Private static String TAG = "com. example. servicetest. MainActivity" ;
/* *启动服务*/
Private ButtonmBtnStart;
/* *绑定服务*/
Private ButtonmBtnBind;
@ Override
Protected void onCreate( Bundle savedInstanceState) {
super. onCreate( savedInstanceState) ;
setContentView( R. layout. activity_main) ;
initView( ) ;
}
/* *
 * init theView
 */
Private void initView( ) {
mBtnStart = ( Button) findViewById( R. id. startservice) ;
mBtnBind = ( Button) findViewById( R. id. bindservice) ;
mBtnStart. setOnClickListener( this) ;
mBtnBind. setOnClickListener( this) ;
}
@ Override
Public void onClick( View view) {
switch( view. getId( ) ) {
//启动服务的方式
caseR. id. startservice:
startService( new Intent( MyService. ACTION) ) ;
break;
//绑定服务的方式
caseR. id. bindservice;
bindService( new Intent( MyService. ACTION) ,conn,BIND_AUTO_CREATE) ;
break;
default;
```

```
break;
}
}
Service Connectionconn = new ServiceConnection(){
Public void onServiceConnected(ComponentNamename,IBinderservice){
Log.v(TAG,"onServiceConnected");
}
Public void onServiceDisconnected(ComponentNamename){
Log.v(TAG,"onServiceDisconnected");
}
};
@Override
Protected void onDestroy(){
super.onDestroy();
System.out.println("--------onDestroy()--");
stopService(newIntent(MyService.ACTION));
unbindService(conn);
}
}
```

(2)MyService.java 类的代码如下:

```
Package com.example.servicetest.service;
import android.app.Service;
import android.content.Intent;
import android.os.Binder;
import android.os.IBinder;
import android.util.Log;
public class MyService extends Service{
/**标志位*/
Private static String TAG = "com.example.servicetest.service.MyService";
/**行为*/
Public static final String ACTION = "com.example.servicetest.service.MyService";
@Override
Public void onCreate(){
```

```
super. onCreate( ) ;
System. out. println( " - - - - - onCreate( ) - - - " ) ;
}
@ Override
Public void onStart( Intentintent, intstartId) {
super. onStart( intent, startId) ;
System. out. println( " - - - - - onStart( ) - - - " ) ;
}
@ Override
Public int onStartCommand( Intentintent, intflags, intstartId) {
System. out. println( " - - - - - onStartCommand( ) - - - " ) ;
Return super. onStartCommand( intent, flags, startId) ;
}
@ Override
Public IBinder onBind( Intentarg0) {
System. out. println( " - - - - - onBind( ) - - - " ) ;
Return null;
}
@ Override
Public void onRebind( Intentintent) {
System. out. println( " - - - - - onRebind( ) - - - " ) ;
super. onRebind( intent) ;
}
@ Override
Public Boolean onUnbind( Intentintent) {
System. out. println( " - - - - - onUnbind( ) - - - " ) ;
Return super. onUnbind( intent) ;
}
@ Override
Public void onDestroy( ) {
System. out. println( " - - - - - onDestroy( ) - - - " ) ;
super. onDestroy( ) ;
}
```

```
}
<! - - 注册 - - >
<service android:name = "com. example. servicetest. service. MyService" >
<intent - filter >
<! - - 用来启动服务的 Intent - - >
<action android:name = "com. example. servicetest. service. MyService"/ >
<category android:name = "android. intent. category. default"/ >
</intent - filter >
</service >
```

6.1.3　Service 的进程通信

本节介绍 service 的进程通信中用到的 Binder 和 Service Manager。

Binder 被用于 service 的进程通信。Binder 是 Android 系统中一个重要的“设备”，之所以加引号，是因为它是虚拟出来的，类似于 Linux 中的块设备，因此它也是基于 IO 的，而且 Binder 是 Parcelable 的，通过 Transaction 与它的代理端即 BinderServer 端交互。

binder 通信是一种 client - server 的通信结构。从表面上看，是 client 通过获得一个 server 的代理接口，对 server 进行直接调用。实际上，代理接口中定义的方法与 server 中定义的方法是一一对应的。client 调用某个代理接口中的方法时，代理接口的方法会将 client 传递的参数打包成 Parcel 对象，代理接口将该 Parcel 发送给内核中的 binder driver，server 会读取 binder driver 中的请求数据，如果是发送给自己的，则解包 Parcel 对象，处理并将结果返回。整个的调用过程是一个同步过程，在 server 处理时，client 会处于 block 状态。

service 的进程通信用到的另一个进程是 Service Manager。Service Manager 是一个 linux 级的进程，顾名思义，就是 service 的管理器。此处的 service 的概念和 init 过程中 init. rc 中的 service 是不同，init. rc 中的 service 都是 linux 进程，但是这里的 service 并不一定是一个进程，也就是说可能有一个或多个 service 属于同一个 linux 进程。

任何 service 在被使用之前，均要向 SM(Service Manager)注册，同时客户端需要访问某个 service 时，应该首先向 SM 查询是否存在该服务。如果 SM 存在这个 service，那么会将该 service 的 handle 返回给 client，handle 是每个 service 的唯一标识符。这个进程的主要工作如下：初始化 binder，打开/dev/binder 设备，在内存中为 binder 映射 128 KB 空间；指定 SM 对应的代理 binder 的 handle 为 0，当 client 尝试与 SM 通信时，需要创建一个 handle 为 0 的代理 binder，这里的代理 binder 其实就是第一节中描述的代理接口；通知 binder driver (BD)，使 SM 成为 BD 的 context manager；维护一个死循环，在这个死循环中不停地去读内核中 binder driver，查看是否有可读的内容，即是否有对 service 的操作要求，如果有，则调用

svcmgr_handler 回调来处理请求的操作；SM 维护了一个 svclist 列表来存储 service 的信息，这里需要声明一下，当 service 在向 SM 注册时，该 service 就是一个 client，SM 则作为 server，而某个进程需要与 service 通信时，这个进程为 client，service 才作为 server，因此 service 不一定为 server，有时它也是作为 client 存在的。

6.1.4 Service 与 Activity 交互

将 Service 用作后台下载，其生命周期不依赖于启动它的组件，且能够与它的组件相互通信。通常，使用 Service 的场景如下：若是前台 Service，一般是用来做类似于音乐播放器的；若是后台 Service，则通常是用来和服务器进行交互（数据下载）或是其他不需要用户参与的操作，同一进程中启动 Service，若直接与服务器交互，则很容易引起 ANR（Application Not Responding）。由于 Service 是由 mainThread 创建出来，因此此时 Service 是运行在 UI 主线程的，如果需要联网下载，则需要开启一个 Thread，然后在子线程中运行。在 Service 中创建/使用线程与在 Activity 中一样，无区别。

1. 数据交互

组件通常是 Activity，可以通过 bindService，当成功绑定时，可以获取 Service 中定义后的一个 IBinder 接口，可以通过这个接口返回该 Service 对象，从而直接访问该 Service 中的公有方法。

当 Service 想要把数据传递给某个组件时，最简单、最好的办法就是通过 Broadcast 在 Intent 中带上数据，广播给组件即可（BroadcastReceiver 中，由于 onReceive 的生命周期为 10 s，因此 onReceive 运行太久也会导致 ANR 问题）。

2. Service 刷新带有进度条的状态栏

通常采用发 Notification 到系统状态栏上的方式提醒用户做一些事情，但是如果仔细看 Notification 的参数，就会发现里面有一个 RemoteViews 类型的成员，如 Widget 应用 RemoteViews：RemoteViews 可以自定义一个 View，里面放一些小的控件，支持自定义一个带有 ProgressBar 的 layout，然后绑定到 Notification 对象上，并通过 NotificationManager 来通知更新即可。

编写 Android Service 需要基础 Service 类，并实现其中的 onBind 方法：

```
/* *
 * Android Service 示例
 */
public class ServiceDemo extends Service {
  private static final String TAG = "ServiceDemo" ;
  public static final String ACTION = "com.lql.service.ServiceDemo";
```

```
@ Override
public IBinder onBind(Intent intent) {
Log. v(TAG, "ServiceDemo onBind");
 return null;
 }
 @ Override
 public void onCreate() {
 Log. v(TAG, "ServiceDemo onCreate");
 super. onCreate();
 }
 @ Override
 public void onStart(Intent intent, int startId) {
 Log. v(TAG, "ServiceDemo onStart");
 super. onStart(intent, startId);
 }
 @ Override
 public int onStartCommand(Intent intent, int flags, int startId) {
 Log. v(TAG, "ServiceDemo onStartCommand");
 return super. onStartCommand(intent, flags, startId);
 }
}
```

在 AndroidManifest. xml 文件中声明 Service 组件:

```
<service android:name = "com. lql. service. ServiceDemo" >
  <intent - filter >
      <action android:name = "com. lql. service. ServiceDemo"/ >
  </intent - filter >
</service >
```

其中,intent - filter 中定义的 action 是用来启动服务的 Intent。

在需要 service 的地方通过 Context. startService(Intent)方法启动 service 或者 Context. bindService 方法来绑定 service:

```
public class ServiceDemoActivity extends Activity {
  private static final String TAG = "ServiceDemoActivity";
  Button bindBtn;
```

```
Button startBtn;
    @ Override
    public void onCreate( Bundle savedInstanceState) {
        super. onCreate( savedInstanceState) ;
        setContentView( R. layout. main) ;
        bindBtn = (Button)findViewById( R. id. bindBtn) ;
        startBtn = (Button)findViewById( R. id. startBtn) ;
        bindBtn. setOnClickListener( new OnClickListener( ) {
  public void onClick( View v) {
  bindService( new Intent( ServiceDemo. ACTION) , conn, BIND_AUTO_CREATE) ;
  }
  }) ;
            startBtn. setOnClickListener( new OnClickListener( ) {
  public void onClick( View v) {
  startService( new Intent( ServiceDemo. ACTION) ) ;
  }
  }) ;
      }
   ServiceConnection conn = new ServiceConnection( ) {
        public void onServiceConnected( ComponentName name, IBinder service) {
  Log. v( TAG, "onServiceConnected" ) ;
  }
  public void onServiceDisconnected( ComponentName name) {
  Log. v( TAG, "onServiceDisconnected" ) ;
  }
  };
  @ Override
  protected void onDestroy( ) {
  Log. v( TAG, "onDestroy unbindService" ) ;
  unbindService( conn) ;
  super. onDestroy( ) ;
  };
}
```

如图 6 - 2 所示为点击绑定服务时输出的内容。可以看出，只调用了 onCreate 方法和 onBind 方法，当重复点击绑定服务时，没有再输出任何日志，并且不报错。onCreate 方法是在第一次创建 Service 时调用的，而且只调用一次。另外，在绑定服务时给定了参数 BIND_AUTO_CREATE，即当服务不存在时自动创建，如果服务已经启动了或者创建了，那么只会调用 onBind 方法。

Application	Tag	Text
com.lql.service.demo	ServiceDemo	ServiceDemo onCreate
com.lql.service.demo	ServiceDemo	ServiceDemo onBind

图 6 - 2　点击绑定服务时输出的内容

如图 6 - 3 所示为多次点击启动服务时输出的内容。可以看出，在第一次点击时，因为 Service 还未创建，所以调用了 onCreate 方法，紧接着调用了 onStartCommand 和 onStart 方法。当再次点击启动服务时，仍然调用了 onStartCommand 和 onStart 方法，所以在 Service 中做任务处理时需要注意这点，因为一个 Service 可以被重复启动。

Application	Tag	Text
com.lql.service.demo	ServiceDemo	ServiceDemo onCreate
com.lql.service.demo	ServiceDemo	ServiceDemo onStartCommand
com.lql.service.demo	ServiceDemo	ServiceDemo onStart
com.lql.service.demo	ServiceDemo	ServiceDemo onStartCommand
com.lql.service.demo	ServiceDemo	ServiceDemo onStart
com.lql.service.demo	ServiceDemo	ServiceDemo onStartCommand
com.lql.service.demo	ServiceDemo	ServiceDemo onStart

图 6 - 3　多次点击启动服务时输出的内容

平常使用多的是 startService 方法，可以把一些耗时的任务放到后台去处理，当处理完成后，可以通过广播来通知前台。而 onBind 方法更多的是结合 AIDL 来使用，这样一个应用可以通过绑定服务获得的 IBinder 来拿到后台的接口，进而调用 AIDL 中定义的方法，进行数据交换等。

6.1.5　Service 中的 ANR

在 Android 上，如果应用程序有一段时间响应不够灵敏，系统会向用户显示一个对话框，这个对话框称为应用程序无响应（Application Not Responding，ANR）对话框。用户可以选择“等待”而让程序继续运行，也可以选择“强制关闭”。一个流畅的合理的应用程序中不能出现 ANR，而让用户每次都要处理这个对话框。因此，在程序里对响应性能的设计很重要，这样系统不会显示 ANR 给用户。

默认情况下，在 android 中 Activity 的最长执行时间是 5 s，BroadcastReceiver 的最长执行

时间则是 10 s。

1. 引发 ANR 的条件

在 Android 里,应用程序的响应性是由 Activity Manager 和 WindowManager 系统服务监视的 。当它监测到以下情况中的一个时,Android 就会针对特定的应用程序显示 ANR:

(1)在 5 s 内没有响应输入的事件(如按键按下、屏幕触摸);

(2)BroadcastReceiver 在 10 s 内没有执行完毕。

造成以上两点的原因有很多,如在主线程中做了非常耗时的操作、下载、IO 异常等。

潜在的耗时操作,如网络或数据库操作,或者高耗时的计算如改变位图尺寸,应该在子线程里(或者以数据库操作为例,通过异步请求的方式)来完成。然而,不是主线程阻塞在那里等待子线程的完成,也不是调用 Thread. wait( )或是 Thread. sleep( )。替代的方法是,主线程应该为子线程提供一个 Handler,以便完成时能够提交给主线程。以这种方式设计应用程序,将能保证主线程保持对输入的响应性并能避免因 5 s 输入事件的超时而引发的 ANR 对话框。

2. 避免 ANR

(1)运行在主线程里的任何方法都尽可能少做事情,特别是 Activity 应该在它的关键生命周期方法(如 onCreate( )和 onResume( ))里尽可能少地去做创建操作(可以采用重新开启子线程的方式,然后使用 Handler + Message 的方式做一些操作,如更新主线程中的 UI 等)。

(2)应用程序应该避免在 BroadcastReceiver 里做耗时的操作或计算,但不再是在子线程里做这些任务(因为 BroadcastReceiver 的生命周期短)。如果响应 Intent 广播需要执行一个耗时的动作,则应用程序应该启动一个 Service(此处需要注意的是,可以在广播接受者中启动 Service,但是却不可以在 Service 中启动 broadcasereciver,原因后续会有介绍)。

(3)避免在 Intent Receiver 里启动一个 Activity,因为它会创建一个新的画面,并从当前用户正在运行的程序上抢夺焦点。如果应用程序在响应 Intent 广播时需要向用户展示什么,则应该使用 Notification Manager 来实现。

ANR 异常也是在程序中经常遇到的问题,主要的解决办法就是不要在主线程中做耗时的操作,而应放在子线程中来实现,如采用 Handler + mesage 的方式,或者是需要做一些和网络相互交互的耗时操作时采用 asyntask 异步任务的方式(它的底层与 Handler + mesage 有所区别的是它是线程池)等,在主线程中更新 UI。

6.2 Service 启动方式比较

startService 是可以独立与调用程序运行的,也就是说启动它的程序消亡了,该 service 还是可以继续运行的。bindService 是允许其他的组件(如 Activities)绑定到其上面,可以发送请求,也可以接受请求,甚至可以进行进程间的通信。当创建一个能提供绑定功能的服务时,就必须要提供一个 IBinder 对象,客户端就可以使用 IBinder 对象来与服务进行交互,在 Android 系统中,有三种方式可以创建 IBinder:扩展 Binder 类,当 service 只给当前的程序用而且在后台执行时,采用这种方式比较好,优点是不需要跨进程间通信;使用 Message 机制,消息机制相对于 Binder 的方式来说就比较复杂,它支持跨进程间调用(这种方式的基础也是 AIDL),这种情况下,service 会定义一个 handle 来处理不同的 object 的服务请求,这里的 IBinder 对所有的客户端来说是共享的,当然客户端也可以定义自己的 handle 来处理和 service 之间来进行交互,消息机制是一种实现进程间通信的最简单的方式,因为所有的请求都会放到一个消息队列当中去进行轮询处理,每次只处理一个服务请求,这样就不需要保证设计的 service 是线程安全状态的;使用安卓接口定义语言(Android interface definition language,AIDL),这种方式是最难的一种方式了,它会把所有的工作都会分解成最原始的语义,从而使系统能够理解该工作目的,然后进行进程间的通信,前面说过 message 采用的是 AIDL 的架构基础的,当需要同时处理多个请求,而不是放在队列里面一个一个处理时就可以采用这种方式了,使用这种方式必须保证 service 能够支持多线程并且保证其线程安全状态,一般情况下会先创建一个. aldl 文件来定义程序的接口,系统会自动生成抽象类以及 IPC 的处理,然后就可以在 service 中进行 extend,实现相关功能。

创建绑定服务要素是必须得实现 onBind( )函数,然后返回一个 IBinder 的接口,IBinder 定义了与 service 通信的接口,其他应用程序通过调用 bindService( )来绑定到该 service 上并获取接口以及调用 service 的方法。service 生存的唯一理由是为了绑定它的应用程序服务,因此应用程序如果消失了,它也将消失。创建一个 bound service 的过程是首先定义一个客户端如何与 service 通信的接口,该接口必须是实现了 IBinder 的接口,该接口是从 onBind( )函数的回调方法中返回来的。一旦客户端收到了 IBinder 接口,就可以和 service 间进行通信了。多个客户端可以与 service 绑定一次,当客户端与 service 交互结束之后,将会调用 unbindService( )来解除绑定,一旦整个系统中没有客户端与 service 进行绑定了,那么系统将会 destory 该 service。下面例子实现一个 service 为客户端提供一个随机产生的数字,并将随机数显示在界面上。

(1)LocalService. java 类继承自 service,代码如下:

```
package com. android. localboundservice;
import java. util. Random;
import android. app. Service;
import android. content. Intent;
import android. os. Binder;
import android. os. IBinder;
```

```
import android. util. Log;
public class LocalService extends Service {
public IBinder localBinder = new MyLocalService( );
public Random m_generator = new Random( );
public static final int num = 2000;
private static final String TAG = "LocalService";
public class MyLocalService extends Binder{
public LocalService getService( ) {
Log. d(TAG, " * * * * * * * * getService" );
return LocalService. this;
}
}
@ Override
public IBinder onBind( Intent arg0) {
// TODO Auto - generated method stub
Log. d(TAG, " * * * * * * * return IBinder interface" );
return localBinder;
}
public int generatorInt( ) {
Log. d(TAG, " * * * * * * get random generator!" );
return m_generator. nextInt( num) ;    }
}
```

(2) LocalBoundServiceActivity. java 继承自 Activity 作为客户端,代码如下:

```
package com. android. huawei. localboundservice;
import com. android. localboundservice. LocalService. MyLocalService;
import android. app. Activity;
import android. os. Bundle;
import android. os. IBinder;
import android. widget. Button;
import android. widget. Toast;
import android. util. Log;
import android. view.    ;
import android. content. ComponentName;
```

```
import android.content.Context;
import android.content.Intent;
import android.content.ServiceConnection;
public class LocalBoundServiceActivity extends Activity {
private static final String TAG = "LocalBoundServiceActivity";
private Button mBtnService = null;
private boolean isConn = false; //该标记位主要用于判断当前是连接状态还是断开状态
private LocalService recSer = null;//定义一个 LocalService 变量,该变量继承自Binder 类(实现了 IBinder 接口)
    @Override
    public void onCreate(Bundle savedInstanceState) {
        super.onCreate(savedInstanceState);
        setContentView(R.layout.main);
        mBtnService = (Button)findViewById(R.id.bindService);
                // 该按钮响应函数用于调用 service 中的随机数生成器。
                mBtnService.setOnClickListener(new Button.OnClickListener(){
                public void onClick(View v){
            if(isConn == true){
            int num = recSer.generatorInt();
 Toast.makeText(LocalBoundServiceActivity.this, "生成数为:" + num, Toast.LENGTH_LONG).show();            }
            }
        });
    }
@Override
    protected void onStart() {
      // TODO Auto-generated method stub
      super.onStart();
      Intent intent = new Intent(LocalBoundServiceActivity.this,
      LocalService.class);
bindService(intent,mcoon,Context.BIND_AUTO_CREATE);//绑定服务
    }
```

```
@ Override
   protected void onStop( ) {
      // TODO Auto - generated method stub
      super. onStop( ) ;
if( isConn) {
unbindService( mcoon) ;
isConn  =  true;
}
   }
private ServiceConnection mcoon  =  new ServiceConnection( ) {
@ Override
public void onServiceDisconnected( ComponentName name) {
// TODO Auto - generated method stub
isConn  =  false;
Log. d( TAG, " service disconnected!!!" ) ;
}
@ Override
public void onServiceConnected( ComponentName name, IBinder service) {
// TODO Auto - generated method stub
MyLocalService bindSer  =  ( MyLocalService) service;
recSer  =  bindSer. getService( ) ;
isConn  =  true;
Log. d( TAG, " service connected!!!" ) ;
}
} ;
   }
```

在 manifest 文件中加上 service 的注册:LocalBoundService Manifest. xml,程序代码如下:

```
< ? xml version = "1.0"  encoding = " utf - 8" ?  >
< manifest xmlns:android = " http://schemas. android. com/apk/res/android"
     package = " com. android. huawei. localboundservice"
     android:versionCode = " 1"
     android:versionName = "1.0" >
   < uses - sdk android:minSdkVersion = "8"  / >
```

```
    <application android:icon = "@drawable/icon"
    android:label = "@string/app_name" >
        <activity android:name = ".LocalBoundServiceActivity"
                  android:label = "@string/app_name" >
            <intent-filter>
                <action android:name = "android.intent.action.MAIN" />
               <category android:name = "android.intent.category.LAUNCHER"  >
            </intent-filter>
        </activity>
<service android:name = ".LocalService" >
</service>
</application>
</manifest>
```

获取随机数并显示 service 的程序运行界面如图 6 -4 所示。

图 6 -4　获取随机数并显示 service 的程序运行界面

6.3　AIDL

Android 系统中的进程之间不能共享内存,因此需要提供一些机制在不同进程之间进行数据通信。为使其他的应用程序也可以访问本应用程序提供的服务,Android 系统采用了远程过程调用(Remote Procedure Call,RPC)方式来实现。与很多其他的基于 RPC 的解决方案一样,Android 使用一种接口定义语言(Interface Definition Language,IDL)来公开服务的接口。四个 Android 应用程序组件中的三个(Activity、BroadcastReceiver 和 ContentProvider)都可以进行跨进程访问,另外一个 Android 应用程序组件 Service 同样可以。因此,可以将这种可以跨进程访问的服务称为 AIDL(Android Interface Definition Language)服务。

建立 AIDL 服务要比建立普通的服务复杂一些,具体步骤如下:

(1)在 Eclipse Android 工程的 Java 包目录中建立一个扩展名为. aidl 的文件,该文件的语法类似于 Java 代码,但会稍有不同;

(2)如果 aidl 文件的内容是正确的,ADT 会自动生成一个 Java 接口文件(*. java);

(3)建立一个服务类(Service 的子类);

(4)实现由. aidl 文件生成的 Java 接口;

(5)在 AndroidManifest. xml 文件中配置 AIDL 服务,尤其要注意的是,<action>标签中 android:name 的属性值就是客户端要引用该服务的 ID,也就是 Intent 类的参数值。

在 Android 中,每个应用(Application)执行在它自己的进程中,无法直接调用到其他应用的资源,这也符合沙箱原理。所谓沙箱原理,一般来说用在移动电话业务中,简单地说旨在部分或全部地隔离应用程序。因此,在 Android 中,当一个应用被执行时,一些操作是被限制的,如访问内存、访问传感器等,这样做可以最大化地保护系统。

AIDL 是 IPC 的一个轻量级实现,遵循 Java 语法。Android 也提供了一个工具,可以自动创建 Stub。当需要在应用间通信时,需要有以下几个步骤:

(1)定义一个 AIDL 接口;

(2)为远程服务(Service)实现对应 Stub;

(3)将服务"暴露"给客户程序使用。

AIDL 的语法类似于 Java 的接口(Interface),只需要定义方法的签名。

AIDL 支持的数据类型与 Java 接口支持的数据类型有些不同:

(1)所有基础类型(int、char 等);

(2)String、List、Map、CharSequence 等类;

(3)其他 AIDL 接口类型;

(4)所有 Parcelable 的类。

为了更好地展示 AIDL 的用法,举一个很简单的例子:两数相加。

新建一个名为 com. android. hellosumaidl 的包,接着创建一个 HelloSumAidlActivity。在 com. android. hellosumaidl 包中,新建一个普通文件(New→File),取名为 LAdditionService. aidl。在这个文件中输入以下代码:

```
package com. android. hellosumaidl;
// Interface declaration
interface IAdditionService {
    // You can pass the value of in, out or inout
    // The primitive types (int, boolean, etc) are only passed by in
    int add(in int value1, in int value2);
}
```

一旦文件被保存,Android 的 AIDL 工具就会在 gen/com/android/hellosumaidl 这个文件夹里自动生成对应的 IAdditionService. java 文件。因为是自动生成的,所以无需改动。这个文件里就包含了 Stub,接下来要为远程服务实现这个 Stub。

实现远程服务,首先在 com. android. hellosumaidl 包中新建一个类,取名为 AdditionService. java。为实现服务,需要让这个类中的 onBind 方法返回一个 IBinder 类的对象,这个 IBinder 类的对象就代表了远程服务的实现,要用到自动生成的子类 IAdditionService. Stub。其中,必须实现之前在 AIDL 文件中定义的 add()函数。下面是远程服务的代码:

```
package com. android. hellosumaidl;
import android. app. Service;
import android. content. Intent;
import android. os. IBinder;
import android. os. RemoteException;
/
  This class exposes the service to client
  /
public class AdditionService extends Service {
    @ Override
    public void onCreate() {
        super. onCreate();
    }
    @ Override
    public IBinder onBind(Intent intent) {
        return new IAdditionService. Stub() {
            /
                Implement com. android. hellosumaidl. IAdditionService. add(int,
int)
```

```
                /
            @ Override
            public int add( int value1, int value2) throws RemoteException {
                return value1 + value2;
            }
        };
    }
    @ Override
    public void onDestroy( ) {
        super. onDestroy( );
    }
}
```

为暴露服务,实现服务中的 onBind 方法后,就可以把客户程序(HelloSumAidlActivity. java)与服务连接起来了。为建立这样的一个链接,需要实现 ServiceConnection 类。在 HelloSumAidlActivity. java 中创建一个内部类 AdditionServiceConnection,这个类继承 ServiceConnection 类,并且重写了它的两个方法:onServiceConnected 和 onServiceDisconnected。下面给出内部类的代码:

```
/*
 * This inner class is used to connect to the service
 */
class AdditionServiceConnection implements ServiceConnection {
  public void onServiceConnected( ComponentName name, IBinder boundService) {
      service = IAdditionService. Stub. asInterface( ( IBinder) boundService) ;
      Toast. makeText( HelloSumAidlActivity. this, "Service connected",
      Toast. LENGTH_LONG) . show( ) ;
  }
  public void onServiceDisconnected( ComponentName name) {
      service = null;
      Toast. makeText( HelloSumAidlActivity. this, "Service disconnected",
      Toast. LENGTH_LONG) . show( ) ;
  }
}
```

这个方法接收一个远程服务的实现作为参数。使用 IAdditionService. Stub. asInterface

((IBinder)boundService)转换(cast)为自己的 AIDL 的实现。

为完成测试项目,首先改写 main.xml(主界面的格局文件)和 string.xml(字符串定义文件):

```
<?xml version="1.0" encoding="utf-8"?>
<LinearLayout xmlns:android="http://schemas.android.com/apk/res/android"
  android:layout_width="match_parent"
  android:layout_height="match_parent"
  android:orientation="vertical" >
   <TextView
        android:layout_width="fill_parent"
        android:layout_height="wrap_content"
        android:text="@string/hello"
         android:textSize="22sp" />
    <EditText
         android:id="@+id/value1"
         android:layout_width="wrap_content"
         android:layout_height="wrap_content"
         android:hint="@string/hint1" >
    </EditText>
    <TextView
         android:id="@+id/TextView01"
         android:layout_width="wrap_content"
         android:layout_height="wrap_content"
         android:text="@string/plus"
         android:textSize="36sp" />
    <EditText
         android:id="@+id/value2"
         android:layout_width="wrap_content"
         android:layout_height="wrap_content"
         android:hint="@string/hint2" >
    </EditText>
    <Button
         android:id="@+id/buttonCalc"
```

```
            android:layout_width = "wrap_content"
            android:layout_height = "wrap_content"
            android:hint = "@ string/equal"  >
        </Button >
        < TextView
            android:id = "@ + id/result"
            android:layout_width = "wrap_content"
            android:layout_height = "wrap_content"
            android:text = "@ string/result"
            android:textSize = "36sp"  / >
    </LinearLayout >
    string. xml
    <? xml version = "1.0"  encoding = "utf - 8"?  >
    < resources >
        < string name = "app_name" > HelloSumAIDL </string >
        < string name = "hello" > Hello Sum AIDL </string >
        < string name = "result" > Result </string >
        < string name = "plus" > + </string >
        < string name = "equal" > = </string >
        < string name = "hint1" > Value 1 </string >
        < string name = "hint2" > Value 2 </string >
    </resources >
```

最后，HelloSumAidlActivity. java 程序如下：

```
package com. android. hellosumaidl;
import android. os. Bundle;
import android. os. IBinder;
import android. os. RemoteException;
import android. view. View;
import android. view. View. OnClickListener;
import android. widget. Button;
import android. widget. EditText;
import android. widget. TextView;
import android. widget. Toast;
```

```
import android.app.Activity;
import android.content.ComponentName;
import android.content.Context;
import android.content.Intent;
import android.content.ServiceConnection;
public class HelloSumAidlActivity extends Activity {
    IAdditionService service;
    AdditionServiceConnection connection;
    @Override
    public void onCreate(Bundle savedInstanceState) {
        super.onCreate(savedInstanceState);
        setContentView(R.layout.main);
        initService();
        Button buttonCalc = (Button)findViewById(R.id.buttonCalc);
        buttonCalc.setOnClickListener(new OnClickListener() {
            TextView result = (TextView)findViewById(R.id.result);
            EditText value1 = (EditText)findViewById(R.id.value1);
            EditText value2 = (EditText)findViewById(R.id.value2);
            @Override
            public void onClick(View v) {
                int v1, v2, res = -1;
                v1 = Integer.parseInt(value1.getText().toString());
                v2 = Integer.parseInt(value2.getText().toString());
                try {
                    res = service.add(v1, v2);
                } catch (RemoteException e) {
                    e.printStackTrace();
                }
                result.setText(Integer.valueOf(res).toString());
            }
        });
    }
    @Override
```

```
    protected void onDestroy() {
        super.onDestroy();
        releaseService();
    }
    /
       This inner class is used to connect to the service
      /
    class AdditionServiceConnection implements ServiceConnection {
        public void onServiceConnected(ComponentName name, IBinder boundService) {
            service = IAdditionService.Stub.asInterface((IBinder)boundService);
                Toast.makeText(HelloSumAidlActivity.this, "Service connected",
Toast.LENGTH_LONG).show();
        }
        public void onServiceDisconnected(ComponentName name) {
            service = null;
            Toast.makeText(HelloSumAidlActivity.this, "Service disconnected",
             Toast.LENGTH_LONG).show();
        }
    }
    /*
     * This function connects the Activity to the service
     */
    private void initService() {
        connection = new AdditionServiceConnection();
        Intent i = new Intent();
        i.setClassName("com.android.hellosumaidl",
        com.android.hellosumaidl.AdditionService.class.getName());
          boolean ret = bindService(i, connection, Context.BIND_AUTO_CREATE);
    }
    /*
```

```
     * This function disconnects the Activity from the service
     */
    private void releaseService() {
        unbindService(connection);
        connection = null;
    }
}
```

第 7 章　Android 数据存储

数据存储在 Android 开发中是使用最频繁的,在这里主要介绍 Android 平台中实现数据存储的五种方式:

(1)使用 SharedPreferences 存储数据;

(2)文件存储数据,按照存储的位置不同,又可细分为内部存储和外部存储;

(3)SQLite 数据库存储数据;

(4)使用 ContentProvider 存储数据;

(5)网络存储数据。

7.1　SharedPreferences 数据存储

使用 SharedPerferences 可在不同的应用程序之间共享数据,其效果与在同一个 Activity 中获取数据的方式是一样的。使用 SharedPreferences 存储数据适用于保存少量的数据,且这些数据的格式非常简单(字符串型、基本类型的值),如应用程序的各种配置信息(是否打开音效、是否使用震动效果、小游戏的玩家积分等)、解锁口令密码等。

SharedPerferences 的核心原理是保存基于 XML 文件存储的 key - value 键值对数据,通常用来存储一些简单的配置信息。通过 DDMS 的 File Explorer 面板展开文件浏览树,可以看到 SharedPreferences 数据总是存储在/data/data/ <package name>/shared_prefs 目录下。SharedPreferences 对象本身只能获取数据而不支持存储和修改,存储修改通过 SharedPreferences. edit( )获取的内部接口 Editor 对象实现。SharedPreferences 本身是一个接口,程序无法直接创建 SharedPreferences 实例,只能通过 Context 提供的 getSharedPreferences(String name, int mode)方法来获取 SharedPreferences 实例,该方法中 name 表示要操作的 xml 文件名,第二个参数具体如下:Context. MODE_PRIVATE,指定该 SharedPreferences 数据只能被本应用程序读、写;Context. MODE_WORLD_READABLE,指定该 SharedPreferences 数据能被其他应用程序读,但不能写;Context. MODE_WORLD_WRITEABLE,指定该 SharedPreferences 数据能被其他应用程序读、写。Editor 有如下主要重要方法:SharedPreferences. Editor clear( ),清空 SharedPreferences 里所有数据;SharedPreferences. Editor put × × ×(String key , × × ×value),向 SharedPreferences 存入指定 key 对应的数据,其中 × × × 可以是 boolean,float,int 等各种基本类型据;SharedPreferences. Editor remove( ),删除 SharedPreferences 中指定 key 对应的数

据项;boolean commit(),当 Editor 编辑完成后,使用该方法提交修改。

SharedPreferences 是 Android 中最容易理解的数据存储技术,实际上 SharedPreferences 处理的就是一个 key - value(键值对)SharedPreferences,常用来存储一些轻量级的数据。

1. 使用 SharedPreferences 保存数据方法

```
//实例化 SharedPreferences 对象
SharedPreferences mySharedPreferences = getSharedPreferences("test",
Activity.MODE_PRIVATE);
//实例化 SharedPreferences.Editor 对象
SharedPreferences.Editor editor = mySharedPreferences.edit();
//用 putString 的方法保存数据
editor.putString("name", "Karl");
editor.putString("habit", "sleep");
//提交当前数据
editor.commit();
//使用 toast 信息提示框提示成功写入数据
Toast.makeText(this, "数据成功写入 SharedPreferences!" ,
Toast.LENGTH_LONG).show();
```

执行以上代码,SharedPreferences 将会把这些数据保存在 test.xml 文件中,可以在 File Explorer 的 data/data/相应的包名/test.xml 下导出该文件并查看。

2. 使用 SharedPreferences 读取数据方法

同样,在读取 SharedPreferences 数据前要实例化出一个 SharedPreferences 对象。

```
SharedPreferencessharedPreferences = getSharedPreferences("test",
Activity.MODE_PRIVATE);
// 使用 getString 方法获得 value,注意第 2 个参数是 value 的默认值
String name =sharedPreferences.getString("name", "");
String habit =sharedPreferences.getString("habit", "");
//使用 toast 信息提示框显示信息
Toast.makeText(this, "读取数据如下:" + "\n" + "name:"
 + name + "\n" + "habit:" + habit,
Toast.LENGTH_LONG).show();
```

读取 PreferenceWriteTest 工程写入的 value1 值的代码如 PreferenceReadTest:

```
public class PreferenceReadTest extends Activity {
    private TextView tv;
```

```
        @ Override
        public void onCreate( Bundle savedInstanceState) {
            super. onCreate( savedInstanceState) ;
            setContentView( R. layout. main) ;
            SharedPreferences sp = getSharedPreferences( "shared_filename" ,
            MODE_WORLD_READABLE) ;
            tv = (TextView) findViewById( R. id. hello) ;
             tv. setText( "value1 =" + sp. getString( "value1" , "default" ) ) ;
        }
    }
```

用于写入 SharedPerferences 到 xml 文件中的代码,如 PreferenceWriteTest:

```
public class PreferenceWriteTest extends Activity {
private Button btn;
    @ Override
    public void onCreate( Bundle savedInstanceState) {
         super. onCreate( savedInstanceState) ;
         setContentView( R. layout. main) ;
         SharedPreferences sp = getSharedPreferences( "shared_filename" ,
         MODE_WORLD_WRITEABLE) ;
         Editor e = sp. edit( ) ;
           e. putString( "value1" , "54321" ) ;
           e. commit( ) ;
    }
```

基于 SharedPreferences 的输入框如图 7 - 1 所示,可以输入密码口令、设置密码口令、获取密码口令。

图 7 - 1　基于 SharedPreferences 的输入框

这里只提供了两个按钮和一个输入文本框,布局简单,在此不给出界面布局文件了,程序核心代码如下:

```
class ViewOcl implements View. OnClickListener{
        @ Override
        public void onClick( View v) {
            switch( v. getId( ) ) {
            case R. id. btnSet:
                //获取输入值
                String code = txtCode. getText( ). toString( ). trim( );
                //创建一个 SharedPreferences. Editor 接口对象,lock 表示要写入
                //的 XML 文件名,MODE_WORLD_WRITEABLE 写操作
                 SharedPreferences. Editor editor = getSharedPreferences( "lock",
                 MODE_WORLD_WRITEABLE). edit( );
                 //将获取过来的值放入文件
                 editor. putString( "code", code);
                 //提交
                 editor. commit( );
                 Toast. makeText( getApplicationContext( ), "口令设置成功",
                 Toast. LENGTH_LONG). show( );
                 break;
             case R. id. btnGet:
                 //创建一个 SharedPreferences 接口对象
                 SharedPreferences read = getSharedPreferences( "lock",
                 MODE_WORLD_READABLE);
                 //获取文件中的值
                 String value = read. getString( "code", "" );
                 Toast. makeText( getApplicationContext( ), "口令为:" + value,
                 Toast. LENGTH_LONG). show( );
                 break;
             }
         }
     }
```

读写其他应用的 SharedPreferences，步骤如下：

（1）在创建 SharedPreferences 时，指定 MODE_WORLD_READABLE 模式，表明该 SharedPreferences 数据可以被其他程序读取；

（2）创建其他应用程序对应的 Context：Context pvCount = createPackageContext("com.tony.app", Context.CONTEXT_IGNORE_SECURITY)，这里的 com.tony.app 就是其他程序的包名；

（3）使用其他程序的 Context 获取对应的 SharedPreferences SharedPreferences read = pvCount.getSharedPreferences("lock", Context.MODE_WORLD_READABLE)；

（4）如果是写入数据，使用 Editor 接口即可，其他操作均与前面一致。

SharedPreferences 对象与 SQLite 数据库相比，免去了创建数据库、创建表、写 SQL 语句等诸多操作，相对而言更加方便、简洁。但是 SharedPreferences 也有其自身缺陷，如其只能存储 boolean、int、float、long 和 String 五种简单的数据类型，无法进行条件查询，等等。无论 SharedPreferences 的数据存储操作如何简单，它也只能是存储方式的一种补充，而无法完全替代如 SQLite 数据库这样的其他数据存储方式。

7.2 内部存储

内部存储是指将文件存储于 Android 手机本地存储空间。Android 中，可以通过绝对路径以 Java 传统方式访问内部存储空间，但是以这种方式创建的文件是私有的，创建它的应用程序对该文件是可读可写，但是别的应用程序并不能直接访问它。不是所有的内部存储空间应用程序都可以访问，默认情况下只能访问“/data/data/应用程序的包名”路径下的文件。

更好的方法是使用 Context 对象的 openFileOutput() 和 openFileInput() 来进行数据持久化存储，数据文件将存储在内部存储空间的/data/data/应用程序的包名/files/目录下，无法指定更深一级的目录，而且默认是 Context.MODE_PRIVATE 模式，即别的应用程序不能访问它。可以使用 openFileOutput() 的 int mode 参数来让别的应用程序也能访问文件。

注意：保存在/data/data/应用程序的包名目录中的文件会在卸载应用程序时被删除掉。

使用 Context 对象访问 Android 内部存储的核心原理：Context 提供了两个方法来打开数据文件里的文件 IO 流，即 FileInputStream openFileInput(String name) 和 FileOutputStream (String name , int mode)。这两个方法第一个参数用于指定文件名，第二个参数用于指定打开文件的模式。具体有以下值可选：MODE_PRIVATE 为默认操作模式，代表该文件是私有数据，只能被应用本身访问，在该模式下，写入的内容会覆盖原文件的内容，如果想把新写入的内容追加到原文件中，可以使用 Context.MODE_APPEND；MODE_APPEND 模式会检查文件是否存在，若存在就往文件追加内容，否则就创建新文件；MODE_WORLD_READABLE 表示当前文件可以被其他应用读取；MODE_WORLD_WRITEABLE 表示当前文件可以被其他应用写入。除此之外，Context 还提供了以下几个重要的方法：getDir(String name , int mode)为在应用程序的数据文件夹下获取或者创建 name 对应的子目录；File getFilesDir()为获取该应用程序的数据文件夹的绝对路径；String[] fileList()为返回该应用数据文件夹的全部文件。

仍然以图 7－1 为例，核心代码如下：

```
public String read() {
        try {
            FileInputStream inStream = this.openFileInput("message.txt");
            byte[] buffer = new byte[1024];
            int hasRead = 0;
            StringBuilder sb = new StringBuilder();
            while ((hasRead = inStream.read(buffer)) != -1) {
                sb.append(new String(buffer, 0, hasRead));
            }
             inStream.close();
             return sb.toString();
        } catch (Exception e) {
             e.printStackTrace();
        }
        return null;
    }
    public void write(String msg){
        //获取输入值
        if(msg == null) return;
        try {
            //创建一个 FileOutputStream 对象,MODE_APPEND 追加模式
            FileOutputStream fos = openFileOutput("message.txt",
                        MODE_APPEND);
            //将获取过来的值放入文件
            fos.write(msg.getBytes());
            //关闭数据流
```

```
                fos. close( ) ;
            } catch (Exception e) {
                e. printStackTrace( ) ;
            }
        }
```

openFileOutput( )方法的第一参数用于指定文件名称,不能包含路径分隔符“/”,如果文件不存在,Android 会自动创建它。创建的文件保存在/data/data/ < package name >/files 目录中,如/data/data/cn. tony. app/files/message. txt。

7.3 外部存储

外部存储即读写 sdcard 上的文件,读写按以下步骤进行。

(1)调用 Environment 的 getExternalStorageState( )方法判断手机上是否插了 SD 卡,且应用程序具有读写 SD 卡的权限,以下代码将返回 true:

Environment. getExternalStorageState( ). equals(Environment. MEDIA_MOUNTED)

(2)调用 Environment. getExternalStorageDirectory( )方法来获取外部存储器,也就是 SD 卡的目录,或者使用"/mnt/sdcard/"目录。

(3)使用 IO 流操作 SD 卡上的文件。

注意:手机应该已插入 SD 卡,对于模拟器而言,可通过 mksdcard 命令来创建虚拟存储卡必须在 AndroidManifest. xml 上配置读写 SD 卡的权限:

< uses - permission android:name = " android. permission. MOUNT_UNMOUNT_FILESYSTEMS"/ >

< uses - permission android:name = " android. permission. WRITE_EXTERNAL_STORAGE"/ >

示例代码如下:

```
private void write(String content) {
        if (Environment. getExternalStorageState( ). equals(
                    Environment. MEDIA_MOUNTED)) { // 如果 sdcard 存在
            File file = new File( Environment. getExternalStorageDirectory( )
                        . toString( )
                        + File. separator
                        + DIR
                        + File. separator
```

```
                + FILENAME); // 定义 File 类对象
        if (! file.getParentFile().exists()) { // 父文件夹不存在
            file.getParentFile().mkdirs(); // 创建文件夹
        }
        PrintStream out = null; // 打印流对象用于输出
        try {
            out = new PrintStream(new FileOutputStream(file, true));
            // 追加文件
            out.println(content);
        } catch (Exception e) {
            e.printStackTrace();
        } finally {
            if (out ! = null) {
                out.close(); // 关闭打印流
            }
        }
    } else { // SDCard 不存在,使用 Toast 提示用户
        Toast.makeText(this, "保存失败,SD 卡不存在!",
        Toast.LENGTH_LONG).show();
    }
}
// 文件读操作函数
private String read() {
    if (Environment.getExternalStorageState().equals(
            Environment.MEDIA_MOUNTED)) { // 如果 sdcard 存在
        File file = new File(Environment.getExternalStorageDirectory()
                .toString()
                + File.separator
                + DIR
                + File.separator
                + FILENAME); // 定义 File 类对象
        if (! file.getParentFile().exists()) { // 父文件夹不存在
            file.getParentFile().mkdirs(); // 创建文件夹
```

```
            }
            Scanner scan = null; // 扫描输入
            StringBuilder sb = new StringBuilder();
            try {
                scan = new Scanner(new FileInputStream(file)); // 实例化 Scanner
                while (scan.hasNext()) { // 循环读取
                    sb.append(scan.next() + "\n"); // 设置文本
                }
                return sb.toString();
            } catch (Exception e) {
                e.printStackTrace();
            } finally {
                if (scan != null) {
                    scan.close(); // 关闭打印流
                }
            }
        } else { // SDCard 不存在,使用 Toast 提示用户
            Toast.makeText(this, "读取失败,SD 卡不存在!",
            Toast.LENGTH_LONG).show();
        }
        return null;
    }
```

Android 实现 SD 卡和实现内存文件存储的做法基本是一样的,只是取得文件路径的方法不一样,基本上与 Java 的文件操作是一致的。外部存储配置取的位置和实现有所不同。对 SD 卡操作常用 Environment 类和 StatFs 类。

1. Environment 类

Environment 是一个提供访问环境变量的类。

Environment 包含常量如下。

(1)MEDIA_BAD_REMOVAL。

解释:返回 getExternalStorageState() ,表明 SDCard 被卸载前已被移除。

(2)MEDIA_CHECKING。

解释:返回 getExternalStorageState() ,表明对象正在磁盘检查。

(3)MEDIA_MOUNTED。

解释:返回 getExternalStorageState(),表明对象是否存在并具有读/写权限。

(4)MEDIA_MOUNTED_READ_ONLY。

解释:返回 getExternalStorageState(),表明对象权限为只读。

(5)MEDIA_NOFS。

解释:返回 getExternalStorageState(),表明对象为空白或正在使用不受支持的文件系统。

(6)MEDIA_REMOVED。

解释:返回 getExternalStorageState(),如果不存在 SDCard 则返回。

(7)MEDIA_SHARED。

解释:返回 getExternalStorageState(),如果 SDCard 未安装,则通过 USB 大容量存储共享返回。

(8)MEDIA_UNMOUNTABLE。

解释:返回 getExternalStorageState(),返回 SDCard 不可以被安装。也就是说,即使 SDCard 是存在的,也不可以被安装。

(9)MEDIA_UNMOUNTED。

解释:返回 getExternalStorageState(),返回 SDCard 已卸掉。也就是说,即使 SDCard 是存在的,也没有被安装。

Environment 常用方法如下。

(1)getDataDirectory()。

解释:返回 File,获取 Android 数据目录。

(2)getDownloadCacheDirectory()。

解释:返回 File,获取 Android 下载/缓存内容目录。

(3)getExternalStorageDirectory()。

解释:返回 File,获取外部存储目录即 SDCard。

(4)getExternalStoragePublicDirectory(String type)。

解释:返回 File,取一个高端的公用外部存储器目录来摆放某些类型的文件。

(5)getExternalStorageState()。

解释:返回 File,获取外部存储设备的当前状态。

(6)getRootDirectory()。

解释:返回 File,获取 Android 的根目录。

2. StatFs 类

StatFs 是一个模拟 linux 的 df 命令的一个类,用来获得 SD 卡和手机内存的使用情况。

StatFs 常用方法如下。

(1)getAvailableBlocks()。

解释:返回 Int ,获取当前可用的存储空间。

(2)getBlockCount()。

解释:返回 Int ,获取该区域可用的文件系统数。

(3)getBlockSize()。

解释:返回 Int ,以字节为单位,一个文件系统。

(4)getFreeBlocks()。

解释:返回 Int ,该块区域剩余的空间。

(5)restat(String path)。

解释:执行一个由该对象所引用的文件系统。

SDCard 存储卡在 Android 手机上是可以随时插拔的,每次的动作都对引起操作系统进行 ACTION_BROADCAST,本例将使用上面学到的方法,计算出 SDCard 的剩余容量和总容量,代码如下:

```
package com.terry;
import java.io.File;
import java.text.DecimalFormat;
import android.R.integer;
import android.app.Activity;
import android.os.Bundle;
import android.os.Environment;
import android.os.StatFs;
import android.view.View;
import android.view.View.OnClickListener;
import android.widget.Button;
import android.widget.ProgressBar;
import android.widget.TextView;
import android.widget.Toast;
public class getStorageActivity extends Activity {
private Button myButton;
/** 在 activity 被第一次创建时调用 */
@Override
public
```

```
void onCreate(Bundle savedInstanceState) {
        super.onCreate(savedInstanceState);
          setContentView(R.layout.main);
          findView();
          viewHolder.myButton.setOnClickListener(new OnClickListener() {
            @Override
            public void onClick(View arg0) {
                getSize();
            }
          });
 }
 void findView() {
          viewHolder.myButton = (Button)findViewById(R.id.Button01);
          viewHolder.myBar = (ProgressBar)findViewById(R.id.myProgressBar);
          viewHolder.myTextView = (TextView)findViewById(R.id.myTextView);
 }
 void getSize() {
          viewHolder.myTextView.setText("");
          viewHolder.myBar.setProgress(0);
        //判断是否有插入存储卡
if(Environment.getExternalStorageState().equals
(Environment.MEDIA_MOUNTED)) {
            File path  = Environment.getExternalStorageDirectory();
            //取得 sdcard 文件路径
           StatFs statfs = new StatFs(path.getPath());
            //获取 block 的 SIZE
 long blocSize = statfs.getBlockSize();
            //获取 BLOCK 数量
long totalBlocks = statfs.getBlockCount();
            //已使用的 Block 的数量
long availaBlock = statfs.getAvailableBlocks();
            String[] total = filesize(totalBlocks * blocSize);
            String[] availale = filesize(availaBlock * blocSize);
```

```
                //设置进度条的最大值
int maxValue = Integer. parseInt(availale[0])
                * viewHolder. myBar. getMax()/Integer. parseInt(total[0]);
                viewHolder. myBar. setProgress(maxValue);
                String Text = "总共:" + total[0] + total[1] + "/n"
 + "可用:" + availale[0] + availale[1];
                viewHolder. myTextView. setText(Text);
            } else if(Environment. getExternalStorageState().
            equals(Environment. MEDIA_REMOVED)) {
                Toast. makeText(getStorageActivity. this, "没有 sdCard",1000). show();
            }
}
//返回数组,下标 1 代表大小,下标 2 代表单位 KB/MB
String[] filesize(long size) {
            String str = "";
          if(size >= 1024) {
                str = "KB";
                size/ = 1024;
                if(size >= 1024) {
                    str = "MB";
                    size/ = 1024;
                }
            }
            DecimalFormat formatter = new DecimalFormat();
            formatter. setGroupingSize(3);
            String result[] = new String[2];
            result[0] = formatter. format(size);
            result[1] = str;
          return result;
}
}
```

7.4　数据库存储

SQLite 是轻量级嵌入式数据库引擎,它支持 SQL 语言,并且只利用很少的内存就有很好的性能。现在的主流移动设备像 Android、iPhone 等都使用 SQLite 作为复杂数据的存储引擎,在为移动设备开发应用程序时,也许就要使用 SQLite 来存储大量的数据,所以需要掌握移动设备上的 SQLite 开发技巧。SQLiteDatabase 类提供了很多种方法,辅助完成添加、更新和删除。

在 SQLite 中执行 SQL 语句:

db. executeSQL(String sql);

db. executeSQL(String sql, Object[] bindArgs);//sql 语句中使用占位符,然后第二个参数是实际的参数集

对 SQLite 的插入、更新、删除:

db. insert(String table, String nullColumnHack, ContentValues values);

db. update(String table, Contentvalues values, String whereClause, String whereArgs);

db. delete(String table, String whereClause, String whereArgs);

以上三种方法的第一个参数都表示要操作的表名;insert 中的第二个参数表示如果插入的数据每一列都为空,则需要指定此行中某一列的名称,系统将此列设置为 NULL,不至于出现错误;insert 中的第三个参数是 ContentValues 类型的变量,是键值对组成的 Map,key 代表列名,value 代表该列要插入的值;update 的第二个参数也很类似,只不过它是更新该字段 key 为最新的 value 值;update 的第三个参数 whereClause 表示 WHERE 表达式,如"age > ? and age < ?"等;最后的 whereArgs 参数是占位符的实际参数值;delete 方法的参数也是一样。

7.4.1　SQLite 的增删改查操作

在 SQLite 中对数据的操作包括添加、删除、修改、查找。

1. insert 方法

```
ContentValues cv = new ContentValues();//实例化一个 ContentValues 用来装载待插入的数据
cv.put("title","you are beautiful");//添加 title
cv.put("weather","sun"); //添加 weather
cv.put("context","xxxx"); //添加 context
String publish = new SimpleDateFormat("yyyy-MM-dd HH:mm:ss").format(new Date());
```

```
cv.put("publish",publish); //添加 publish
db.insert("diary",null,cv);//执行插入操作
```

使用 execSQL 方式来实现:

```
String sql = "insert into user(username,password) values ('Jack Johnson','iLovePopMuisc');//插入操作的 SQL 语句
db.execSQL(sql);//执行 SQL 语句
```

2. 数据的删除

有两种方式可以实现数据的删除。

(1)使用 delete 函数方式的实现。

```
String whereClause = "username=?";//删除的条件
String[] whereArgs = {"Jack Johnson"};//删除的条件参数
db.delete("user",whereClause,whereArgs);//执行删除
```

(2)使用 execSQL 方式的实现。

```
String sql = "delete from user where username='Jack Johnson'";//删除操作的 SQL 语句
db.execSQL(sql);//执行删除操作
```

3. 数据的修改

有两种方式可以实现数据的修改。

(1)使用 update 函数方式的实现。

```
ContentValues cv = new ContentValues();//实例化 ContentValues
cv.put("password","iHatePopMusic");//添加要更改的字段及内容
String whereClause = "username=?";//修改条件
String[] whereArgs = {"Jack Johnson"};//修改条件的参数
db.update("user",cv,whereClause,whereArgs);//执行修改
```

(2)使用 execSQL 方式的实现。

```
String sql = "update user set password = "iHatePopMusic" where username = "Jack Johnson";//修改的 SQL 语句
db.execSQL(sql);//执行修改
```

4. 数据查询

查询操作相对于上面的几种操作要复杂些,因为经常要面对着各种各样的查询条件,所以系统也考虑到这种复杂性,为 Android 提供了较为丰富的查询形式:

```
db.rawQuery(String sql, String[] selectionArgs);
db.query(String table, String[] columns, String selection, String[] selectionArgs, String
```

groupBy, String having, String orderBy);

db.query(String table, String[ ] columns, String selection, String[ ] selectionArgs, String groupBy, String having, String orderBy, String limit);

db.query(String distinct, String table, String[ ] columns, String selection, String[ ] selectionArgs, String groupBy, String having, String orderBy, String limit);

上面几种都是常用的查询方法,第一种最为简单,将所有的 SQL 语句都组织到一个字符串中,使用占位符代替实际参数,selectionArgs 就是占位符实际参数集。各参数说明如下。

table:表名称。

colums:表示要查询的列所有名称集。

selection:表示 WHERE 之后的条件语句,可以使用占位符。

selectionArgs:条件语句的参数数组。

groupBy:指定分组的列名。

having:指定分组条件,配合 groupBy 使用。

orderBy:指定排序的列名。

limit:指定分页参数。

distinct:指定"true"或"false"表示要不要过滤重复值。

Cursor:返回值,相当于结果集 ResultSet。

最后,它们同时返回一个 Cursor 对象,代表数据集的游标,有点类似于 JavaSE 中的 ResultSet。

下面是 Cursor 对象的常用方法:

```
c.move(int offset); //以当前位置为参考,移动到指定行
c.moveToFirst();      //移动到第一行
c.moveToLast();       //移动到最后一行
c.moveToPosition(int position); //移动到指定行
c.moveToPrevious(); //移动到前一行
c.moveToNext();       //移动到下一行
c.isFirst();          //是否指向第一条
c.isLast();      //是否指向最后一条
c.isBeforeFirst();  //是否指向第一条之前
c.isAfterLast();     //是否指向最后一条之后
c.isNull(int columnIndex);  //指定列是否为空(列基数为 0)
c.isClosed();          //游标是否已关闭
c.getCount();          //总数据项数
```

```
c. getPosition( );        //返回当前游标所指向的行数
c. getColumnIndex( String columnName);//返回某列名对应的列索引值
c. getString( int columnIndex);    //返回当前行指定列的值
```

游标实现代码如下：

```
String[ ] params = {12345,123456};
Cursor cursor = db. query( "user",columns,"ID = ?",params,null,null,null);//查询并获得游标
if( cursor. moveToFirst( ) ){//判断游标是否为空
    for( int i =0;i < cursor. getCount( );i + + ){
        cursor. move( i);//移动到指定记录
        String username = cursor. getString( cursor. getColumnIndex( "username");
        String password = cursor. getString( cursor. getColumnIndex( "password"));
    }
}
//通过 rawQuery 实现的带参数查询
Cursor result = db. rawQuery( "SELECT ID, name, inventory FROM mytable");
//Cursor c = db. rawQuery( "s name, inventory FROM mytable where ID = ?",new Stirng[ ]{"123456"});
result. moveToFirst( );
while (! result. isAfterLast( )) {
    int id = result. getInt(0);
    String name = result. getString(1);
    int inventory = result. getInt(2);
    // do something useful with these
    result. moveToNext( );
}
result. close( );
```

在上面的代码示例中已经用到了这几个常用方法，更多的信息可以参考官方文档中的说明。最后，当完成了对数据库的操作后，记得调用 SQLiteDatabase 的 close( )方法释放数据库连接，否则容易出现 SQLiteException。上面就是 SQLite 的基本应用，但在实际开发中，为更好地管理和维护数据库，会封装一个继承自 SQLiteOpenHelper 类的数据库操作类，然后以这个类为基础，再封装业务逻辑方法。

7.4.2　SQLiteOpenHelper 类介绍

SQLiteOpenHelper 是 SQLiteDatabase 的一个帮助类,用来管理数据库的创建和版本的更新。一般是建立一个类继承它,并实现它的 onCreate 和 onUpgrade 方法。SQLiteOpenHelper 常用函数见表 7 - 1。

表 7 - 1　SQLiteOpenHelper 常用函数

方法名	方法描述
SQLiteOpenHelper (Context context, String name, SQLiteDatabase. CursorFactory factory, int version)	构造方法 context:程序上下文环境,即 × × × Activity. this name :数据库名字 factory:游标工厂,默认为 null,即为使用默认工厂 version:数据库版本号
onCreate(SQLiteDatabase db)	创建数据库时调用
onUpgrade(SQLiteDatabase db, int oldVersion , int newVersion)	版本更新时调用
getReadableDatabase( )	创建或打开一个只读数据库
getWritableDatabase( )	创建或打开一个读写数据库

基于 SQLiteOpenHelper 实现对数据的操作如图 7 - 2 所示。首先创建数据库类,然后用一个 Dao 来封装所有的业务方法,实现 Dao 中调用的 getWritableDatabase( ) 和 getReadable-Database( ),最后使用这些数据操作方法来显示数据。

图 7 - 2　基于 SQLiteOpenHelper 实现对数据的操作

(1)创建数据库类。

```
import android.content.Context;
import android.database.sqlite.SQLiteDatabase;
import android.database.sqlite.SQLiteDatabase.CursorFactory;
import android.database.sqlite.SQLiteOpenHelper;
public class SqliteDBHelper extends SQLiteOpenHelper {
      //设置常数参量
      private static final String DATABASE_NAME = "diary_db";
    private static final int VERSION = 1;
    private static final String TABLE_NAME = "diary";
    //重载构造方法
    public SqliteDBHelper(Context context) {
         super(context, DATABASE_NAME, null, VERSION);
    }
    public SqliteDBHelper(Context context, String name, CursorFactory factory,
              int version) {
         super(context, name, factory, version);
    }
    //数据库第一次被创建时,onCreate()会被调用
    @Override
    public void onCreate(SQLiteDatabase db) {
         //数据库表的创建
         String strSQL = "create table "
                   + TABLE_NAME
                   + "(tid integer primary key autoincrement,title varchar(20),
                 weather varchar(10),context text,publish date)";
         //使用参数 db,创建对象
         db.execSQL(strSQL);
    }
    //数据库版本变化时,会调用 onUpgrade()
    @Override
    public void onUpgrade(SQLiteDatabase arg0, int arg1, int arg2) {
  }
```

```
}
```

正如上面所述，数据库第一次创建时，onCreate 方法会被调用，可以执行创建表的语句。当系统发现版本变化之后，会调用 onUpgrade 方法，可以执行修改表结构等语句。

（2）用一个 Dao 来封装所有的业务方法。

```
import android.content.Context;
import android.database.Cursor;
import android.database.sqlite.SQLiteDatabase;
import com.chinasoft.dbhelper.SqliteDBHelper;
public class DiaryDao {
private SqliteDBHelper sqliteDBHelper;
    private SQLiteDatabase db;
     // 重写构造方法
    public DiaryDao(Context context) {
            this.sqliteDBHelper = new SqliteDBHelper(context);
            db = sqliteDBHelper.getWritableDatabase();
     }
     // 读操作
      public String execQuery(final String strSQL) {
          try {
              System.out.println("strSQL>" + strSQL);
                // Cursor 相当于 JDBC 中的 ResultSet
                Cursor cursor = db.rawQuery(strSQL, null);
                // 始终让 cursor 指向数据库表的第 1 行记录
                cursor.moveToFirst();
                // 定义一个 StringBuffer 的对象,用于动态拼接字符串
                StringBuffer sb = new StringBuffer();
                // 循环游标,如果不是最后一项记录
                while (! cursor.isAfterLast()) {
                  sb.append(cursor.getInt(0) + "/" + cursor.getString(1) + "/"
                         + cursor.getString(2) + "/" + cursor.getString(3) + "/"
                         + cursor.getString(4) + "#");
                   //cursor 游标移动
                   cursor.moveToNext();
```

```
                }
                db.close();
                return sb.deleteCharAt(sb.length() - 1).toString();
            } catch (RuntimeException e) {
                e.printStackTrace();
                return null;
            }
        }
    // 写操作
    public boolean execOther(final String strSQL) {
            db.beginTransaction();  //开始事务
            try {
                System.out.println("strSQL" + strSQL);
                db.execSQL(strSQL);
                db.setTransactionSuccessful();  //设置事务成功完成
                db.close();
                return true;
            } catch (RuntimeException e) {
                e.printStackTrace();
                return false;
            }finally {
                db.endTransaction();      //结束事务
            }
        }
}
```

在 Dao 构造方法中实例化 sqliteDBHelper 并获取一个 SQLiteDatabase 对象，作为整个应用的数据库实例；在增删改信息时，采用事务处理，确保数据完整性；最后要注意释放数据库资源 db.close()，这一步骤在整个应用关闭时执行，这个环节容易被忘记，所以需要注意。

在 Dao 中获取数据库实例时使用了 getWritableDatabase()方法，下面分析 getWritableDatabase()和 getReadableDatabase()。

(3) SQLiteOpenHelper 中的 getReadableDatabase()方法实现。

```
public synchronized SQLiteDatabase getReadableDatabase() {
        if (mDatabase != null && mDatabase.isOpen()) {
```

```
        // 如果发现 mDatabase 不为空并且已经打开,则直接返回
        return mDatabase;
    }
    if (mIsInitializing) {
        // 如果正在初始化,则抛出异常
        throw new IllegalStateException("getReadableDatabase called recursively");
}
    // 开始实例化数据库 mDatabase
    try {
        // 注意这里是调用了 getWritableDatabase()方法
        return getWritableDatabase();
    } catch (SQLiteException e) {
        if (mName == null)
            {Throw e;}
        Log.e(TAG, "Couldn't open " + mName +
        " for writing (will try read-only):", e);
    }
    // 如果无法以可读写模式打开数据库 则以只读方式打开
    SQLiteDatabase db = null;
    try {
        mIsInitializing = true;
        String path = mContext.getDatabasePath(mName).getPath();
        // 以只读方式打开数据库
        db = SQLiteDatabase.openDatabase(path, mFactory,
        SQLiteDatabase.OPEN_READONLY);
        if (db.getVersion() != mNewVersion) {
            throw new SQLiteException
            ("Can't upgrade read-only database from version " +
            db.getVersion() + " to "
                    + mNewVersion + ": " + path);
        }
        onOpen(db);
        Log.w(TAG, "Opened " + mName + " in read-only mode");
```

```
            mDatabase = db;// 为 mDatabase 指定新打开的数据库
            return mDatabase;// 返回打开的数据库
        } finally {
            mIsInitializing = false;
            if (db ! = null && db ! = mDatabase)
                db. close( );
        }
    }
```

在 getReadableDatabase( )方法中，首先判断是否已存在数据库实例并且是打开状态。如果是，则直接返回该实例；否则，试图获取一个可读写模式的数据库实例。如果遇到磁盘空间已满等情况获取失败，再以只读模式打开数据库，获取数据库实例并返回，然后为 mDatabase 赋值为最新打开的数据库实例。

(4) getWritableDatabase( )方法的实现。

```
public synchronized SQLiteDatabase getWritableDatabase( ) {
    if (mDatabase ! = null && mDatabase. isOpen( ) && ! mDatabase. isReadOnly
        ( )) {
        // 如果 mDatabase 不为空已打开并且不是只读模式 则返回该实例
        return mDatabase;
    }
    if (mIsInitializing) {
        throw new IllegalStateException(" getWritableDatabase called recursively" );
    }
/
    boolean success = false;
    SQLiteDatabase db = null;
    // 如果 mDatabase 不为空，则加锁阻止其他的操作
    if (mDatabase ! = null)
        mDatabase. lock( );
    try {
        mIsInitializing = true;
        if (mName = = null) {
            db = SQLiteDatabase. create( null) ;
        } else {
```

```
            // 打开或创建数据库
            db = mContext.openOrCreateDatabase(mName, 0, mFactory);
        }
        // 获取数据库版本(如果刚创建的数据库,版本为 0)
        int version = db.getVersion();
        // 比较版本(代码中的版本 mNewVersion 为 1)
        if (version != mNewVersion) {
            db.beginTransaction();// 开始事务
            try {
                if (version == 0) {
                    // 执行 onCreate 方法
                    onCreate(db);
                } else {
                    onUpgrade(db, version, mNewVersion);
                }
                db.setVersion(mNewVersion);// 设置最新版本
                db.setTransactionSuccessful();// 设置事务成功
            } finally {
                db.endTransaction();// 结束事务
            }
        }
          onOpen(db);
        success = true;
        return db;// 返回可读写模式的数据库实例
    } finally {
        mIsInitializing = false;
        if (success) {
            // 打开成功
            if (mDatabase != null) {
                // 如果 mDatabase 有值则先关闭
                try {
                    mDatabase.close();
                } catch (Exception e) {
```

```
                    }
                    mDatabase. unlock( );// 解锁
                }
                mDatabase = db;// 赋值给 mDatabase
            } else {
                // 打开失败的情况:解锁、关闭
                if (mDatabase ! = null)
                    mDatabase. unlock( );
                if (db ! = null)
                    db. close( );
            }
        }
    }
```

首先判断 mDatabase 如果不为空,已打开并且不是只读模式,则直接返回;否则,如果 mDatabase 不为空,则加锁,然后开始打开或创建数据库比较版本,根据版本号来调用相应的方法为数据库设置新版本号,最后释放旧的不为空的 mDatabase 并解锁,把新打开的数据库实例赋予 mDatabase,并返回最新实例。在遇到磁盘空间不满的情况,getReadableDatabase( )一般都会返回与 getWritableDatabase( )一样的数据库实例,所以在 DBManager 构造方法中使用 getWritableDatabase( )获取整个应用所使用的数据库实例是可行的。如果担心磁盘空间已经满的情况会发生,那么可以先用 getWritableDatabase( )获取数据实例,如果遇到异常,再试图用 getReadableDatabase( )获取实例,当然这时获取的实例只能读不能写。

(5)使用这些数据操作方法来显示数据。

```
public class SQLiteActivity extends Activity {
    public DiaryDao diaryDao;
     //因为 getWritableDatabase 内部调用了 mContext. openOrCreateDatabase(mName, 0, mFactory);
    //所以要确保 context 已初始化,可以把实例化 Dao 的步骤放在 Activity 的 onCreate 里
    @ Override
    protected void onCreate( Bundle savedInstanceState) {
        diaryDao = new DiaryDao( SQLiteActivity. this);
        initDatabase( );
    }
```

```
    class ViewOcl implements View.OnClickListener {
        @Override
        public void onClick(View v) {
            String strSQL;
            boolean flag;
            String message;
            switch (v.getId()) {
            case R.id.btnAdd:
                String title = txtTitle.getText().toString().trim();
                String weather = txtWeather.getText().toString().trim();;
                String context = txtContext.getText().toString().trim();;
                String publish = new SimpleDateFormat("yyyy-MM-dd HH:mm:ss")
                        .format(new Date());
                // 动态组件 SQL 语句
                strSQL = "insert into diary values(null,'" + title + "','"
                        + weather + "','" + context + "','" + publish + "')";
                flag = diaryDao.execOther(strSQL);
                //返回信息
                message = flag?"添加成功":"添加失败";
                    Toast.makeText(getApplicationContext(), message, Toast.LENGTH_LONG).show();
                break;
            case R.id.btnDelete:
                strSQL = "delete from diary where tid = 1";
                flag = diaryDao.execOther(strSQL);
                //返回信息
                message = flag?"删除成功":"删除失败";
                    Toast.makeText(getApplicationContext(), message, Toast.LENGTH_LONG).show();
                break;
            case R.id.btnQuery:
                strSQL = "select * from diary order by publish desc";
```

```
                    String data = diaryDao.execQuery(strSQL);
                    Toast.makeText(getApplicationContext(), data, Toast.LENGTH_LONG).show();
                    break;
                case R.id.btnUpdate:
                    strSQL = "update diary set title = SK'测试标题 1-1' where tid = 1";
                    flag = diaryDao.execOther(strSQL);
                    //返回信息
                    message = flag?"更新成功":"更新失败";
                    Toast.makeText(getApplicationContext(), message, Toast.LENGTH_LONG).show();
                    break;
            }
        }
    }
    private void initDatabase() {
        // 创建数据库对象
        SqliteDBHelper sqliteDBHelper = new SqliteDBHelper(SQLiteActivity.this);
        sqliteDBHelper.getWritableDatabase();
        System.out.println("数据库创建成功");
    }
}
```

Android sqlite3 数据库管理工具 Android SDK 的 tools 目录下提供了一个 sqlite3.exe 工具,这是一个简单的 sqlite 数据库管理工具。开发者可以方便地使用它对 sqlite 数据库进行命令行的操作。程序运行生成的 *.db 文件一般位于"/data/data/项目名(包括所处包名)/databases/*.db",因此要对数据库文件进行操作,需要先找到数据库文件。

1. 进入 shell 命令

```
adb shell
```

2. 找到数据库文件

```
#cd data/data
#ls                    --列出所有项目
```

```
#cd project_name        - -进入所需项目名
#cd databases
#ls                     - -列出现寸的数据库文件
```

3. 进入数据库

```
#sqlite3 test_db    - -进入所需数据库
```

会出现类似如下字样:

```
SQLite version 3.6.22
Enter ".help" for instructions
Enter SQL statements terminated with a ";"
sqlite >
```

至此,可对数据库进行 sql 操作。

4. sqlite 常用命令

```
>.databases            - -查看当前数据库
>.tables               - -查看当前数据库中的表
>.help                 - -sqlite3 帮助
>.schema               - -各个表的生成语句
```

SQLiteSpy 是一个快速和紧凑的图形用户界面的 SQLite 数据库管理工具,可以实现与 sqlite3.exe 相同的功能。

SQLiteSpy 主要特点如下:

(1)树状显示所有的架构,包括表、列、索引和触发器在数据库中包含的项目;

(2)本机的 SQL 数据类型显示不同的背景颜色来帮助检测类型错误;

(3)SQLiteSpy 完全支持 SQLite 的 Unicode 的能力;

(4)支持正则表达式,并增加了完整的 Perl 的正则表达式语法;

(5)支持加密。

7.5　网络存储

Android 提供了通过网络来实现数据的存储和获取的方法,可以调用 WebService 返回的数据或是解析 HTTP 协议实现网络数据交互。需要熟悉 java.net. 和 Android.net. 这两个包的内容,详细的类与方法的说明请参考 SDK。

基于网络存储实现将数据发送到电子邮件中备份,要发送电子邮件首先需要在电子邮件中配置电子邮件账户。Android 中发送电子邮件通过 startActivity 方法来调用要发送的邮件数据的 Intent。可以通过 putExtra 方法来设置邮件的主题、内容、附件等。当点击返回按

钮 - - - - > back 时，就会出现发送邮件，发送邮件的界面如图 7 - 3 所示，当点击发送的时候就会发送给设置的邮箱一封邮件。

图 7 - 3　发送邮件的界面

发送邮件界面的示例代码如下：

```
package test.datastore;
import android.app.Activity;
import android.content.Intent;
import android.net.Uri;
import android.os.Bundle;
import android.view.KeyEvent;
public class Activity01 extends Activity {
  private int miCount = 0;
  @Override
   public void onCreate(Bundle savedInstanceState) {
   super.onCreate(savedInstanceState);
   setContentView(R.layout.main);
   miCount = 1000;
   }
   public boolean onKeyDown(int keyCode, KeyEvent event) {
   if (keyCode == KeyEvent.KEYCODE_BACK) {
```

```
// 退出应用程序时保存数据
/* 发送邮件的地址 */
Uri uri = Uri.parse("mailto:yongjinquanli@gmail.com");
// 创建 Intent
Intent it = new Intent(Intent.ACTION_SENDTO, uri);
// 设置邮件的主题
it.putExtra(android.content.Intent.EXTRA_SUBJECT, "数据备份");
// 设置邮件的内容
it.putExtra(android.content.Intent.EXTRA_TEXT, "本次计数:"
+ miCount);
// 开启
startActivity(it);
return true;
}
return super.onKeyDown(keyCode, event);
}
}
```

下面演示通过网络来读取一个文件的内容,然后将其显示在定义好的 TextView 上。将文件 xh.txt 放置在 tomcat 服务器上,文件的内容为“欢迎疯狂热爱 android 开发的朋友加入”。

由于在程序中访问了外部网络,因此需要在 AndroidManifest.xml 文件中给予权限,其代码如下:

```
<uses-permission android:name="android.permission.INTERNET" />
package test.datastore;
import java.io.BufferedInputStream;
import java.io.InputStream;
import java.net.URL;
import java.net.URLConnection;
import android.app.Activity;
import android.graphics.Color;
import android.os.Bundle;
import android.widget.TextView;
public class Activity01 extends Activity {
    @Override
```

```
public void onCreate(Bundle savedInstanceState) {
super. onCreate(savedInstanceState);
setContentView(R. layout. main);
TextView tv = new TextView(this);
String myString = null;
try {
/* 定义要访问的地址 url */
URL uri = new URL("http://192. 168. 0. 100:8080/examples/xh. txt");
/* 打开这个 url */
URLConnection uConnection = uri. openConnection();
// 从上面的链接中取得 InputStream
InputStream is = uConnection. getInputStream();
// new 一个带缓冲区的输入流
BufferedInputStream bis = new BufferedInputStream(is);
/* 解决中文乱码 */
byte[] bytearray = new byte[1024];
int current = -1;
int i = 0;
while ((current = bis. read()) != -1) {
bytearray[i] = (byte) current;
i++;
}
myString = new String(bytearray, "GB2312");
} catch (Exception e) {
// 获取异常信息
myString = e. getMessage();
}
// 设置到 TextView 颜色
tv. setTextColor(Color. RED);
// 设置字体
tv. setTextSize(20. 0f);
tv. setText(myString);
// 将 TextView 显示到屏幕上
```

```
        this.setContentView(tv);
    }
}
```

7.6　数据共享

Android 官方指出的 Android 的数据存储方式总共有五种,分别是 Shared Preferences、内部(文件)存储、外部储存、SQLite、网络存储。但是这些存储都只是在单独的一个应用程序之中达到一个数据的共享,有时需要操作其他应用程序的一些数据,如需要操作系统里的媒体库、通讯录等,这时就可能通过 ContentProvider 来满足需求了。ContentProvider 为存储和获取数据提供统一的接口,可以在不同的应用程序之间共享数据。Android 已经为常见的一些数据提供了默认的 ContentProvider。

1. ContentProvider 使用表的形式来组织数据

无论数据的来源是什么,ContentProvider 都会认为是一种表,然后把数据组织成表格。

2. ContentProvider 提供的方法

query:查询。

insert:插入。

update:更新。

delete:删除。

getType:得到数据类型。

onCreate:创建数据时调用的回调函数。

3. 公共 UPI

每个 ContentProvider 都有一个公共的 URI,这个 URI 用于表示这个 ContentProvider 所提供的数据。Android 提供的 ContentProvider 都存放在 android.provider 包中。

7.6.1　ContentProvider 的内部原理

Android 为常见的一些数据提供了默认的 ContentProvider(包括音频、视频、图片和通讯录等)。ContentProvider 为存储和获取数据提供了统一的接口。ContentProvide 对数据进行封装,不用关心数据存储的细节,使用表的形式来组织数据。使用 ContentProvider 可以在不同的应用程序之间共享数据。

总的来说,使用 ContentProvider 对外共享数据的好处是统一了数据的访问方式,因为系统的每一个资源给其一个名字,如通话记录。

(1)每一个 ContentProvider 都拥有一个公共的 URI,用于表示这个 ContentProvider 所提供的数据。

(2)Android 提供的 ContentProvider 都存放在 android. provider 包中,将其分为 A、B、C、D 四个部分。

A:标准前缀,用来说明一个 Content Provider 控制这些数据是无法改变的。

B:URI 的标识,用于唯一标识这个 ContentProvider,外部调用者可以根据这个标识来找到它。它定义了是哪个 Content Provider 提供这些数据。对于第三方应用程序,为保证 URI 标识的唯一性,它必须是一个完整的、小写的类名。这个标识在元素的 authorities 属性中说明,用于定义该 ContentProvider 的包、类的名称。

C:路径(path),通俗地讲就是要操作的数据库中表的名字,也可以自己定义,记得在使用的时候保持一致就可以。

D:如果 URI 中包含表示需要获取的记录的 ID,就返回该 id 对应的数据;如果没有 ID,就表示返回全部。

路径(path)可以用来表示要操作的数据,路径的构建应根据业务而定。

要操作 person 表中 id 为 10 的记录,可以构建这样的路径:/person/10。

要操作 person 表中 id 为 10 的记录的 name 字段,可以构建这样的路径:person/10/name。

要操作 person 表中的所有记录,可以构建这样的路径:/person。

要操作×××表中的记录,可以构建这样的路径:/×××。

当然,要操作的数据不一定来自数据库,也可以是文件、xml 或网络等其他存储方式。

要操作 xml 文件中 person 节点下的 name 节点,可以构建这样的路径:/person/name。

如果要把一个字符串转换成 Uri,可以使用 Uri 类中的 parse()方法:Uri uri = Uri. parse("content://com. bing. provider. personprovider/person")。

举例,自定义一个 ContentProvider,实现对其的操作,定义一个 CONTENT_URI 常量(里面的字符串必须是唯一):

Public static final Uri CONTENT_URI = Uri. parse("content://com. test. MyContentprovider");

如果有子表,URI 如下:

```
Public static final Uri CONTENT_URI =
Uri. parse("content://com. test. MyContentProvider/users");
Public class MyContentProvider extends ContentProvider
package com. test. cp;
import java. util. HashMap;
```

```
import com.test.cp.MyContentProviderMetaData.UserTableMetaData;
import com.test.data.DatabaseHelp;
import android.content.ContentProvider;
import android.content.ContentUris;
import android.content.ContentValues;
import android.content.UriMatcher;
import android.database.Cursor;
import android.database.sqlite.SQLiteDatabase;
import android.database.sqlite.SQLiteQueryBuilder;
import android.net.Uri;
import android.text.TextUtils;
public class MyContentProvider extends ContentProvider {
 //访问表的所有列
 public static final int INCOMING_USER_COLLECTION = 1;
 //访问单独的列
public static final int INCOMING_USER_SINGLE = 2;
//操作 URI 的类
public static final UriMatcher uriMatcher;
//为 UriMatcher 添加自定义的 URI
static{
uriMatcher = new UriMatcher(UriMatcher.NO_MATCH);
uriMatcher.addURI(MyContentProviderMetaData.AUTHORITIES,"/user",
INCOMING_USER_COLLECTION);
uriMatcher.addURI(MyContentProviderMetaData.AUTHORITIES,"/user/#",
INCOMING_USER_SINGLE);
    }
private DatabaseHelp dh;
//为数据库表字段起别名
public static HashMap userProjectionMap;
static
{
userProjectionMap = new HashMap();
userProjectionMap.put(UserTableMetaData._ID,UserTableMetaData._ID);
```

```
userProjectionMap. put( UserTableMetaData. USER_NAME,
UserTableMetaData. USER_NAME) ;
}
// 删除表数据
    @ Override
public int delete( Uri uri, String selection, String[ ] selectionArgs) {
System. out. println( "delete" ) ;
//得到一个可写的数据库
SQLiteDatabase db = dh. getWritableDatabase( ) ;
//执行删除,得到删除的行数
int count = db. delete( UserTableMetaData. TABLE_NAME, selection,
selectionArgs) ;
return count;
}
    // 数据库访问类型
    @ Override
public String getType( Uri uri) {
System. out. println( "getType" ) ;
//根据用户请求,得到数据类型
switch (uriMatcher. match( uri) ) {
case INCOMING_USER_COLLECTION:
return MyContentProviderMetaData.
UserTableMetaData. CONTENT_TYPE;
case INCOMING_USER_SINGLE:
return MyContentProviderMetaData. UserTableMetaData.
CONTENT_TYPE_ITEM;
default:
throw new IllegalArgumentException( "UnKnown URI" + uri) ;    }    }
// 插入数据
    @ Override
public Uri insert( Uri uri, ContentValues values) {
//得到一个可写的数据库
SQLiteDatabase db = dh. getWritableDatabase( ) ;
```

```
//向指定的表插入数据,得到返回的 Id
long rowId = db.insert(UserTableMetaData.TABLE_NAME, null, values);
if(rowId > 0){//判断插入是否执行成功
//如果添加成功,利用新添加的 Id 和
Uri insertedUserUri = ContentUris.withAppendedId
(UserTableMetaData.CONTENT_URI, rowId);
//通知监听器,数据已经改变
getContext().getContentResolver().notifyChange(insertedUserUri, null);
return insertedUserUri;
}
return uri;
}
        // 创建 ContentProvider 时调用的回调函数
        @Override
public boolean onCreate() {
System.out.println("onCreate");
//得到数据库帮助类
dh = new DatabaseHelp(getContext(),
MyContentProviderMetaData.DATABASE_NAME);
return false;
}
//查询数据库
        @Override
public Cursor query(Uri uri, String[] projection, String selection,
String[] selectionArgs, String sortOrder) {
//创建一个执行查询的 Sqlite
SQLiteQueryBuilder qb = new SQLiteQueryBuilder();
//判断用户请求,查询所有还是单个
switch(uriMatcher.match(uri)){
case INCOMING_USER_COLLECTION:
//设置要查询的表名
qb.setTables(UserTableMetaData.TABLE_NAME);
//设置表字段的别名
```

```
qb. setProjectionMap( userProjectionMap) ;
break;
case INCOMING_USER_SINGLE:
qb. setTables( UserTableMetaData. TABLE_NAME) ;
qb. setProjectionMap( userProjectionMap) ;
qb. appendWhere( UserTableMetaData. _ID + " =" +
uri. getPathSegments( ). get(1) ) ;
break; }
//设置排序
String orderBy;
if( TextUtils. isEmpty( sortOrder) ) {
orderBy = UserTableMetaData. DEFAULT_SORT_ORDER; }
else{
orderBy = sortOrder;
}
//得到一个可读的数据库
SQLiteDatabase db = dh. getReadableDatabase( ) ;
//执行查询,把输入传入
Cursor c = qb. query( db, projection, selection,
selectionArgs, null, null, orderBy) ;
//设置监听
c. setNotificationUri( getContext( ). getContentResolver( ) , uri) ;
return c; }
//更新数据库
@ Override
public int update( Uri uri, ContentValues values, String selection,
String[ ] selectionArgs) {
System. out. println( " update" ) ;
//得到一个可写的数据库
SQLiteDatabase db = dh. getWritableDatabase( ) ;
//执行更新语句,得到更新的条数
int count = db. update( UserTableMetaData.
TABLE_NAME, values, selection, selectionArgs) ;
```

```
return count;
}
    }
```

在 AndroidManifest. xml 中进行声明:

```
android:name = ". cp. MyContentProvider"
android:authorities = "com. test. cp. MyContentProvider"
/>
```

name 所对应的项为(contentProvider(数据存储))的具体操作的类。

authorities(授权)即访问这个. MyContentProvider 类的权限,说明 - - - com. test. cp. MyContentProvider 是可以访问的,别的类可以通过。

Uri = Uri. parse("content://" + AUTHORITY);

public static final String AUTHORITY = "com. test. cp. MyContentProvider";

对这个数据库进行直接的增删改查的操作,如果这个数据库有多个表,则这个 Uri 需要加上对应的表名,如 Uri = Uri. parse("content://" + AUTHORITY + "/User"); - - - User 为其中一个表。

下面示例为 ContentProvider 提供一个常量类 MyContentProviderMetaData. java:

```
package com. test. cp;
import android. net. Uri;
import android. provider. BaseColumns;
public class MyContentProviderMetaData {
 //URI 的指定,此处的字符串必须和声明的 authorities 一致
 public static final String AUTHORITIES =
 "com. wangweida. cp. MyContentProvider";
 //数据库名称
 public static final String DATABASE_NAME = "myContentProvider. db";
  //数据库的版本
  public static final int DATABASE_VERSION = 1;
  //表名
  public static final String USERS_TABLE_NAME = "user";
  public static final class UserTableMetaData implements BaseColumns{
  //表名
  public static final String TABLE_NAME = "user";
  //访问该 ContentProvider 的 URI
```

```
public static final Uri CONTENT_URI = Uri.parse("content://" + AUTHORITIES + "/user");
//该 ContentProvider 所返回的数据类型的定义
public static final String CONTENT_TYPE =
"vnd.android.cursor.dir/vnd.myprovider.user";
public static final String CONTENT_TYPE_ITEM =
"vnd.android.cursor.item/vnd.myprovider.user";
//列名
public static final String USER_NAME = "name";
//默认的排序方法
public static final String DEFAULT_SORT_ORDER = "_id desc";
}}
```

最后整个应用被编译成 apk。安装之后,该应用中的 contentProvider 就可以被其他应用访问了。对于 Provider 使用者来说,如果特定 Provider 有 permission 要求,则要在自己的 Androidmanifest.xml 中添加指定 Permission 引用, 如:

```
<uses-permission android:name="com.example.demos.permission.READ_WORDS"/>
<uses-permission android:name="com.example.demos.permission.WRITE_WORDS"/>
```

Android 提供了 Context 级别的 ContentResolver 对象来对 Content Provider 进行操作。正是因为有了 ContentResolver, 使用者才不用关心 Provider 到底是哪个应用或哪个类实现的, 只要知道它的 uri 就能访问。ContentResolver 对象存在于每个 Context 中,几乎所有对象都有自己的 Context,使用 getContext().getContentResovler()可以获取 Context。

有些情况下,ContentProvider 使用者想监听数据的变化,可以注册一个 Observer:

```
Class MyContentObServer extends ContentObserver{
Public MyConentObServer(Handler handler){
Super(handler);
}
Public void onChange(boolean selfNotify){…}
}
getContext().getContentResolver().registerContentObserver(uri, true, new MycontentObserver(new Handler()));
```

7.6.2　UriMatcher 类和 ContentUris 类使用介绍

ContentProvider 向外界提供了一个标准的，也是唯一的用于查询的接口：

public final Cursor query(Uri uri, String[] projection, String selection, String[] selectionArgs, String sortOrder);

其中 uri 用于指定哪一个数据源，当一个数据源含有多个内容(如多个表)时，就需要用不同的 Uri 进行区分，例如：

public static final Uri CONTENT_URI_A = Uri.parse("content://" + AUTHORITY + "/" + TABLE_A);

public static final Uri CONTENT_URI_B = Uri.parse("content://" + AUTHORITY + "/" + TABLE_B);

这时，使用 UriMatcher 就可以方便地过滤到 TableA 或 TableB，然后进行下一步查询。如果不用 UriMatcher，就需要手动过滤字符串，用起来有点麻烦，可维护性也不好。

Uri 代表了要操作的数据，所以经常需要解析 Uri，并从 Uri 中获取数据。Android 系统提供了两个用于操作 Uri 的工具类，分别为 UriMatcher 和 ContentUris 。掌握它们的使用，会便于开发工作。

UriMatcher 类使用介绍：UriMatcher 类用于匹配 Uri，它的用法如下。

首先把需要匹配 Uri 路径全部注册，如下：

```
//常量 UriMatcher.NO_MATCH 表示不匹配任何路径的返回码
UriMatcher sMatcher = new UriMatcher(UriMatcher.NO_MATCH);
//如果 match()方法匹配 content://com.bing.procvide.personprovider/person 路径，返回匹配码为 1
sMatcher.addURI("com.bing.procvide.personprovider", "person", 1);
//添加需要匹配 uri，如果匹配就会返回匹配码
//如果 match()方法匹配 content://com.bing.provider.personprovider/person/230 路径，返回匹配码为 2
sMatcher.addURI("com.bing.provider.personprovider", "person/#", 2);
//#号为通配符
switch (sMatcher.match(Uri.parse("content://com.ljq.provider.personprovider/person/10"))) {
case 1      break;
case 2      break;
default://不匹配
```

```
break;
}
```

注册完需要匹配的 Uri 后,就可以使用 sMatcher. match(uri)方法对输入的 Uri 进行匹配,如果匹配就返回匹配码,匹配码是调用 addURI()方法传入的第三个参数,假设匹配 content://com. ljq. provider. personprovider/person 路径,返回的匹配码为 1。

ContentUris 类用于操作 Uri 路径后面的 ID 部分,它有以下两个比较实用的方法。

withAppendedId(uri, id)用于为路径加上 ID 部分:

```
Uri uri = Uri.parse("content://com.bing.provider.personprovider/person")
Uri resultUri = ContentUris.withAppendedId(uri, 10);
//生成后的 Uri 为:content://com.bing.provider.personprovider/person/10
```

parseId(uri)方法用于从路径中获取 ID 部分:

```
Uri uri = Uri.parse("content://com.ljq.provider.personprovider/person/10")
long personid = ContentUris.parseId(uri);//获取的结果为 10
```

第 8 章　位置服务与地图应用

8.1　位 置 服 务

位置服务(Location - Based Services,LBS)又称定位服务或基于位置的服务,融合了 GPS 定位、移动通信、导航等多种技术,提供与空间位置相关的综合应用服务。近年来,基于位置的服务发展非常迅速,涉及商务、医疗、工作和生活的各个方面,为用户提供定位、追踪和敏感区域警告等一系列服务。

8.1.1　Android 平台的位置服务 API

Android 平台支持提供位置服务的 API,在开发过程中主要用到 LocationManager 对象。LocationManager 可以用来获取当前的位置、追踪设备的移动路线或设定敏感区域,在进入或离开敏感区域时设备会发出特定警报。

为使开发的程序能够提供位置服务,首先的问题是如何获取 LocationManager。获取 LocationManager 可以通过调用 android. app. Activity. getSystemService( ) 函数获取,代码如下:

```
1   String serviceString = Context. LOCATION_SERVICE;
2   LocationManager LocationManager = (LocationManager) getSystemService(serviceString);
```

在上述代码中,第 1 行的 Context. LOCATION_SERVICE 指明获取的是位置服务;第 2 行的 getSystemService( ) 函数,可以根据服务名称获取 Android 提供的系统级服务。

Android 支持的系统级服务见表 8 - 1。

表 8 - 1　Android 支持的系统级服务

Context 类的静态常量	返回对象	说明
LOCATION_SERVICE	LocationManager	控制位置等设备的更新
WINDOW_SERVICE	WindowManager	最顶层的窗口管理器
LAYOUT_INFLATER_SERVICE	LayoutInflater	将 XML 资源实例化为 View
POWER_SERVICE	PowerManager	电源管理
ALARM_SERVICE	AlarmManager	在指定时间接受 Intent

续表 8－1

Context 类的静态常量	返回对象	说明
NOTIFICATION_SERVICE	NotificationManager	后台事件通知
KEYGUARD_SERVICE	KeyguardManager	锁定或解锁键盘
SEARCH_SERVICE	SearchManager	访问系统的搜索服务
VIBRATOR_SERVICE	Vibrator	访问支持振动的硬件
CONNECTIVITY_SERVICE	ConnectivityManager	网络连接管理
WIFI_SERVICE	WifiManager	Wi－Fi 连接管理
INPUT_METHOD_SERVICE	InputMethodManager	输入法管理

在获取到 LocationManager 后，还需要指定 LocationManager 的定位方法，然后才能够调用 LocationManager. getLastKnowLocation( )方法获取当前位置。目前 LocationManager 中主要两种定位方法，LocationManager 支持的定位方法见表 8－2。

表 8－2　LocationManager 支持的定位方法

LocationManager 类的静态常量	说明
GPS_PROVIDER	使用 GPS 定位，利用卫星提供精确的位置信息，但定位速度和质量受卫星数量和环境情况的影响，此外需要用户权限： android. permissions. ACCESS_FINE_LOCATION 用户权限
NETWORK_PROVIDER	使用网络定位，利用基站或 Wi－Fi 访问提供近似的位置信息，但速度较 GPS 定位要迅速，此外需要用户权限： android. permission. ACCESS_COARSE_LOCATION 或 android. permission. ACCESS_FINE_LOCATION

在指定 LocationManager 的定位方法后，可以调用 getLastKnownLocation( )方法获取当前的位置信息。以使用 GPS 定位为例，获取位置信息的代码如下：

```
1  String provider = LocationManager.GPS_PROVIDER;
2  Location location = locationManager.getLastKnownLocation(provider);
```

代码第 2 行返回的 Location 对象中包含了可以确定位置的信息，如经度、纬度和速度等。

通过调用 Location 中的 getLatitude( )和 getLonggitude( )方法可以分别获取位置信息中的纬度和经度，示例代码如下：

```
double lat = location.getLatitude();
double lng = location.getLongitude();
```

在很多提供定位服务的应用程序中,不仅需要获取当前的位置信息,还需要监视位置的变化,在位置改变时调用特定的处理方法。LocationManager 提供了一种便捷、高效的位置监视方法 requestLocationUpdates(),可以根据位置的距离变化和时间间隔设定,产生位置改变事件的条件,这样可以避免因微小的距离变化而产生大量的位置改变事件。LocationManager 中设定监听位置变化的代码如下:

```
locationManager.requestLocationUpdates(provider, 1000, 5, locationListener);
```

在上面的代码中,第 1 个参数是定位的方法,GPS 定位或网络定位;第 2 个参数是产生位置改变事件的时间间隔,单位为 μs;第 3 个参数是距离条件,单位为 m;第 4 个参数是回调函数,用于处理位置改变事件。代码将产生位置改变事件的条件设定为距离改变 5 m,时间间隔为 1 s。

实现 locationListener 的代码如下:

```
1  LocationListener locationListener = new LocationListener(){
2  public void onLocationChanged(Location location) {
3  }
4  public void onProviderDisabled(String provider) {
5  }
6  public void onProviderEnabled(String provider) {
7  }
8  public void onStatusChanged(String provider, int status, Bundle extras) {
9  }
10 };
```

在上面的代码中,第 2 行代码 onLocationChanged()在位置改变时被调用,第 4 行的 onProviderDisabled()在用户禁用具有定位功能的硬件时被调用,第 6 行的 onProviderEnabled()在用户启用具有定位功能的硬件时被调用,第 8 行的 onStatusChanged()在定位功能硬件状态改变时被调用(如从不可获取位置信息状态到可以获取位置信息的状态,反之亦然)。

为使 GPS 定位功能生效,还需要在 AndroidManifest.xml 文件中加入用户许可,实现代码如下:

```
<uses-permission
android:name="android.permission.ACCESS_FINE_LOCATION"/>
```

8.1.2 CurrentLocationTest 实例

CurrentLocationTest 是一个提供基本位置服务的实例,可以显示当前位置信息,并能够监视设备的位置变化，CurrentLocationTest 实例用户界面如图 8 - 1 所示。

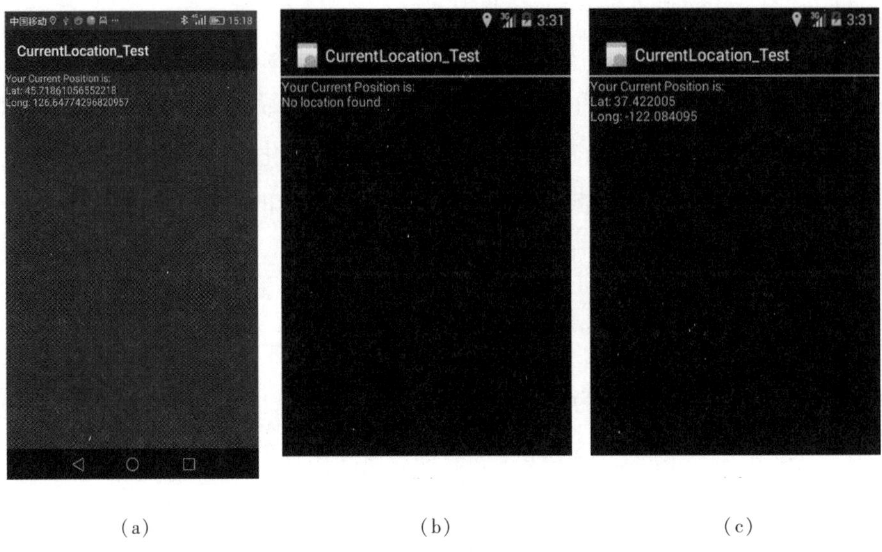

(a)　　　　(b)　　　　(c)

图 8 - 1　CurrentLocationTest **实例用户界面**

调试位置服务实例最理想的方式是在真机上运行，如图 8 - 1(a) 所示。但在没有真机的情况下，也可以采用 Android 模拟器运行程序，不过由于 Android 模拟器不支持硬件模拟，因此需要使用 Android 模拟器的控制器来模拟设备的位置变化：首先打开 DDMS 中的模拟器控制器，在 Location Controls 中的 Longitude 和 Latitude 部分输入设备当前的经度和纬度；然后点击 Send 按钮，就将虚拟的位置信息发送到 Android 模拟器中。模拟器控制器如图 8 - 2 所示。

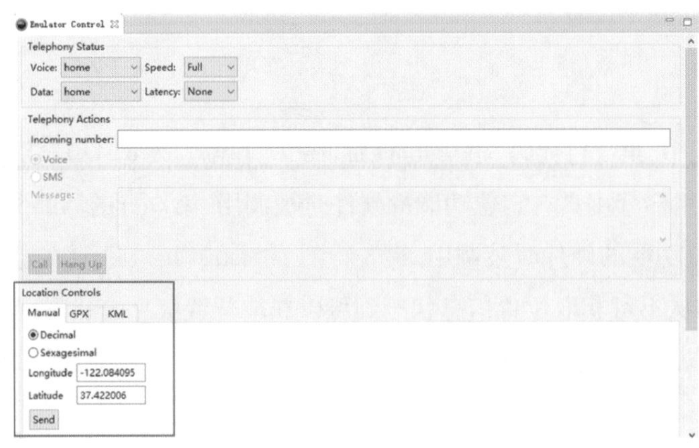

图 8 - 2　模拟器控制器

在程序运行过程中，可以在模拟器控制器中改变经度和纬度坐标值，程序在检测到位置的变化后会将最新的位置信息显示在界面上，如图 8 - 1(c) 所示。下面给出 CurrentLocationTest 实例中主活动的 Java 代码：

```
public class CurrentLocationTestActivity extends Activity {
  public void onCreate(Bundle savedInstanceState) {
      super.onCreate(savedInstanceState);
      setContentView(R.layout.main);
      String serviceString = Context.LOCATION_SERVICE;
      LocationManager locationManager = (LocationManager)getSystemService(serviceString);
      String provider = LocationManager.GPS_PROVIDER;
      Location location = locationManager.getLastKnownLocation(provider);
      getLocationInfo(location);
       locationManager.requestLocationUpdates(provider, 1000, 0, locationListener);
   }
   private void getLocationInfo(Location location) {
     String latLongInfo;
     TextView locationText = (TextView)findViewById(R.id.label);
     if (location != null) {
     double lat = location.getLatitude();
     double lng = location.getLongitude();
     latLongInfo = "Lat: " + lat + "\nLong: " + lng;
     }
     else{
     latLongInfo = "No location found";
     }
     locationText.setText("Your Current Position is:\n" + latLongInfo);
   }
   private final LocationListener locationListener = new LocationListener(){
 public void onLocationChanged(Location location) {
 getLocationInfo(location);
 }
 public void onProviderDisabled(String provider) {
 getLocationInfo(null);
 }
```

```
32  public void onProviderEnabled(String provider) {
33  getLocationInfo(null);
34  }
35  public void onStatusChanged(String provider, int status, Bundle extras) {
36  }
37      };
38}
```

LocationManager. GPS_PROVIDER 精度较高,但是慢且消耗电力,而且可能因为天气原因或者障碍物而无法获取卫星信息,另外设备可能没有 GPS 模块。LocationManager. NETWORK_PROVIDER 通过网络获取定位信息,精度低,耗电少,获取信息速度较快,不依赖 GPS 模块。为了程序的通用性,希望动态选择 location provider。在 Android 手机上实际提供的定位服务采用的定位模式如图 8 -3(a)所示,可以看出其可以根据用户的选择而动态地采取不同的 location provider。

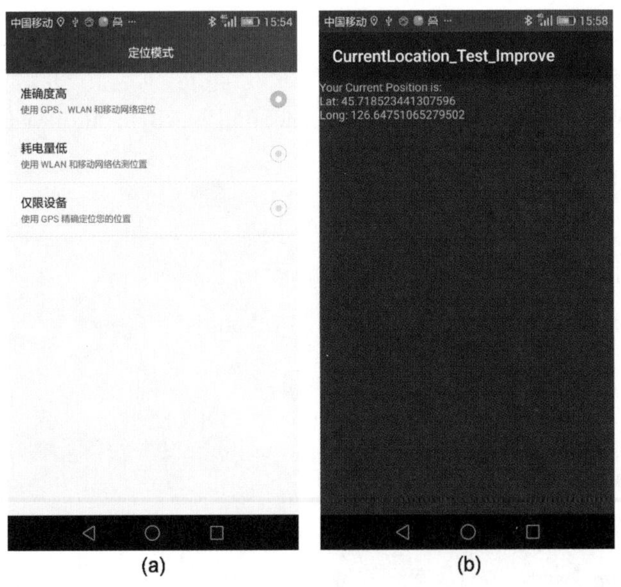

图 8 -3　用户自定义选择 location provider

下面给出改进实现的关键代码。这里使用到了 Criteria 类,可根据当前设备情况自动选择哪种 location provider。使用下面的 7 行代码将上面例子中的第 7、第 8 两行代码进行替换,程序执行结果如图 8 -3(b)所示,程序代码如下:

```
1  Criteria criteria = new Criteria();
2  criteria. setAccuracy(Criteria. ACCURACY_FINE); //设置为最大精度
3  criteria. setAltitudeRequired(false); //不要求海拔信息
```

```
criteria. setBearingRequired( false) ; //不要求方位信息
criteria. setCostAllowed( true) ; //设置是否允许运营商收费
criteria. setPowerRequirement( Criteria. POWER_LOW) ; //对电量的要求
location =
locationManager. getLastKnownLocation
locationManager. getBestProvider( criteria, true) ;
```

8.2　百度地图应用

开发者可利用 SDK 提供的接口,使用百度提供的基础地图数据。目前百度地图 SDK 所提供的地图等级为 3 ~ 19 级,所包含的信息有建筑物、道路、河流、学校、公园等内容。百度地图上 1 cm 代表实地距离依次为{“20 m”,“50 m”,“100 m”,“200 m”,“500 m”,“1 km”,“2 km”,“5 km”,“10 km”,“20 km”,“25 km”,“50 km”,“100 km”,“200 km”,“500 km”,“1 000 km”,“2 000 km”},其中缩放级别 3 对应 2 000 km。

所有叠加或覆盖到地图的内容统称为地图覆盖物,如标注、矢量图形元素(包括折线、多边形和圆等)、定位图标等。覆盖物拥有自己的地理坐标,当拖动或缩放地图时,它们会相应地移动。

百度地图 SDK 为广大开发者提供的基础地图和上面的各种覆盖物元素具有一定的层级压盖关系,具体如下(从下至上的顺序):

(1)基础底图(包括底图、底图道路、卫星图等);

(2)地形图图层(GroundOverlay);

(3)热力图图层(HeatMap);

(4)实时路况图图层(BaiduMap. setTrafficEnabled(true);) ;

(5)百度城市热力图(BaiduMap. setBaiduHeatMapEnabled(true);) ;

(6)底图标注(指的是底图上面自带的那些 POI 元素);

(7)几何图形图层(点、折线、弧线、圆、多边形);

(8)标注图层(Marker),文字绘制图层(Text);

(9)指南针图层(当地图发生旋转和视角变化时,默认出现在左上角的指南针);

(10)定位图层(BaiduMap. setMyLocationEnabled(true);) ;

(11)弹出窗图层(InfoWindow);

(12)自定义 View(MapView. addView(View);)。

8.2.1　申请秘钥

申请 Key,地址为 http://lbsyun. baidu. com/apiconsole/key。百度地图开发平台页面如

图 8－4 所示，百度地图申请 Key 的页面如图 8－5 所示。

图 8－4　百度地图开发平台页面

图 8－5　百度地图申请 Key 的页面

安全码的组成规则为 Android 签名证书的 sha1 值 +“;”+ packagename（即数字签名 + 分号 + 包名），例如：

36:97:A9:17:5A:B3:A2:AF:A2:12:EB:9A:71:35:29:97:B3:48:73:AF;com.chao.hellobaidumap

获取 Android 签名证书的 sha1 值，可以在 eclipse 中直接查看：windows→preferance→android→build。获取数字签名如图 8－6 所示。

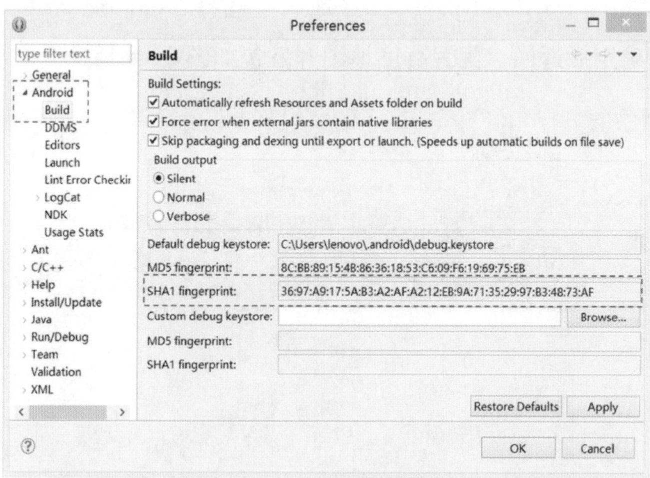

图 8－6　获取数字签名

8.2.2　显示百度地图 Hello BaiduMap

百度地图 SDK v3.4.0 为开发者提供了便捷的显示百度地图数据的接口，通过以下几步操作，即可在应用中使用百度地图数据。

1. 创建并配置工程（集成开发环境 ADT－Bundle－Windows－x86－　－20140321）

（1）在新创建的工程 HelloBaiduMap 中找到 libs 根目录，将开发包 Android SDK v3.4.0 中的 BaiduLBS_Android. jar 拷贝到 libs 根目录下，在 libs 根目录下新建 armeabi 文件夹，将 libBaiduMapSDK_v3_4_0_15. so 拷贝到 libs\armeabi 子目录下，拷贝完成后的百度地图应用程序工程目录如图 8－7 所示。

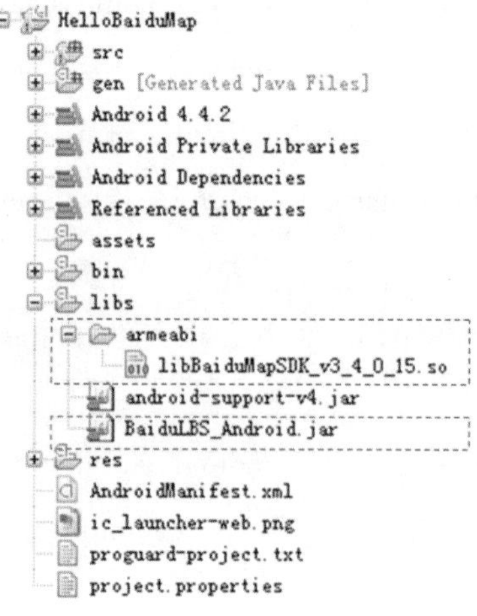

图 8－7　百度地图应用程序工程目录

(2)在工程属性→Java Build Path→Libraries 中选择“Add External JARs”，选定 BaiduLBS_Android.jar，确定后返回。配置百度地图开发包如图 8－8 所示。

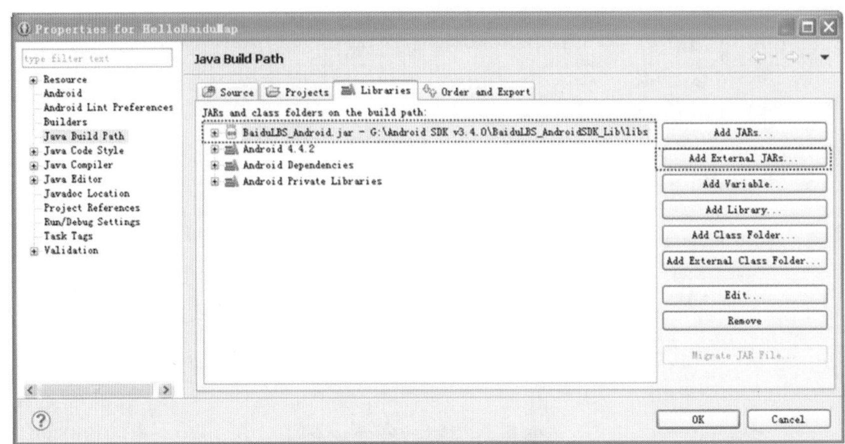

图 8－8　配置百度地图开发包

2. 在 AndroidManifest.xml 中添加开发密钥、所需权限等信息

(1)在 application 中添加开发密钥，代码如下：

```
<application>
<meta-data
android:name="com.baidu.lbsapi.API_KEY"
android:value="开发者 key" />
</application>
```

(2)添加所需权限，代码如下：

```
<uses-permission android:name="android.permission.GET_ACCOUNTS" />
<uses-permission android:name="android.permission.USE_CREDENTIALS" />
<uses-permission android:name="android.permission.MANAGE_ACCOUNTS" />
<uses-permission android:name="android.permission.AUTHENTICATE_ACCOUNTS" />
<uses-permission android:name="android.permission.ACCESS_NETWORK_STATE" />
<uses-permission android:name="android.permission.INTERNET" />
<uses-permission android:name="com.android.launcher.permission.READ_SETTINGS" />
<uses-permission android:name="android.permission.CHANGE_WIFI_STATE" />
<uses-permission android:name="android.permission.ACCESS_WIFI_STATE" />
```

```
<uses - permission android:name = "android. permission. READ_PHONE_STATE" / >
<uses - permission android: name = " android. permission. WRITE _ EXTERNAL _ STORAGE" / >
<uses - permission android:name = "android. permission. BROADCAST_STICKY" / >
<uses - permission android:name = "android. permission. WRITE_SETTINGS" / >
```

3. 在布局文件 activity_main. xml 中添加地图控件 MapView

代码如下:

```
<com. baidu. mapapi. map. MapView
android:id = "@ + id/bmapView"
android:layout_width = "fill_parent"
android:layout_height = "fill_parent"
android:clickable = "true" / >
```

4. 在应用程序创建时初始化 SDK 引用的 Context 全局变量

具体做法是在 SDK 各功能组件使用之前都需要执行 SDKInitializer. initialize( ) 函数,代码如下:

```
public class MainActivity extends Activity {
   protected void onCreate( Bundle savedInstanceState) {
        super. onCreate( savedInstanceState) ;
        //在使用 SDK 各组件之前初始化 context 信息,传入 ApplicationContext
        //注意该方法要在 setContentView 方法之前实现
        SDKInitializer. initialize( getApplicationContext( ) ) ;
        setContentView( R. layout. activity_main) ;
   }
}
```

5. 创建地图 Activity,管理地图生命周期

代码如下:

```
public class MainActivity extends Activity {
   protected void onCreate( Bundle savedInstanceState) {
        super. onCreate( savedInstanceState) ;
        SDKInitializer. initialize( getApplicationContext( ) ) ;
        setContentView( R. layout. activity_main) ;
        //获取地图控件引用
```

```
7          mMapView  =  (MapView) findViewById(R. id. bmapView);
8      }
9      protected void onDestroy() {
10             super. onDestroy();
11             mMapView. onDestroy();
12       }
13       protected void onResume() {
14             super. onResume();
15             mMapView. onResume();
16             }
17       protected void onPause() {
18             super. onPause();
19             mMapView. onPause();
20             }
21   }
```

代码第 11 行的含义是在 activity 执行 onDestroy 时执行 mMapView. onDestroy( ),实现地图生命周期管理;代码第 15 行的含义是在 activity 执行 onResume 时执行 mMapView. onResume ( ),实现地图生命周期管理;代码第 19 行的含义是在 activity 执行 onPause 时执行 mMapView. onPause ( ),实现地图生命周期管理。

完成以上步骤后运行程序,即可在应用中显示地图。

8.2.3 基础地图

1. 地图类型

百度地图 Android SDK 提供了两种类型的地图资源(普通矢量地图和卫星图),开发者可以利用 BaiduMap 类中的 mapType( )方法来设置地图类型。核心代码如下:

```
mMapView  =  (MapView) findViewById(R. id. bmapView);
mBaiduMap  =  mMapView. getMap();
mBaiduMap. setMapType(BaiduMap. MAP_TYPE_NORMAL);   //普通地图
mBaiduMap. setMapType(BaiduMap. MAP_TYPE_SATELLITE); //卫星地图
```

运行程序,即可在应用中显示地图。

2. 实时交通图

当前,全国范围内已支持多个城市实时路况查询,且会陆续开通其他城市。在地图上打开实时路况的核心代码如下:

```
mMapView = (MapView) findViewById(R.id.bmapView);
mBaiduMap = mMapView.getMap();
mBaiduMap.setTrafficEnabled(true); //开启实时路况图
```

运行程序，即可在应用中显示地图。

3. 百度城市热力图

百度地图 SDK 继为广大开发者开放热力图本地绘制能力之后，再次进一步开放百度自有数据的城市热力图层，帮助开发者构建形式更加多样的移动端应用。百度城市热力图的性质及使用与实时交通图类似，只需要简单的接口调用，即可在地图上展现样式丰富的百度城市热力图。在地图上开启百度城市热力图的核心代码如下：

```
mMapView = (MapView) findViewById(R.id.bmapView);
mBaiduMap = mMapView.getMap();
mBaiduMap.setBaiduHeatMapEnabled(true); //开启百度城市热力图
```

运行程序，即可在应用中显示地图。其中，“非常舒适”表示少于 10 人/100 m^2；“舒适”表示 10～20 人/100 m^2；“一般”表示 20～40 人/100 m^2；“拥挤”表示 40～60 人/100 m^2；“非常拥挤”表示大于 60 人/100 m^2。

4. 标注覆盖物

开发者可根据自己实际的业务需求，利用标注覆盖物，在地图指定的位置上添加标注信息。具体实现方法如下：

```
LatLng point = new LatLng(45.718484, 126.647336); //定义 Maker 坐标点
//构建 Marker 图标
BitmapDescriptor bitmap = BitmapDescriptorFactory.fromResource(R.drawable.icon_marka);
//构建 MarkerOption，用于在地图上添加 Marker
OverlayOptions option = new MarkerOptions().position(point).icon(bitmap);
mBaiduMap.addOverlay(option); //在地图上添加 Marker，并显示
```

运行程序，即可在应用中显示地图。

针对已经添加在地图上的标注，可采用如下方式进行手势拖拽。

(1)设置可拖拽。

程序代码如下：

```
OverlayOptions options = new MarkerOptions()
    .position(llA)  //设置 marker 的位置
    .icon(bdA)  //设置 marker 图标
    .zIndex(9)  //设置 marker 所在层级
```

```
5       .draggable(true);  //设置手势拖拽
6   marker = (Marker) (mBaiduMap.addOverlay(options)); //将 marker 添加到地图上
```

在上述代码中,第 2 行中参数 llA 和第 3 行中参数 bdA 均需要在前面通过以下语句进行声明:

```
LatLng llA = new LatLng(39.963175, 116.400244);
BitmapDescriptor bdA =
BitmapDescriptorFactory.fromResource(R.drawable.icon_marka);
```

(2)设置监听方法:

```
//调用 BaiduMap 对象的 setOnMarkerDragListener 方法设置 marker 拖拽的监听
mBaiduMap.setOnMarkerDragListener(new OnMarkerDragListener() {
    public void onMarkerDrag(Marker marker) {//拖拽中
    }
    public void onMarkerDragEnd(Marker marker) {//拖拽结束
Toast.makeText(
                MainActivity.this,"拖拽结束,新位置:" +
               marker.getPosition().latitude + ", " + marker.getPosition().longitude,
Toast.LENGTH_LONG).show();
      }
      public void onMarkerDragStart(Marker marker) { //开始拖拽
      }
});
```

针对已添加在地图上的标注覆盖物,可利用如下方法进行修改和删除操作:

```
marker.remove();   //调用 Marker 对象的 remove 方法实现指定 marker 的删除
```

第 9 章　Android 多线程

多线程编程是 Android 程序开发人员必须掌握的技能之一，当将用户界面显示和数据处理分开时等情况均需要多线程技术。因此，本章主要介绍 Android 平台下的多线程技术。

9.1　Android 下的线程

线程是进程中的一个实体，是被系统独立调度和分配的基本单位。在一个进程中可以创建几个线程来提高程序的执行效率，同一个进程中的多个线程之间可以并发执行。在 Android 平台下，多线程编程为充分利用系统资源提供了便利，同时也为设计复杂用户界面和耗时操作提供了途径，提升了 Android 用户的使用体验。

Android 系统启动某个应用后，将会创建一个线程来运行该应用，这个线程称为“主”线程。主线程非常重要，这是因为它要负责消息的分发，给界面上相应的 UI 组件分发事件，包括绘图事件，这也是应用可以和 UI 组件（为 android. widget 和 android. view 中定义的组件）发生直接交互的线程。因此，主线程也通常称为用户界面线程（UI 线程）。UI 线程只有一个，因此应用可以说是单线程（Single - threaded）的。

在 Android 平台下，为实现“线程安全”，Android 规定只有 UI 线程才能更新用户界面和接受用户的按钮及触摸事件。因此，需要遵循以下两个规则：

（1）永远不要阻塞 UI 线程；

（2）不要在非 UI 线程中操作 UI 组件。

由于 Android 使用单线程工作模式，因此不阻塞 UI 线程对于应用程序的响应性能至关重要。如果在应用中包含一些不是一瞬间就能完成的操作，就应该使用额外的线程（即辅助线程）来执行这些操作。

为让辅助线程和 UI 线程顺利的进行通信，Android 提出了循环者 - 消息机制（Looper - Message 机制）。

9.2　循环者 - 消息机制

循环者 - 消息机制（Looper - Message 机制）是指线程间可以通过该消息队列并结合处理者（Handler）和循环者（Looper）组件来进行信息交换。

1. Message(消息)

Message 是线程间交流的信息。辅助线程若需要更新界面,则发送内含一些数据的消息给 UI 线程。

2. Handler(处理者)

Handler 直接继承自 Object,一个 Handler 允许发送和处理 Message 或者 Runnable 对象,并且会关联到 UI 线程的 MessageQueue 中。当实例化一个 Handler 时,这个 Handler 可以把 Message 或 Runnable 压入消息队列,并且从消息队列中取出 Message 或 Runnable,进而操作它们。Handler 主要有以下两个作用:

(1)在辅助线程中发送消息;

(2)在 UI 线程中获取、处理消息。

Handler 本身没有开辟一个新线程,而更像是 UI 线程的秘书,是一个触发器,负责管理从辅助线程中得到更新的数据,然后在 UI 线程中更新界面。辅助线程通过 Handler 的 sendMessage()方法发送一个消息后,Handler 就会回调 Handler 的 HandlerMessage 方法来处理消息。

3. MessageQueue(消息队列)

MessageQueue 用来存放通过 Handler 发送的消息,按照先进先出执行,每个消息队列都会有一个对应的 Handler。Handler 通过 sendMessage 将消息发送到消息队列,插在消息队列队尾并按先进先出执行,这个消息会被 Handler 的 handleMessage 函数处理。

Android 没有全局的消息队列,而 Android 会自动为 UI 线程建立消息队列,但在子线程里并没有建立消息队列,所以调用 Looper. getMainLooper()得到 UI 线程的循环者。UI 线程的循环者不会为 NULL。

4. Looper(循环者)

Looper 是每条线程里的消息队列的管家,是处理者和消息队列之间的通信桥梁,程序组件首先通过处理者把消息传递给循环者,循环者则把消息放入队列。对于 UI 线程,系统已经给它建立了消息队列和循环者,但要想向 UI 线程发消息和处理消息,用户必须在 UI 里建立自己的 Handler 对象,那么 Handler 是属于 UI 线程的,从而是可以与 UI 线程交互的。

UI 线程的 Looper 一直在进行 Loop 操作,在 MessageQueue 中读取符合要求的 Message 给属于它的 Handler 来处理。因此,只要在辅助线程中将最新的数据放到 Handler 所关联的 Looper 的 MessageQueue 中,Looper 一直在 loop 操作,一旦有符合要求的 Message,就将 Message 交给该 Message 的 Handler 来处理,最终被从属于 UI 线程的 Handler 的 handlMessag(Message msg)方法调用。下面通过实例进一步说明这一多线程使用的机制。

编辑主活动代码,在 UI 线程中控制辅助线程的开启和关闭,使用 Handler 将辅助线程中传递过来的数据信息提取出来,并以此来更新 UI 线程的界面控件,代码如下:

```
public class MainActivity extends Activity {
  private Button btn_StartThread;
  private Button btn_StopThread;
  private TextView threadOutputInfo;
  private TextView threadStateOutputInfo;
  private MyTaskThread myThread = null;
  private Handler mHandler;
    public void onCreate(Bundle savedInstanceState) {
        super.onCreate(savedInstanceState);
         setContentView(R.layout.activity_main);
         threadOutputInfo = (TextView)findViewById(R.id.ThreadOuputInfo);
   threadStateOutputInfo =
   (TextView)findViewById(R.id.ThreadStateOuputInfo);
         threadStateOutputInfo.setText("线程未运行");
         mHandler = new Handler() {
             public void handleMessage(Message msg) {
              switch (msg.what) {
                case MyTaskThread.MSG_REFRESHINFO:
                threadOutputInfo.setText((String)(msg.obj));
                break;
                default:
             break;
             }
           }
       };
      btn_StartThread = (Button)findViewById(R.id.startThread);//开始运行线程
      btn_StartThread.setOnClickListener(new OnClickListener(){
          public void onClick(View v) {
             myThread = new MyTaskThread(mHandler);  // 创建一个线程
             myThread.start();  // 启动线程
 setButtonAvailable();
             threadStateOutputInfo.setText("线程运行中");
           }
```

```
        });
        btn_StopThread = (Button) findViewById(R.id.stopThread);//中止线程运行
        btn_StopThread.setOnClickListener(new OnClickListener() {
            public void onClick(View v) {
                if (myThread != null && myThread.isAlive()) {
                myThread.stopRun();
threadStateOutputInfo.setText("线程已中止");
}
                try {
                if (myThread != null) {
                myThread.join();    // 等待线程运行结束
                myThread = null;
                }
                } catch (InterruptedException e) {
                // 空语句块,表示忽略强行中止异常
                }
setButtonAvailable();
            }
        });
setButtonAvailable();
    }
    private void setButtonAvailable()   // 新增函数,用于设置各按钮的可选性
    {
        btn_StartThread.setEnabled(myThread == null);
        btn_StopThread.setEnabled(myThread != null);
    }
}
```

辅助线程的代码如下,在辅助线程中实现简单的计数功能:

```
public class MyTaskThread extends Thread {
private static final int stepTime = 600; //每一步执行时间(单位:ms)
private volatile boolean isEnded; //线程是否运行的标记,用于终止线程的运行
private Handler mainHandler;  // 用于发送消息的处理者
public static final int MSG_REFRESHINFO = 1;  // 更新界面的消息
```

```
public MyTaskThread(Handler mh) {// 定义构造函数
super();// 调用父类的构建器创建对象
isEnded = false;
mainHandler = mh;
 }
 public void run(){// 在线程体 run 方法中书写运行代码
 Message msg ;
 for (int i = 0; ! isEnded; i++) {
 try {
 Thread.sleep(stepTime);  // 让线程的每一步睡眠指定时间
            String s = "完成第" + i +"步";
            msg = new Message();
            msg.what = MSG_REFRESHINFO; // 定义消息类型
            msg.obj = s; //给消息附带数据
            mainHandler.sendMessage(msg); // 发送消息
 } catch (InterruptedException e) {
 e.printStackTrace();
 }
 }
 }
   public void stopRun(){// 停止线程的运行的控制函数
     isEnded = true;
   }
 }
```

程序执行过程示意图如图 9－1 所示。

(a)　　(b)　　(c)

图 9－1　程序执行过程示意图

9.3 AsyncTask

除使用循环者 - 消息(Looper - Message)机制来实现辅助线程与 UI 线程的通信外,还可以使用一种称为异步任务(AsyncTask)的类来实现通信。Android 的 AsyncTask 比 Handler 更轻量级一些,适用于简单的异步处理。AsyncTask 的一般使用框架如下。

AsyncTask 是抽象类,定义了三种泛型类型:Params、Progress 和 Result。

①Params 启动任务执行的输入参数,如 HTTP 请求的 URL。

②Progress 后台任务执行的百分比。

③Result 后台执行任务最终返回的结果,如 String、Integer 等。

AsyncTask 的执行分为四个步骤,每一步都对应一个回调方法,开发者需要实现这些方法。

(1)onPreExecute()。该方法将在执行实际的后台处理工作前被 UI 线程调用,用于做一些准备工作,如在界面上显示一条进度条或者一些控件的实例化。该方法不是必须实现的。

(2)doInBackground(Params...)。该方法将在 onPreExecute()方法执行后立即执行。该方法运行在辅助线程中,它将主要负责执行那些很耗时的后台处理工作。可以调用 publishProgress()方法来更新实时的任务进度。该方法是抽象方法,子类必须实现。

(3)onProgressUpdate(Progress...)。在 publishProgress()方法被调用之后,UI 线程将调用这个方法在界面上展示任务的进展情况,如通过一个进度条进行显示。

(4)onPostExecute(Result)。在 doInBackground()方法执行后,onPostExecute()方法将被 UI 线程调用,后台的计算结果将通过该方法传递到 UI 线程,并且在用户界面上显示给用户。

此外,还有 onCancelled()方法,该方法在用户取消线程操作时调用。

为正确地使用 AsyncTask 类,必须遵守以下的准则:

①Task 的实例必须在 UI 线程中创建;

②execute 方法必须在 UI 线程中调用;

③不要手动调用 onPreExecute()、onPostExecute(Result)、doInBackground(Params...)、onProgressUpdate(Progress...)这四个方法,需要在 UI 线程中实例化的 task 中来调用;

④该 task 只能被执行一次,否则多次调用时将会出现异常。

doInBackground()方法和 onPostExecute()方法的参数必须对应,这两个参数在 AsyncTask声明的泛型参数列表中指定,第一个为 doInBackground()方法接受的参数,第二个为显示进度的参数,第三个为 doInBackground()方法返回和 onPostExecute()方法传入的参数。

下面通过实例进一步说明 AsyncTask 使用的方式,代码如下:

```
public class MainActivity extends Activity {
private Button button;
private ImageView imageView;
private ProgressDialog progressDialog;
private final String IMAGE_PATH =
"http://192.168.1.108:8080/URLResource/urlc.jpg";
protected void onCreate(Bundle savedInstanceState){
  super.onCreate(savedInstanceState);
  setContentView(R.layout.activity_main);
   button = (Button)findViewById(R.id.button);
   imageView = (ImageView)findViewById(R.id.imageView);
   //弹出要给 ProgressDialog
   progressDialog = new ProgressDialog(MainActivity.this);
   progressDialog.setTitle("提示信息");
   progressDialog.setMessage("正在下载中,请稍后……");
   //设置 setCancelable(false);表示不能取消这个弹出框,
   //等下载完成后再让弹出框消失
   progressDialog.setCancelable(false);
   //设置 ProgressDialog 样式为水平的样式
   progressDialog.setProgressStyle(ProgressDialog.STYLE_HORIZONTAL);
   button.setOnClickListener(new View.OnClickListener(){
     public void onClick(View v) {
       new MyAsyncTask().execute(IMAGE_PATH);
     }
   });
 }
 /*定义一个类,让其继承 AsyncTask 这个类
 * Params: String 类型,表示传递给异步任务的参数类型是 String,通常指定的是 URL 路径
 * Progress: Integer 类型,进度条的单位通常都是 Integer 类型
 * Result:byte[]类型,表示下载好的图片以字节数组返回 */
 public class MyAsyncTask extends AsyncTask<String, Integer, byte[]>{
```

```
    protected void onPreExecute(){
        super.onPreExecute();
        //在 onPreExecute()中让 ProgressDialog 显示出来
        progressDialog.show();
    }
protected byte[] doInBackground(String... params){
    //通过 Apache 的 HttpClient 来访问请求网络中的一张图片
    HttpClient httpClient = new DefaultHttpClient();
    HttpGet httpGet = new HttpGet(params[0]);
    byte[] image = new byte[]{};
    try{
        HttpResponse httpResponse = httpClient.execute(httpGet);
        HttpEntity httpEntity = httpResponse.getEntity();
        InputStream inputStream = null;
        ByteArrayOutputStream byteArrayOutputStream =
        new ByteArrayOutputStream();
        if(httpEntity != null &&
        httpResponse.getStatusLine().getStatusCode() ==
        HttpStatus.SC_OK){
            //得到文件的总长度
            long file_length = httpEntity.getContentLength();
            //每次读取后累加的长度
            long total_length = 0;
            int length = 0;
            byte[] data = new byte[1024]; //每次读取 1024 个字节
            inputStream = httpEntity.getContent();
            while(-1 != (length = inputStream.read(data))){
                //每读一次,就将 total_length 累加起来
                total_length += length;
                byteArrayOutputStream.write(data, 0, length);
                //得到当前图片下载的进度
                int progress = ((int)(total_length/(float)file_length) * 100);
                publishProgress(progress);
```

```
                }
            }
            image = byteArrayOutputStream.toByteArray();
            inputStream.close();
            byteArrayOutputStream.close();
        }
        catch (Exception e) {
            e.printStackTrace();
        }
        finally{
            httpClient.getConnectionManager().shutdown();
        }
        return image;
    }
    protected void onProgressUpdate(Integer... values) {
        super.onProgressUpdate(values);
        progressDialog.setProgress(values[0]); //更新 ProgressDialog 的进度条
    }
    protected void onPostExecute(byte[] result) {
        super.onPostExecute(result);
        //将 doInBackground 方法返回的 byte[]解码成要给 Bitmap
        Bitmap bitmap = BitmapFactory.decodeByteArray(result, 0, result.length);
        imageView.setImageBitmap(bitmap); //更新我们的 ImageView 控件
        progressDialog.dismiss();//使 ProgressDialog 框消失
    }
  }
}
```

程序执行过程示意图如图 9 – 2 所示。

(a)　　(b)　　(c)

图 9 – 2　程序执行过程示意图

第 10 章　Android 网络通信开发

Android 平台为网络通信提供了丰富的 API,包括:Java 标准平台的 java. net、javax. net、javax. net. ssl 包;Apache 旗下的 Http 通信相关的 org. apache. http 包;android. net、android. net. http 包。java. net 包中主要类/接口说明见表 10－1,org. apache. http 包中主要类/接口说明见表 10－2。

表 10－1　java. net 包中主要类/接口说明

类/接口	说明
IntetAddress	表示 IP 地址
UnkownHostException	主机位置异常
HttpURLConnection	用于管理 Http 链接的资源连接管理器
URL	用于指定互联网上一个资源的位置信息
ServerSocket	表示用于等待客户端连接的服务方的套接字
Scoket	提供一个客户端的 TCP 套接字
DatagramSocket	实现一个用于发送和接收数据报的 UDP 套接字
DatagramPacket	数据包

表 10－2　org. apache. http 包中主要类/接口说明

类/接口	说明
DefaultHttpClient	表示一个 Http 客户端默认实现接口
HttpGet /HttpPost	表示 Http 的 Get 和 Post 访问方式
HttpResponse	一个 Http 响应
StatusLine	状态行
Header	表示 Http 头部字段
HeaderElement	Http 头部值中的一个元素
NameValuesPair	封装了属性－值对的类
HttpEntity	一个可以同 Http 消息进行接收或发送的实体

10.1 HTTP 网络通信

HTTP 协议即超文本传送协议(Hypertext Transfer Protocol),是 Web 网络通信的基础,也是手机网络通信常用的协议之一,是建立在 TCP 协议之上的一种应用。HTTP 连接最显著的特点是客户端发送的每次请求都需要服务器回送响应,在请求结束后会主动释放连接。从建立连接到关闭连接的过程称为“一次连接”。

由于 HTTP 在每次请求结束后都会主动释放连接,因此 HTTP 连接是一种无状态的短连接,要保持客户端程序的在线状态,需要不断地向服务器发起连接请求。通常的做法是即使不需要获得任何数据,客户端也保持每隔一段固定的时间向服务器发送一次“保持连接”的请求,服务器在收到该请求后对客户端进行回复,表明知道客户端“在线”。若服务器长时间无法收到客户端的请求,则认为客户端“下线”;若客户端长时间无法收到服务器的回复,则认为网络已经断开。

Android 应用经常会和服务器端交互,这就需要 Android 客户端发送网络请求。Android 对于 Http 网络通信,提供了标准的 java 接口即 httpURLConnection 接口,以及 apache 的接口即 httpclient 接口。同时,http 通信也分为 post 方式和 get 的方式,两种方式的区别如下。

(1)post 请求可以向服务器传送数据,而且数据放在 HTML HEADER 内一起传送到服务端 URL 地址,数据对用户不可见。而 get 是把参数数据队列加到提交的 URL 中,值和表单内各个字段一一对应,如“http://192.168.1.108:8080/MyHTTP/msg? message = HelloWorld! (get)”。

(2)get 传送的数据量较小,不能大于 2 KB。post 传送的数据量较大,一般被默认为不受限制。

(3)get 安全性非常低,post 安全性较高。

10.1.1 基于 HTTP 协议的网络程序设计

HTTP 协议是一个基于请求与响应模式的应用层协议,是基于 TCP 的连接方式,默认端口为 80。大多数的 Web 应用服务都是构建在 HTTP 协议之上的。

HTTP 协议的工作原理为由 HTTP 的客户端发起请求,建立一个到服务器指定端口的 TCP 连接,HTTP 服务器则在那个端口监听客户端发送过来的请求。一旦收到请求,服务器(向客户端)即发回一个响应信息,如“HTTP/1.1 200 OK”。

1. 请求方式

HTTP 协议有 GET 方式和 POST 方式两种请求方式。

（1）GET 方式。请求获取 Request - URI 所标识的资源，浏览器采用 GET 方式向服务器获取资源。

（2）POST 方式。在 Request - URI 所标识的资源后附加新的数据，常用于提交表单。

2. 请求信息

HTTP 请求报文由三个部分组成，分别是请求行、消息报头、请求正文。HTTP 请求信息示例如下：

GET /hello. htm HTTP/1. 1。　　　　　　　//请求行，请求信息的标志

Accept：*/*

Accept - Language：zh - cn

Accept - Encoding：gzip，deflate

If - Modified - Since：Wed，17 Oct 2007 02:15:55 GMT

If - None - Match：W/"158 - 1192587355000"

User - Agent：Mozilla/4. 0（compatible；MSIE 6. 0；Windows NT 5. 1；SV1）

Host：192. 168. 2. 162:8080

Connection：Keep - Alive

HTTP 协议请求报文的头部参数见表 10 - 3。

表 10 - 3　HTTP 协议请求报文的头部参数

请求参数	参数说明
Accept	介质类型，*/* 表示任何类型
Acccpt - Charsct	声明接收的字符集
Connection	保持连接的持续性，Keep - Alive 为保持连接的时间（单位为 s）
Content - Encoding	压缩方式（gzip 或 deflate）
Content - Language	Web 服务器响应时使用的语言
Host	客户端指定访问 Web 服务器的域名/IP 地址和端口号
GET/HTTP/1. 1	以 GET 方式请求，HTTP 的版本是 1. 1
User - Agent	浏览器类型

3. HTTP 响应信息

HTTP 响应信息的报文示例如下：

HTTP/1. 1 200 OK　　　　　　　//响应信息的标志

Last - Modified：Wed，17 Oct 2007 03:01:41 GMT

Content - Type：text /html

Content - Length:158

Date: Wed, 17 Jul 2012 03:01:59 GMT

Server: Apache - Coyote/1 .1

HTTP 协议响应报文的头部参数见表 10 - 4。

表 10 - 4 HTTP 协议响应报文的头部参数

相应参数	参数说明
Date	当前响应的 GMT 时间
Connection	保持连接的持续性
Content - Length	响应文档的长度,以字节方式存储的十进制数表示
Content - Type	Web 服务器响应文档的 MIME 类型
Cache - Control	缓存的控制权限
Content - Encoding	文档编码的压缩方式(gzip 或 deflate)
Expires	文档已过期时间
HTTP/1.1 200 OK	HTTP 的版本是 1.1,返回 200 表示成功
Server	Web 服务器系统及版本等信息

【例 10 - 1】显示 HTTP 协议报文头部信息。

在该项目的界面设计中设置一个文本编辑框和一个按钮。在按钮的事件中,通过套接字 Socket 建立的输出流向 www.baidu.com 网站发出访问请求信息,通过套接字 Socket 建立的输入流接收网站发来的响应报文,最后将接收到的报文头部信息显示到文本编辑框中。

程序代码如下:

```
package com.example.ex09_03;
import java.io.BufferedReader;
import java.io.InputStream;
import java.io.InputStreamReader;
import java.io.OutputStream;
import java.net.Socket;
import android.os.Bundle;
import android.view.View;
import android.view.View.OnClickListener;
import android.widget.Button;
import android.widget.TextView;
```

```
import android. app. Activity;

public class MainActivity extends Activity
{
TextView text = null;
Button httpBtn;
@ Override
public void onCreate( Bundle savedInstanceState)
{
  super. onCreate( savedInstanceState) ;
  setContentView ( R. layout. activity_main) ;
  text = (TextView)this. findViewById( R. id. textView1) ;
  httpBtn = ( Button) findViewById( R. id. button1) ;
  httpBtn. setonClickListener( new mClick( ) ) ;
}
class mClick implements OnClickListener
{
  @ Override
  public void onClick( View arg0)
  {
     String host = "www. baidu. com" ;
     String url ="/index. html" ;
     String method ="GET" ;
     StringBuffer sb1, sb2;
     String str;
     OutputStream outStream;
     InputStream inStream;
     InputStreamReader inReader;
     BufferedReader buff;
     try {
       Socket socket = new Socket( host,80) ;
       outStream = socket. getOutputStream( ) ;
       sb1 = new StringBuffer( ) ;
```

```
        /* *
         * 请求信息第 1 行:方式,请求的内容,HTTP 协议的版本
         * 用 GET 方式,请求的内容时 url,HTTP 协议的版本为 1.1
        版"HTTP/1.1" * */
        sb1.append(method + " " +url + "HTTP/1.1\r\n");
        /* 请求信息的第 2 行:主机名,格式为"Host:主机"    */
        sb1.append("Host:" + host + "\r\n");
        /* 请求信息的第 3 行:接受的数据类型              */
        sb1.append("Accept: : */* \r\n");
        /* 请求信息的第 4 行:连接设置设定为一直保持连接 */
        sb1.append("Connection: Keep - Alive\r\n");
        /* 请求信息的第 5 行:注意最后一定要有\r\n 回车换行 */
        sb1.append("\r\n");
        outStream.write(sbl.toString().getBytes());
        outStream.flush();   //发送请求报

        inStream = socket.getInputStream();  //建立输入流对象
        inReader = new InputStreamReader(inStream);
        buff = new BufferedReader(inReader);
        sb2 =new StringBuffer();
        while((str = buff.readLine())! = null){

        sb2.append(str + "\n");
        }
        buff.close();
        inReader.close();
        outStream.close();
        inStream.close() ;
        text.setText(sb2 .tostring());  //将读取的内容放到文本框中显示
    } catch (Exception e) {
        System.out.println("套接字连接错误" + e);
    }
}
```

```
}
}
```

在配置文件 AndroidManifest. xml 中加入允许访问网络的权限语句：

< uses - permission android：name = " android. permission. INTERNET" / >

显示 HTTP 协议的头部信息的程序运行结果如图 10 - 1 所示。

图 10 - 1　显示 HTTP 协议的头部信息的程序运行结果

10. 1. 2　HttpURLConnection 接口开发

HttpURLConnection 是 Java 的标准类，继承自 URLConnection 类。URLConnection 与 HttpURLConnection 都是抽象类，因此无法直接实例化对象，其对象主要通过 URL 的 openConnection( ) 方法获得。需要注意的是，此方法只是创建 URLConnection 或 HttpURLConnection 类的实例，而不是真正的连接操作，因此在连接之前可以对其一些属性进行设置。

创建一个项目，项目中包括三个 Activity 子类，通过主 Activity 类的两个按钮和一个文本输入框分别跳转到另外两个 Activity 网页信息显示活动。在每个跳转后的 Activity 活动的生命周期函数 OnCreat( ) 里，调用 HttpURLConnection 网络接口向服务发出请求，获取服务器返回信息。

跳转后的第一个 Activity 页面显示信息是通过 Get 方式携带参数请求服务器返回信息，第二个 Activity 页面显示信息是通过 Post 方式携带参数请求服务器返回信息，返回信息均显示在两个 Activity 布局页面的文本显示框中。

需要在 AndroidManifest. xml 中设置网络权限，关键代码如下：

```
<manifest ... >
<uses-permission android:name="android.permission.INTERNET"/>
</manifest>
```

该项目包含三个活动 MainActivity. java、GetActivity. java、PostActivity. java 文件，分别对应布局文件 activity_main. xml、get. xml、post. xml 布局文件。

布局文件 activity_main. xml 中包含两个 Button 按钮和一个 EditText 文本输入框，点击两个按钮分别跳转到 GetActivity、PostActivity 活动上，两个活动对应的布局文件中各包含一个 TextView 文本显示框。

布局文件 get. xml 中的文本显示框中所显示的内容是携带参数[message = " HelloWorld!"]的 Get 请求方法所获得的目标网页页面信息。

布局文件 post. xml 中的文本显示框中所显示的内容是携带参数[message = str]的 Post 请求方法所获得的目标网页页面信息（其中，str 为 activity_main. xml 布局文件的文本输入框中用户所输入信息）。

活动 MainActivity. java 文件事务处理：点击“通过 get 方式”按钮后，程序会调用 StartActivity( )方法跳转到 GetActivity 活动。在 EditText 中输入所需传递的参数后，点击“Post”按钮，程序会调用 StartActivity( )方法并携带输入信息跳转到 PostActivity 活动。

活动 GetActivity. java 文件事务处理：在此活动的 OnCreate( )方法中，程序会开启一个子线程，在子线程中来访问页面信息。访问的方式很简单，创建一个 HttpUrlConnection 连接，读取流中的内容，完成之后关闭此连接，将获取到的页面信息在文本显示框中显示。

首先使用 HttpUrlConnection 打开连接。创建过程为建立一个 http 目标地址并构造一个 URL 对象，创建 HttpURLConnection 对象打开 url 对象连接，其关键代码如下：

```
httpUrl = "http://192.168.1.108:8080/MyHTTP/msg?message=HelloWorld!(get)";
url = new URL(httpUrl); //携带参数的 get 方式
HttpURLConnection urlConn = (HttpURLConnection)url.openConnection()
```

然后获得读取流中的内容。创建过程为获得读取内容的流，为输出创建一个 BufferReader 对象，并使用循环来读取该对象所获得的数据信息，其关键代码如下：

```
InputStreamReader in = new InputStreamReader(urlConn.getInputStream());
BufferedReader buffer = new BufferedReader(in);
String inputLine = null;
while((inputLine = buffer.readLine())! = null){
resultData += inputLine + "\n";
}
```

```
in. close( ) ;
urlConn. disconnect( ) ;
```

最后将读取到的内容显示在文本显示框中,其信息发送关键代码如下:

```
if( resultData!  = null) {
Bundle b  = new Bundle( ) ;
b. putString( " megpost" , resultData) ;
Message meg  =  new Message( ) ;
meg. setData( b) ;
h_One. sendMessage( meg) ;
}
Handler h_One  =  new Handler( ) { //文本显示框信息更新代码
public void handleMessage( Message msg)  {
 super. handleMessage( msg) ;
 t_One. setText( msg. getData( ). toString( ) ) ;} } ;
```

活动 PostActivity. java 文件事务处理:使用 post 请求的方式获取页面信息,将其返回的页面信息显示在文本显示框中。由于 HttpUrlConnection 默认使用 Get 方式,因此在使用 Post 方式之前需要进行 setRuquestMethod 设置。首先使用 HttpUrlConnection 打开连接,创建过程和关键代码与 Get 方式相同,然后进行 setRuquestMethod 设置,并携带参数向目标页面进行 post 请求,创建过程与代码如下:

```
urlConn. setDoOutput( true) ;
//post 请求需要设置标志指示允许输入输出 UrlConnection
urlConn. setDoInput( true) ;
urlConn. setRequestMethod( " POST" ) ; // 设置以 POST 方式进行 http 请求
urlConn. setUseCaches( false) ; // Post 请求不能使用缓存
urlConn. setInstanceFollowRedirects( true) ;  //设置连接遵循重定向
// 配置本次连接的 Content - type,配置为 application/x - www - form - urlencoded
urlConn. setRequestProperty ( " Content  - Type" ," application/x  - www  - form  - urlencoded" ) ;
/ * urlConn 连接,从 postUrl. openConnection( ) 至此的配置必须要在 connect 之前完成,而 connect 是在 connection. getOutputStream 时隐含的进行的创建 DataOutputStream 流 * /
 DataOutputStream out  =  new DataOutputStream( urlConn. getOutputStream( ) ) ;
 String content  =  " par = "   +  URLEncoder. encode( str,  " gb2312" ) ; // 创建要上传的参数
```

```
out.writeBytes(content); // 将要上传的内容写入流中
out.flush(); // 刷新、关闭
out.close();
```

最后将读取流中的内容并显示在文本显示框中，其创建过程、代码与 Get 方式相同。

HttpUrlConnection 接口开发实例如图 10 –2 所示。

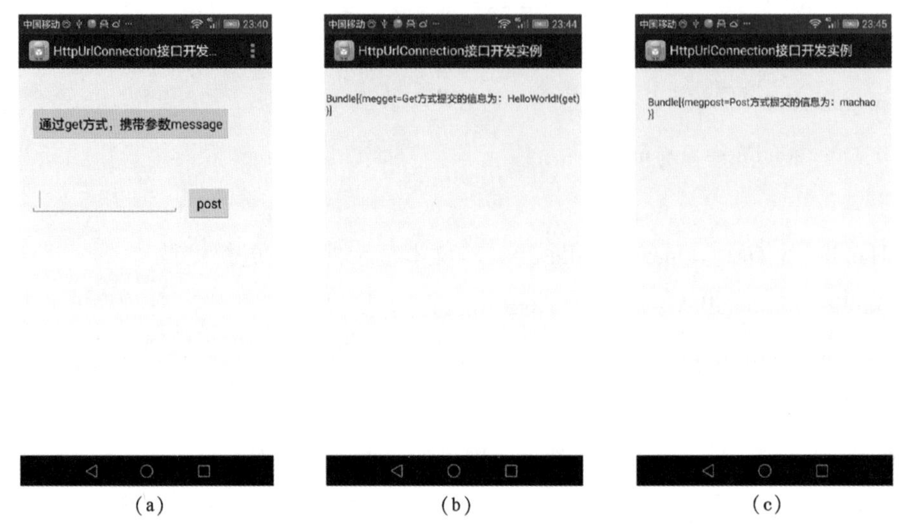

(a)　　(b)　　(c)

图 10 –2　HttpUrlConnection 接口开发实例

10.1.3　HttpClient 接口开发

在 Android 项目实际开发中，HttpClient 接口相比于 HttpURLConnection 接口更常见，因为它比后者更适合运用到更复杂的联网操作。Apache 提供的 HttpClient 接口对 java.net 中的类做了一些封装与抽象，更适合在 Android 平台上开发互联网应用。下面是一些与其相关的常见接口和类。

ClientConnectionManager 为客户端连接管理器接口，为客户端连接提供了如下方法：关闭空闲连接（closeIdleConnections）、释放一个连接（releaseConnection）、请求一个新的连接（requestConnection）、关闭管理器并释放资源（shutdown）。

下面是一些与其相关的常见接口和类。

（1）HttpClient 接口。Http 客户端接口，DefaultHttpClient 是常用于实现 HttpClient 接口的子类。

（2）HttpResponse 接口。Http 响应接口，HttpResponse 提供了一系列 get 方法。

（3）StatusLine 接口。StatusLine 也就是 HTTP 协议中的状态行。HTTP 状态行由三部分组成：HTTP 协议版本、服务器发回的响应状态码、状态码的文本描述。

（4）HttpEntity 接口。HttpEntity 就是 HTTP 消息发送或接收的实体。

(5)NameValuePair 接口。NameValuePair 是一个简单的封闭的键值对,只提供了一个 getName()和 getValue()方法。

(6)HttpGet 类。HttpGet 实现了 HttpRequest、HttpUriRequest 接口。

(7)HttpPost 类。HttpPost 也实现了 HttpRequest、HttpUriRequest 接口。

创建一个项目,项目中包括三个 Activity 子类,通过主 Activity 类的两个按钮和一个文本输入框分别跳转到另外两个 Activity 网页信息显示活动。在每个跳转后的 Activity 活动的生命周期函数 OnCreat()里,调用 HttpClient 网络接口向服务发出请求,获取服务器返回信息。

跳转后的第一个 Activity 页面显示信息是通过 Get 方式请求服务器返回信息不携带参数,第二个 Activity 页面显示信息是通过 Post 方式携带参数请求服务器返回信息。返回信息均显示在两个 Activity 布局页面的文本显示框中。

该项目包含三个活动 MainActivity. java、GetActivity. java、PostActivity. java 文件,分别对应布局文件 activity_main. xml、get. xml、post. xml 布局文件。

activity_main. xml 布局文件中包含两个 Button 按钮和一个 EditText 文本输入框,点击两个按钮分别跳转到 GetActivity、PostActivity 活动上,两个活动对应的布局文件中各包含一个 TextView 文本显示框。

get. xml 布局文件中的文本显示框中所显示的内容是使用 HttpClient 接口中携带参数[par = "HelloWorld!"]的 Get 请求方法所获得的目标网页页面信息。

post. xml 布局文件中的文本显示框中所显示的内容是使用 HttpClient 接口携带参数[par = Str]的 Post 请求方法所获得的目标网页页面信息(其中,Str 为 activity_main. xml 布局文件的文本输入框中用户所输入信息)。

此外,还需要在 AndroidManifest. xml 中设置网络权限,其关键代码如下:

```
<manifest ...  >
<uses-permission android:name="android.permission.INTERNET" />
</manifest>
```

活动 MainActivity. java 文件事务处理:点击“Get 方式”按钮后,程序会调用 StartActivity()方法跳转到 GetActivity 活动并通过 HttpClient 接口使用 Get 方式获取网页信息。在 EditText 中输入所需传递的参数后,点击“Post 方式”按钮,程序会调用 StartActivity()方法跳转到 PostActivity 活动。

活动 GetActivity. java 文件事务处理 :首先需要使用 HttpGet 来构造一个 Get 方式的 Http 请求,然后通过 HttpClient 来执行此请求,当 HttpResponse 接收到此请求后判断请求是否成功,并进行处理,最后在文本显示框中显示请求所获得的网页信息。其创建过程与关键代码如下:

```
String httpUrl =
"http://192.168.1.108:8080/MyHTTP/msg? message = HelloHttpClient!";
HttpClient httpclient = new DefaultHttpClient(); //新建 HttpClient 对象
HttpGet Request = new HttpGet(httpUrl); //使用 Get 方式获取请求
//通过 HttpClient 实例执行请求,获取响应结果
HttpResponse Response = httpclient.execute (Request);
//判断请求是否成功
if(Response.getStatusLine().getStatusCode() = = HttpStatus.SC_OK){
//获取返回的字符串
String Result = EntityUtils.toString (Response.getEntity (), "gb2312");
Bundle b = new Bundle ();
b.putString ("par", Result);
Message mess = new Message ();
mess.setData (b);
h_One.sendMessage (mess); }
```

活动 PostActivity.java 文件事务处理:PostActivity 活动中通过 HttpClient 接口使用 Post 方式获取网页信息,其过程与 Get 方式相类似。需要注意的是,使用 Post 方法进行参数传递时需要使用 NameValuePair 来存储所需传递的参数,同时还需设置所使用的字符集,其关键步骤与代码如下:

```
HttpClient httpclient = new DefaultHttpClient(); //新建 HttpClient 对象
HttpPost Request = new HttpPost(httpUrl); //新建 HttpPost 对象
//使用 NameValuePair 来保存需要传递的 Post 参数
List<NameValuePair> params = new ArrayList<NameValuePair>();
//添加要传递的参数
params.add(new BasicNameValuePair("username","" + username));
params.add(new BasicNameValuePair("password","" +password));
//获取返回的字符集
HttpEntity httpentity = new UrlEncodedFormEntity(params,"gb2312");
Request.setEntity(httpentity);
HttpResponse Response = httpClient.execute(Request); //获取响应结果 if(Re-
sponse.getStatusLine().getStatusCode() = = HttpStatus.SC_OK){
//响应通过
String result = EntityUtils.toString(Response.getEntity(), "gb2312");
```

```
Bundle b = new Bundle ( );
b. putString ( "par" , Result) ;
Message mess = new Message ( ) ;
mess. setData ( b) ;
h_One. sendMessage ( mess) ; }
```

HttpClient 接口开发实例如图 10－3 所示。

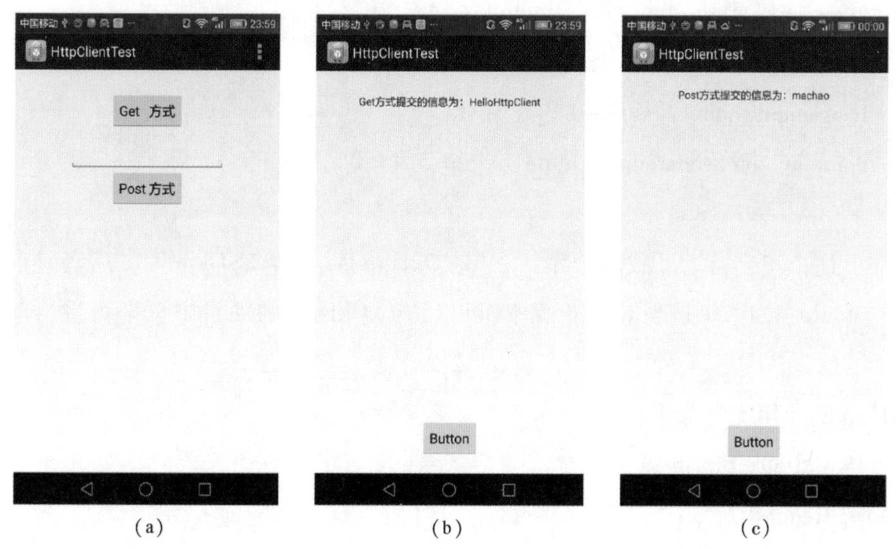

图 10－3　HttpClient 接口开发实例

10.1.4　使用 OkHttp

OkHttp 是由 Square 公司开发的,这个公司在开源事业上面贡献良多,除 OkHttp 外,还开发了像 Picasso、Retrofit 等著名的开源项目。OkHttp 不仅在接口封装上面做得简单易用,就连在底层实现上也是自成一派,比起原生的 HttpURLConnection, 可以说是有过之而无不及,现在已经成为广大 Android 开发者首选的网络通信库。本节主要学习 OkHttp 的用法, OkHttp 的官方网址是 htps:// github. com/square/okhttp。

1. OkHttp 特性

OkHttp 是一个高效的 HTTP 客户端,它有以下默认特性:

①支持 HTTP/2,允许所有同一个主机地址的请求共享同一个 socket 连接;

②连接池减少请求延时;

③透明的 GZIP 压缩减少响应数据的大小;

④缓存响应内容,避免一些完全重复的请求。

当网络出现问题时,OkHttp 依然坚守自己的职责,它会自动恢复一般的连接问题,如果服务有多个 IP 地址，当第一个 IP 请求失败时，OkHttp 会交替尝试所配置的其他 IP，OkHttp 使用现代 TLS 技术（SNI，ALPN）初始化新的连接，当握手失败时会回退到 TLS 1.0。

2. 使用步骤

在使用 OkHttp 之前，需要先在项目中添加 OkHttp 库的依赖。编辑 app/build.gradle 文件，在 dependencies 闭包中添加如下内容：

```
dependencies {
    compile fileTree(dir: 'libs', include: ['*.jar'])
    compile 'com.android.support:appcompat-v7:26.+'
    testCompile 'junit:junit:4.12'
    compile 'com.squareup.okhttp3:okhttp:3.14.0'
}
```

添加上述依赖会自动下载两个库：一个是 OkHttp 库，一个是 Okio 库。后者是前者的通信基础。okhttp:3.14.0 是当前最新版本，可以访问 OkHttp 的项目主页来查看当前最新的版本。

OkHttp 的使用流程如下：

①创建一个 OkHttpClietn 示例；

②构造 Request 对象；

③调用 OkHttpClient 的 newCall()方法创建一个 Call 对象；

④通过 Call#enqueue(Callback)方法来提交异步请求或者是通过 Call#execute()方法提交同步请求。

同步请求的最后一步是通过 Call#execute() 来提交请求，注意这种方式会阻塞调用线程，所以在 Android 中应放在子线程中执行，否则有可能引起 ANR 异常，Android 3.0 以后已经不允许在主线程访问网络了。

【例 10-2】同步请求 Get 方法，设置一个按钮。在按钮的事件中，通过向 www.baidu.com 网站发出访问请求信息，将发送结果以 Toast 方式显示到屏幕上。同步 Get 方法如图 10-4 所示。

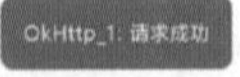

图 10-4　同步 Get 方法

```
public class MainActivity extends AppCompatActivity {

    @Override
    protected void onCreate(Bundle savedInstanceS-
```

```
tate) {
        super.onCreate(savedInstanceState);
        setContentView(R.layout.activity_main);

        Button button = (Button) findViewById(R.id.button);
        button.setOnClickListener(new View.OnClickListener() {
            @Override
            public void onClick(View view) {
                sendRequestWithOkHttp();
            }
        });
    }
    private void sendRequestWithOkHttp()
    {
        OkHttpClient mOkHttpClient = new OkHttpClient();
        final Request request = new Request.Builder()
                .url("http://www.baidu.com")
                .build();

        Call call = mOkHttpClient.newCall(request);
        call.enqueue(new Callback() {
            @Override
            public void onFailure(Call call, IOException e) {
                Toast.makeText(getApplication(), "请求失败",
                        Toast.LENGTH_SHORT).show();
            }

            @Override
            public void onResponse(Call call, Response response) throws
IOException {
                String str = response.body().string();
                runOnUiThread(new Runnable() {
                    @Override
```

```
                    public void run() {
                        Toast.makeText(getApplication(), "请求成功",
                                Toast.LENGTH_SHORT).show();
                    }

                });
            }
        });
    }
}public class MainActivity extends AppCompatActivity {

    @Override
    protected void onCreate(Bundle savedInstanceState) {
        super.onCreate(savedInstanceState);
        setContentView(R.layout.activity_main);

        Button button = (Button) findViewById(R.id.button);
        button.setOnClickListener(new View.OnClickListener() {
            @Override
            public void onClick(View view) {
                sendRequestWithOkHttp();
            }
        });
    }
    private void sendRequestWithOkHttp()
    {
        OkHttpClient mOkHttpClient = new OkHttpClient();
        final Request request = new Request.Builder()
                .url("http://www.baidu.com")
                .build();

        Call call = mOkHttpClient.newCall(request);
        call.enqueue(new Callback() {
```

```
            @ Override
            public void onFailure( Call call, IOException e) {
                Toast. makeText( getApplication( ), "请求失败",
Toast. LENGTH_SHORT). show( );
            }

            @ Override
            public void onResponse( Call call, Response response) throws
IOException {
                String str = response. body( ). string( );
                runOnUiThread( new Runnable( ) {
                    @ Override
                    public void run( ) {
                        Toast. makeText( getApplication( ), "请求成功",
                                    Toast. LENGTH_SHORT). show( );
                    }

                });
            }
        });
    }
}
```

【例 10 – 3】异步请求 Post 方法，设置一个按钮和一个文本框。在按钮的事件中，通过向网站发出访问请求信息，将返回的报文结果显示到文本框中。异步 Post 方法显示结果如图 10 – 5 所示。

```
public class MainActivity extends AppCompatAc-
tivity {

    private TextView textView;
    @ Override
    protected void onCreate( Bundle savedInstanceS-
tate) {
```

图 10 – 5　异步 Post 方法

```
        super. onCreate( savedInstanceState) ;
        setContentView( R. layout. activity_main) ;
        textView = ( TextView) findViewById( R. id. textView) ;
        Button button = ( Button) findViewById( R. id. button) ;
         button. setOnClickListener( new View. OnClickListener( ) {
            @ Override
            public void onClick( View view) {
                sendRequestWithOkHttp( ) ;
            }
        }) ;
    }
    private void sendRequestWithOkHttp( )
    {
        new Thread( new Runnable( ) {
            @ Override
            public void run( ) {
                try {
                    OkHttpClient client = new OkHttpClient( ) ;
                    RequestBody requestBody = new FormBody. Builder( )
                            . add( " username" ," val" )
                            . add( " password" ," 123456" )
                            . add( " email" ,98765431@ qq. com) ;
                            . build( ) ;
                     Request request = new Request. Builder( )
                           . url( " " http://139. 196. 103. 84:8080/HH_war/r
equestServlet" )
                            . post( requestBody)
                            . build( ) ;
                    Response response = client. newCall( request) . execute( ) ;
                    String responseData = response. body( ) . string( ) ;
                    showResponse( responseData) ;
                } catch ( Exception e)
                {
```

```
                    e.printStackTrace();
                }
            }
        });
    }
    private void showResponse(final String response)
    {
        runOnUiThread(new Runnable() {
            @Override
            public void run() {
                textView.setText(response);
            }
        });
    }
public class MainActivity extends AppCompatActivity {

    private TextView textView;
    @Override
    protected void onCreate(Bundle savedInstanceState) {
        super.onCreate(savedInstanceState);
        setContentView(R.layout.activity_main);
        textView = (TextView) findViewById(R.id.textView);
        Button button = (Button) findViewById(R.id.button);
        button.setOnClickListener(new View.OnClickListener() {
            @Override
            public void onClick(View view) {
                sendRequestWithOkHttp();
            }
        });
    }
    private void sendRequestWithOkHttp()
    {
        new Thread(new Runnable() {
```

```
                @Override
                public void run() {
                    try {
                        OkHttpClient client = new OkHttpClient();
                        RequestBody requestBody = new FormBody.Builder()
                                .add("username","val")
                                .add("password","132456")
                                .build();
                        Request request = new Request.Builder()
                                .url("http://www.baidu.com")
                                .post(requestBody)
                                .build();
                        Response response = client.newCall(request).execute();
                        String responseData = response.body().string();
                        showResponse(responseData);
                    } catch (Exception e)
                    {
                        e.printStackTrace();
                    }
                }
            });
    }
    private void showResponse(final String response)
    {
        runOnUiThread(new Runnable() {
            @Override
            public void run() {
                textView.setText(response);
            }
        });
    } }
```

10.1.5 使用 Retrofit2.0

Retrofit 是现在比较火的一个网络请求框架,它的底层是依靠 okhttp 实现的。确切地讲，Retrofit 是对 okhttp 的进一步封装，它功能强大，支持同步和异步，支持多种数据的解析（默认使用 Gson），也支持 RxJava。

1. Retrofit 功能

①基于 OkHttp，遵循 Restful API 设计风格。

②通过注解方法配置网络请求参数。支持同步、异步网络请求。

③支持多种数据的解析 & 序列化格式（Gson、Json、XML、Protobuf）。

④提供对 RxJava 支持。

基于以上功能，Retrofit 有它的优点：功能强大，支持同步 & 异步，支持多种数据的解析 & 序列化格式，支持 RxJava；简洁易用，通过注解方法配置网络请求参数，采用大量设计模式简化使用；可拓展性好，功能模块高度封装，解耦彻底，如自定义 Converters 等。

2. Retrofit 使用流程

①添加 Retrofit 库的依赖；

②创建接收服务器返回数据的类；

③创建用于描述网络请求的接口；

④创建 Retrofit 实例；

⑤创建网络请求接口实例并配置网络请求参数；

⑥发送网络请求（异步 / 同步）。

3. Retrofit 注解类型

Retrofit 网络请求方法及其解释见表 10－5，Retrofit 标记见表 10－6，Retrofit 请求参数见表 10－7。

表 10－5　Retrofit 网络请求方法及其解释

网络请求方法	解释
@ GET	所有方法分别对应 Http 中的网络请求方法：都接收和一个网络地址 URL（也可以不指定，通过@ Http 设置）
@ POST	
@ PUT	
@ DELETE	
@ PATH	
@ HEAD	
@ OPTIONS	
@ HTTP	用于替换以上 7 个注解的作用及更多功能拓展

表 10 -6　Retrofit 标记

标记类	解释
@ FormUrlEncoded	表示请求体是一个 Form 表单
@ Multipart	表示请求体是一个支持文件上传的 Form 表单
@ Streaming	表示返回的数据以流的形式返回 适用于返回数据较大的场景 （如果没有使用该注解，默认把数据全部载入内存，然后获取数据也是从内存中读取）

表 10 -7　Retrofit 请求参数

网络请求参数	解释
@ Headers	添加请求头
@ Header	添加不固定值的 Header
@ Body	用于非表单请求体
@ Field	向 Post 表单传入键值对
@ FieldMap	
@ Part	用于表单字段； 适用于有文件上传的情况
@ PartMap	
@ Query	用于表单字段； 功能同@ Field 与@ FieldMap； （区别在于@ Query 和@ QueryMap 的数据体现在 URL 上，@ Field 与@ FieldMap的数据体现在请求体上，但生成的数据是一致的）
@ QueryMap	
@ Path	URL 缺省值
@ URL	URL 设置

【例 10 -4】用户登录 App，输入个人信息后，向服务器提交用户名、密码和邮箱，返回通过 Retrofit 的 POST 方法的通信内容。Retrofit 用户登录界面实现结果如图 10 -6 所示，其代码如下：

```
//MainActivity 类
package httprequest. example. com. retrofit;

import android. support. v7. app. AppCompatActivity;
import android. os. Bundle;
```

```
import android. util. Log;
import android. view. View;
import android. widget. Button;
import android. widget. EditText;
import android. widget. TextView;
import android. widget. Toast;
import retrofit2. Call;
import retrofit2. Callback;
import retrofit2. Response;
import retrofit2. Retrofit;
import retrofit2. converter. gson. GsonConverterFactory;

public class MainActivity extends AppCompatActivity {

    private EditText uName;
    private EditText uPsd;
    private EditText uEmail;

    private TextView tShow;
    private String saveMsg;
    private String uUrl = " http://139. 196. 103. 84:8080/" ;
    @ Override
    protected void onCreate( Bundle savedInstanceState) {
        super. onCreate( savedInstanceState) ;
        setContentView( R. layout. activity_main) ;

        final EditText uPsd2 = ( EditText) findViewById( R. id. psd2) ;
        uPsd = ( EditText) findViewById( R. id. psd) ;
        uName = ( EditText) findViewById( R. id. name) ;
        uEmail = ( EditText) findViewById( R. id. email) ;
        Button button = ( Button) findViewById( R. id. button) ;
        tShow = ( TextView) findViewById( R. id. show) ;

```

```
            button.setOnClickListener(new View.OnClickListener() {
                @Override
                public void onClick(View view) {
              if(!uPsd.getText().toString().equals(uPsd2.getText().toString()))
                    {
                        Toast.makeText(MainActivity.this,"两次密码不一
                        致",Toast.LENGTH_SHORT).show();
                    }
                    else {
                        send();
                    }
                }
            });
        }

    public void send()
        {
            new Thread(new Runnable() {
                @Override
                public void run() {

                    //创建 Retrofit 实例
                    Retrofit retrofit = new Retrofit.Builder().baseUrl(uUrl)
                        .addConverterFactory(GsonConverterFactory.create())
                        .build();

                    //请求网络接口的实例
        UserMgrService userMgrService = retrofit.create(UserMgrService.class);
                    //对发送请求进行封装
        Call<UserInfoModel> call = userMgrService.getCall(uName.getText()
            .toString(),uPsd.getText().toString(),uEmail.getText().toString());


```

```
                    //发送网络请求
                    call.enqueue(new Callback<UserInfoModel>() {
                        @Override
                        //请求成功回调时
                    public void onResponse(Call<UserInfoModel> call,
Response<UserInfoModel> response) {
                            String data = response.body().show();
                            show(data);

                        }

                        @Override
                        //请求失败回调时
                public void onFailure(Call<UserInfoModel> call, Throwable t) {
                            show("连接失败");
                        }
                    });

                }
            }).start();

    }


    public void show(final String msg)
    {
        runOnUiThread(new Runnable() {
            @Override
            public void run() {
Toast.makeText(MainActivity.this,msg,Toast.LENGTH_SHORT).show();
                tShow.setText(msg);
            }

```

```
            });
        }
}

//UserMgrService 类
/**
 * Created by sang on 2019/6/6 0006.
 * 采用注释描述网络请求参数
 */

public interface UserMgrService {
    @POST("HH_war/requestServlet?")
    Call<UserInfoModel> getCall(@Query("username") String username,
 @Query("password") String pwd, @Query("email") String email);

}

//UserInfoModel 类
/**
 * Created by sang on 2019/6/6 0006.
 * 数据格式说明
 */

public class UserInfoModel {

            public String password;
            public String name;
            public String email;

    public String show()
    {
            String msg = "\n\nname: " + name + "\npassword: "
                            + password + "\nemail: " + email;
```

```
            return msg;
        }
    }
```

GET 方法实现并不复杂,将采用注释描述网络请求参数改成 GET 即可,其代码如下:

```
public interface UserMgrService {
    @GET("HH_war/requestServlet?")
    Call < UserInfoModel >  getCall();

}
```

(a)发送前

(b)发送后

图 10 -6　Retrofit **用户登录界面实现结果**

10.2　Socket 网络通信

10.2.1　Socket 工作机制

要想明白 Socket 连接,先要明白 TCP 连接。手机能够使用联网功能是因为手机底层实现了 TCP/IP 协议,可以使手机终端通过无线网络建立 TCP 连接。TCP 协议可以对上层网络提供接口,使上层网络数据的传输建立在“无差别”的网络之上。建立起一个 TCP 连接需要经过“三次握手”。

(1)第一次握手。客户端发送 syn 包(syn = j)到服务器,并进入 SYN_SEND 状态,等待服务器确认。

(2)第二次握手。服务器收到 syn 包,必须确认客户的 SYN(ack = j + 1),同时自己也发送一个 SYN 包(syn = k),即 SYN + ACK 包,此时服务器进入 SYN_RECV 状态。

(3)第三次握手。客户端收到服务器的 SYN + ACK 包,向服务器发送确认包 ACK(ack = k + 1),此包发送完毕,客户端和服务器进入 ESTABLISHED 状态,完成三次握手。

假设在如图 10 – 7 所示的 Socket 通信原理,主机 A 就需要知道信息是被正确传送到主机 B,而不是被传送到主机 C,基于 TCP/IP 网络中的每一个主机均被赋予了一个唯一的 IP 地址。每一个基于 TCP/IP 网络通信的程序都被赋予了唯一的端口和端口号,端口是一个信息缓冲区,用于保留 Socket 中的输入/输出信息,端口号是一个 16 位无符号整数,范围是 0 ~ 65 535,以区别主机上的每一个程序(端口号就像房屋中的房间号),低于 256 的端口号保留给标准应用程序。

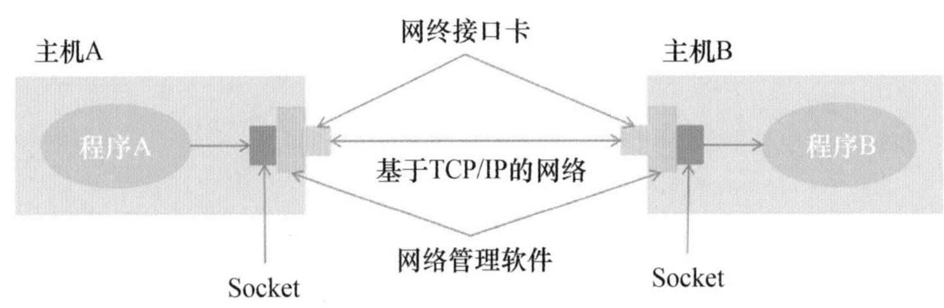

图 10 – 7　Socket 通信原理

10.2.2　Socket 通信开发

网络上的两个程序通过一个双向的通信连接实现数据的交换,这个双向链路的一端称为一个 Socket。Socket 通常用来实现客户方和服务方的连接。Socket 是 TCP/IP 协议的一个十分流行的编程界面,一个 Socket 由一个 IP 地址和一个端口号唯一确定。Socket 和 ServerSocket 类库位于 java . net 包中。ServerSocket 用于服务器端,Socket 是建立网络连接时使用的。在连接成功时,应用程序两端都会产生一个 Socket 实例,操作这个实例,完成所需的会话,Socket 通信建立过程如图 10 – 8 所示。

服务器端以监听端口号为参数实例化 ServerSocket 类,以 accept( )方法接收客户的连接,其代码如下:

```
ServerSocket ss = new ServerSocket(Int port);
```

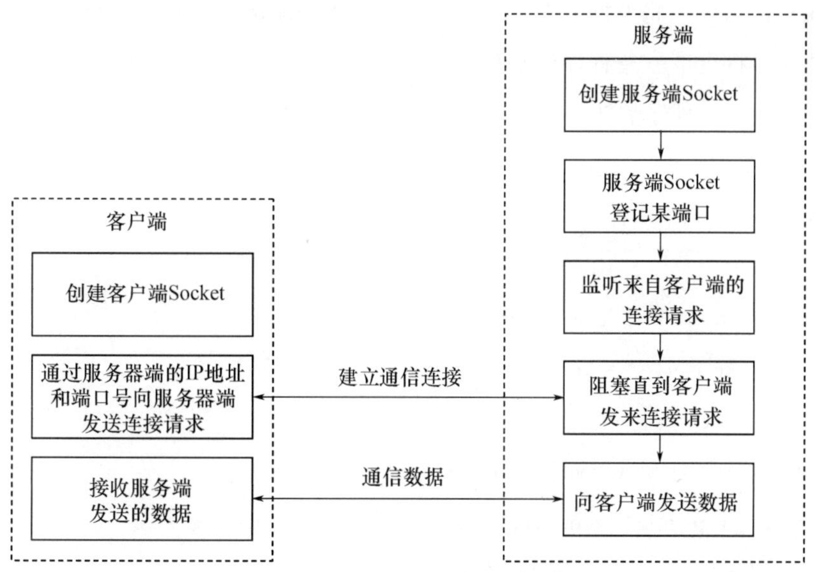

图 10－8　Socket 通信建立过程

```
Socket socket = ss.accept();
```

ss 是声明一个 ServerSocket 对象，ServerSocket()方法创建一个新的 ServerSocket 对象并绑定到给定端口，accept()方法用来接受客户连接。

客户端则直接以服务器的地址和监听端口为参数实例化 Socket 类，连接服务器，其代码如下：

```
Socket socket = Socket(String dstName, int dstPort);
```

当二者建立连接口时，就可以进行网络通信。服务器端和客户端之间通过流的形式进行交互。

服务器端调用 getOutputStream()方法得到输出流，并向其中写入数据信息传递给客户端，其代码如下：

```
PrintWriter out = new PrintWriter(new BufferedWriter( new OutputStreamWriter(socket.getOutputStream())), true);
```

客户端调用 getInputStream()方法得到输入流，接收服务端发送的数据信息，其代码如下：

```
BufferedReader in = new BufferedReader( new InputStreamReader(socket.getInputStream()));
```

Socket 类包含了许多有用的方法，例如：

①getLocalAddress()将返回一个包含客户程序 IP 地址的 InetAddress 子类对象的引用；

②getLocalPort()将返回客户程序的端口号；

③getInetAddress( )将返回一个包含服务器 IP 地址的 InetAddress 子类对象的引用;

④getPort( )将返回服务程序的端口号。

Socket 服务端的程序端执行界面如图 10－9 所示,其关键代码如下:

```
public class Server implements Runnable {
public void run( ) {
try {
System. out. println( "connected. . . . " ) ;
ServerSocket serverSocket = new ServerSocket(55555) ;
while (true) {
Socket client = serverSocket. accept( ) ;
System. out. println( "receiving. . . . . . " ) ;
String clientip = client. getInetAddress( ) . toString( ) ;
System. out. println( "accept:" + clientip) ;
try {
//服务器读取客户端发过来的消息
BufferedReader in = new BufferedReader(
new InputStreamReader( client. getInputStream( ) ) ) ;
String str = in. readLine( ) ;
System. out. println( "read:" + str) ;
//服务器写给客户端的消息
PrintWriter out = new PrintWriter( new BufferedWriter(
new OutputStreamWriter( client. getOutputStream( ) ) ) ,true) ;
out. println( "connection has been created and infromation has
also been received!" ) ;
out. close( ) ;
in. close( ) ;
} catch (Exception e) {
System. out. println( e. getMessage( ) ) ;
e. printStackTrace( ) ;
} finally {
client. close( ) ;
System. out. println( "close" ) ;
}
```

```
}
} catch (Exception e) {
System. out. println( e. getMessage( ) );
}
}
public static void main( String a[ ]) {
Thread desktopServerThread = new Thread( new Server( ) );
desktopServerThread. start( );
}
}
```

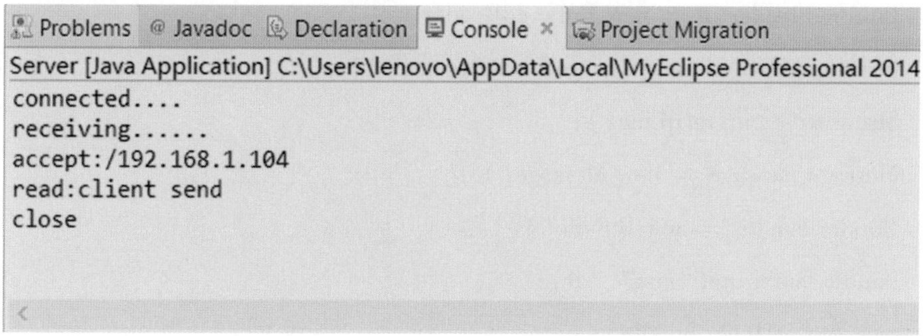

```
Problems  @ Javadoc  Declaration  Console ×  Project Migration
Server [Java Application] C:\Users\lenovo\AppData\Local\MyEclipse Professional 2014
connected....
receiving......
accept:/192.168.1.104
read:client send
close
```

图 10－9　Socket 服务器的程序端执行界面

Socket 客户端的关键代码如下：

```
public class MainActivity extends Activity {
EditText et, etSERVER;
private Thread thread = null;
protected void onCreate( Bundle savedInstanceState) {
super. onCreate( savedInstanceState);
setContentView( R. layout. activity_main);
et = (EditText) findViewById( R. id. et);
etSERVER = (EditText) findViewById( R. id. et1);
Button btn = (Button) findViewById( R. id. btn);
btn. setOnClickListener( new OnClickListener( ) {
public void onClick( View arg0) {
thread = new Thread( new Runnable( ) {
public void run( ) {
```

```
String ip = "192.168.1.108";
int port = 55555;
Socket so = null;
try {
so = new Socket(ip, port);
String msg = et.getText().toString();
PrintWriter out = new PrintWriter(new BufferedWriter(
new OutputStreamWriter(so.getOutputStream())),true);
out.println(msg);
out.flush();
BufferedReader in = new BufferedReader(
new InputStreamReader(so.getInputStream()));
String str = in.readLine();
Message message = new Message();
Bundle bundle = new Bundle();
bundle.putString("msg", str);
message.setData(bundle);
handler.sendMessage(message);
} catch (Exception e) {
e.printStackTrace();
}
}
});
thread.start();
}
});
}
Handler handler = new Handler() {
public void handleMessage(Message msg) {
Bundle bundle = msg.getData();
String returnMsg = bundle.get("msg").toString();
etSERVER.setText(returnMsg);
};
```

```
    };
}
```

Socket 客户端执行结果如图 10－10 所示。

(a) (b)

图 10－10 Socket 客户端执行结果

10.3 URL 通信

URL 对象代表统一资源定位器,它是指向互联网“资源”的指针。资源可以是简单的文件或目录,也可以是对更为复杂的对象引用,如对数据库或搜索引擎的查询。URL 的组成形式如下:

protocol: //host:port/resourceName

其中,protocol 表示协议名;host 表示主机;port 表示端口;resourceName 表示资源。网络访问离不开 URL,URL 是访问 Web 页面的地址。URL 类中提供了多个构造器用于创建 URL 对象,一旦创建了 URL 对象后,就可以调用以下方法来访问该 URL 对应的资源。

①String getFile()。获取此 URL 的资源名。

②String getHost()。获取此 URL 的主机名。

③String getPath()。获取此 URL 的路径部分。

④int getPort()。获取此 URL 的端口。

⑤String getProtocol()。获取此 URL 的协议名称。

⑥String getQuery()。获取此 URL 的查询字符串部分。

⑦URLConnection openConnection()。返回一个 URLConnection 对象,它表示到 URL 所引用的远程对象的连接。

⑧InputStream openStream()。打开与此 URL 的连接,并返回一个用于读取该 URL 资源的 InputStream。

URL 方式是通过 URLConnection 对象请求服务器资源,以此来实现在客户端和服务器之间的通信。URLConnection 类是 java. net 接口中的标准 java 类,其实现过程如下:

(1)根据指定的 URL 网址创建 URL 对象;

(2)调用 URLConnection. openConnection()方法打开连接;

(3)获取输入流;

(4)将网络信息提取显示。

下面给出实例程序,URLConnection 接口执行界面如图 10 – 11 所示。在实现服务器端程序时,将 Tomcat 作为服务器,在其 webapps 目录下新建文件夹 URLResource,这个文件夹用来存放客户端将要获取的资源,将名为 urlc. txt 和 urlc. jpg 的文件放入 URLResource 文件夹下。

(a)　　(b)

图 10 – 11　URLConnection **接口执行界面**

在实现客户端程序时,第一步先完成布局文件的绘制,代码如下:

```
<LinearLayout … >
<TextView
android:id = "@ +id/text"
…/>
<ImageView
```

```
android:id="@+id/image"
.../>
<Button
android:id="@+id/btn"
android:text="获取服务器资源"
…/>
</LinearLayout>
```

第二步,在活动文件中,onCreate()函数中的关键代码如下:

```
if (android.os.Build.VERSION.SDK_INT > 9) {
StrictMode.ThreadPolicy policy = new
StrictMode.ThreadPolicy.Builder().permitAll().build();
StrictMode.setThreadPolicy(policy);
}
Button btn = (Button) findViewById(R.id.btn);
btn.setOnClickListener(new View.OnClickListener() {
public void onClick(View v) {
String txturl = "http://192.168.1.108:8080/URLResource/urlc.txt";
URL mytxtUrl = new URL(txturl);
URLConnection mytxtCon = mytxtUrl.openConnection();
mytxtCon.setDoOutput(false);
InputStream txtin = mytxtCon.getInputStream();
BufferedInputStream bis = new BufferedInputStream(txtin);
ByteArrayBuffer baf = new ByteArrayBuffer(bis.available());
int data = 0;
while ((data = bis.read()) != -1) {
baf.append((byte)data);
}
String msg = EncodingUtils.getString(baf.toByteArray(), "gb2312");
TextView text = (TextView) findViewById(R.id.text);
text.setText(msg);
String jpgurl = "http://192.168.1.108:8080/URLResource/urlc.jpg";
URL myjpgUrl = new URL(jpgurl);
URLConnection myjpgCon = myjpgUrl.openConnection();
```

```
InputStream jpgin = myjpgCon. getInputStream( );
Bitmap bmp = BitmapFactory. decodeStream( jpgin);
ImageView image = ( ImageView) findViewById( R. id. image);
image. setImageBitmap( bmp);
}});
```

第三步,配置文件中添加的权限代码如下:

```
< uses - permission android:name = " android. permission. INTERNET"/ >
```

10.4 WiFi 管理

WiFi 全称 wireless Fidelity,又称 802. 11b 标准,其最大优势是传输速度快,可以达到 11 Mbit/s。另外,它的有效距离也很长,并与各种 802. 11DSSS 设备兼容。WiFi 是一个无线网络通信技术的品牌,由 WiFi 联盟(WiFi Alliance 拥有),目的是改善基于 IEEE 802. 11 标准的无线网络产品间互通性。一般情况下,WiFi 无线电波覆盖范围为 300 英尺左右(约合 100 m),而蓝牙仅为 50 英尺左右(约合 15 m)。

Android. net. wifi 包提供的类管理设备上的无线功能。WiFi API 提供了一种方式,应用程序可以与较低的无线协议栈提供 WiFi 网络连接,几乎获取所有装置信息,包括所连接网络的连接速度、IP 地址、协商状态等信息。其他一些 API 的功能包括扫描、增加、保存、终止和启动无线网络连接。在 Android. net. wifi 包下面主要包括以下几个类和接口。

(1) ScanResult。主要用来描述已经检测出的接入点,包括接入点的地址、接入点的名称、身份认证、频率、信号强度等信息。

(2) WifiConfiguration。WiFi 网络的配置,包括安全设置等。

(3) WifiInfo。WiFi 无线连接的描述,包括接入点、网络连接状态、隐藏的接入点、IP 地址、连接速度、MAC 地址、网络 ID、信号强度等信息。下面简单介绍一下方法。

①getBSSID( ) 。获取 BSSID。

②getDetailedStateOf( ) 。获取客户端的连通性。

③getHiddenSSID( ) 。获得 SSID 是否被隐藏。

④getIpAddress( ) 。获取 IP 地址。

⑤getLinkSpeed( ) 。获得连接的速度。

⑥getMacAddress( ) 。获得 Mac 地址。

⑦getRssi( ) 。获得 802. 11n 网络的信号。

⑧getSSID( ) 。获得 SSID。

⑨getSupplicanState( ) 。返回具体客户端状态的信息。

（4）WifiManager。用来管理 WiFi 连接，这里已经定义好了一些类以供使用。

WiFi 网卡的状态是由一系列的整形常量来表示的。

①WIFI_STATE_DISABLED。WIFI 网卡不可用(1)。

②WIFI_STATE_DISABLING。WIFI 网卡正在关闭(0)。

③WIFI_STATE_ENABLED。WIFI 网卡可用(3)。

④WIFI_STATE_ENABLING。WIFI 网正在打开(2)（WiFi 启动需要一段时间）。

⑤WIFI_STATE_UNKNOWN。未知网卡状态。

在开发 WiFi 应用程序时，首先需要添加权限，在 AndroidManifest. xml 文件中分别添加更改网络状态、更改 WiFi 状态、访问网络状态、访问 WiFi 状态四个权限，其代码如下：

```
< uses - permission android: name = " android. permission. CHANGE _ NETWORK _ STATE" >
</uses - permission >
< uses - permission android: name = " android. permission. CHANGE_WIFI_STATE" > </uses - permission >
< uses - permission android: name = " android. permission. ACCESS _ NETWORK _ STATE" >
</uses - permission >
< uses - permission android: name = " android. permission. ACCESS_WIFI_STATE" > </uses - permission >
```

接着，通过获取系统服务 getSystemService(Context. WIFI_SERVICE)并向下转型为 WifiManager 的方式取得 WifiManager 对象，再通过使用 WifiManager 对象的 getConnectionInfo( )方法创建 WifiInfo 对象，就可以打开、关闭网卡与搜索网络，最后得到扫描结果和配置好的网络。下面给出一个管理 WiFi 的类的实现：

```
public class MainActivity extends Activity {
private Button startButton  = null;
private Button stopButton  = null;
private Button checkButton  = null;
private Button getresultButton  = null;
private TextView ShowresultView  = null;
WifiManager wifiManager  = null;
private List < ScanResult > mWifiList;
public void onCreate( Bundle savedInstanceState) {
        super. onCreate( savedInstanceState) ;
```

```
        setContentView(R. layout. activity_main);
        startButton = (Button) findViewById(R. id. startButton);
        stopButton = (Button) findViewById(R. id. stopButton);
        checkButton = (Button) findViewById(R. id. checkButton);
        getresultButton = (Button) findViewById(R. id. getresultButton);
        showresultView = (TextView) findViewById(R. id. showresultView);
ShowresultView. setMovementMethod(ScrollingMovementMethod
. getInstance());
        startButton. setOnClickListener(new startButtonListener());
        stopButton. setOnClickListener(new stopButtonListener());
        checkButton. setOnClickListener(new checkButtonListener());
        getresultButton. setOnClickListener(new getresultButtonListener());
    }
    class startButtonListener implements OnClickListener {
        public void onClick(View v) {
            //创建 WifiManager 对象
            wifiManager = (WifiManager) MainActivity. this
. getSystemService(Context. WIFI_SERVICE);
            wifiManager. setWifiEnabled(true); //打开 WiFi 网卡
        }     }
    class stopButtonListener implements OnClickListener {
        public void onClick(View v) {
            wifiManager = (WifiManager) MainActivity. this
. getSystemService(Context. WIFI_SERVICE);
            wifiManager. setWifiEnabled(false); //关闭 WiFi 网卡
        }     }
    class checkButtonListener implements OnClickListener {
        public void onClick(View v) {
            wifiManager = (WifiManager) MainActivity. this
. getSystemService(Context. WIFI_SERVICE);
            Toast. makeText(MainActivity. this,"当前网卡状态为:"
 + wifiManager. getWifiState(), Toast. LENGTH_SHORT). show();
        }     }
```

```
    class getresultButtonListener implements OnClickListener {
        public void onClick(View v) {
            wifiManager = (WifiManager) MainActivity.this
.getSystemService(Context.WIFI_SERVICE);
            wifiManager.startScan();
            mWifiList = wifiManager.getScanResults();//得到扫描结果
            StringBuilder stringBuilder = new StringBuilder();
            for (int i = 0; i < mWifiList.size(); i++){
            stringBuilder.append("Index_" + new Integer(i + 1).toString() + ":");
              //将 ScanResult 信息转换成一个字符串包
              stringBuilder.append((mWifiList.get(i)).toString());
              stringBuilder.append("\n");
              ShowresultView.setText(stringBuilder);
            }
        }
    }
}
```

WiFi 管理实例的程序运行结果如图 10－12 所示。

(a) (b)

图 10－12 WiFi 管理实例的程序运行结果

第 11 章　Android 近距离通信开发

11.1　蓝牙 Bluetooth

Android 平台提供了对蓝牙协议栈的支持，它允许一个蓝牙设备和其他的蓝牙设备进行无线的数据交换。应用程序层通过 Android 蓝牙 API 来调用蓝牙相关功能，这些 API 使应用程序无线连接蓝牙设备，并拥有点到点和多点无线连接的特性。使用蓝牙 API，Android 程序能扫描其他蓝牙设备、查询本地已经配对的蓝牙适配器、建立 RFCOMM 通道、通过服务发现并连接其他设备、在设备间传输数据管理多个蓝牙连接。

使用 Android 平台中的蓝牙 API 完成蓝牙通信需要完成四项主要任务：设置蓝牙、查找已配对或区域内可用的蓝牙设备、连接设备、设备间传输数据。所有蓝牙 API 都在 android. bluetooth 包下。android. bluetooth 包中的类和接口见表 11 - 1。

表 11 - 1　android. bluetooth 包中的类和接口

类和接口	说明
BluetoothAdapter	它是本地蓝牙适配器，使用它可以发现其他蓝牙设备，查询已配对的设备列表，实例化一个 BluetoothDevice，创建一个 BluetoothServerSocket
BluetoothDevice	它是一个远程蓝牙设备，使用它可以请求一个与远程设备的 BluetoothSocket 连接，或者查询关于设备名称、地址、类和连接状态设备信息
BluetoothSocket	它是一个蓝牙 socket 的接口，是一个连接点，它允许一个应用与其他蓝牙设备通过 InputStream 和 OutputStream 交换数据
BluetoothServerSocket	它是一个开放的服务器 socket，监听接受的请求，为了连接两台 Android 设备，一个设备必须使用这个类开启一个服务器 socket
BluetoothClass	它用来描述一个蓝牙设备的基本特性和性能，这是一个只读的属性集合，它定义了设备的主要和次要的设备类以及它的服务
BluetoothProfile	它是一个表示蓝牙配置文件的接口
BluetoothProfile. ServiceListener	它是一个接口，当 BluetoothProfile IPC 客户端从服务器上建立连接或断开连接时，它负责通知它们
BluetoothHealth	它表示一个 Health Device Profile 代理，它控制蓝牙服务
BluetoothHealthCallback	它是一个抽象类，可以使用它来实现 BluetoothHealth 的回调函数，必须扩展这个类并实现回调函数方法来接收应用程序的注册状态改变以及蓝牙串口状态的更新

续表 11-1

类和接口	说明
BluetoothHealthApp Configuration	它表示一个应用程序配置,Bluetooth Health 第三方应用程序注册和一个远程 Bluetooth Health 设备通信
BluetoothHeadset	它提供了对移动手机使用的蓝牙耳机的支持
BluetoothA2dp	它定义了高品质的音频如何通过蓝牙连接从一个设备传输到另一个设备

11.1.1 蓝牙的设置与发现

为在应用中使用蓝牙特性,需要至少声明一种蓝牙权限:BLUETOOTH 和BLUETOOTH_ADMIN。为执行任何蓝牙通信,如请求一个连接、接受一个连接以及传输数据,必须请求 BLUETOOTH 权限。

为初始化设备查找或控制蓝牙设置,必须请求 BLUETOOTH_ADMIN 权限。大多数应用需要这个权限,仅仅是为了可以发现本地蓝牙设备。这个权限授权的其他功能不应该被使用,除非该应用是一个“强大的控制器”,来通过用户请求修改蓝牙设置。

注意:如果使用 BLUETOOTH_ADMIN 权限,那么必须拥有 BLUETOOTH 权限,在应用程序清单文件中声明蓝牙权限。

1. 设置蓝牙

在应用程序能够利用蓝牙通道通信之前,需要确认设备是否支持蓝牙通信,可以通过以下两个步骤完成。

(1)获得 BluetoothAdapter 对象。BluetoothAdapter 对象是所有蓝牙活动都需要的,要获得这个对象,就要调用静态的 getDefaultAdapter( )方法。这个方法会返回一个代表设备自己的蓝牙适配器的 BluetoothAdapter 对象。整个系统有一个蓝牙适配器,应用程序能够使用这个对象来进行交互。如果 getDefaultAdapter( )方法返回 null,那么该设备不支持蓝牙,处理也要在此结束。

(2)启用蓝牙功能。开发者需要确保蓝牙是可用的,调用 isEnabled( )方法来检查当前蓝牙是否可用。如果这个方法返回 false,那么蓝牙是被禁用的。要申请启用蓝牙功能,就要调用带有 ACTION_REQUEST_ENABLE 操作意图的 startActivityForResult( )方法。它会给系统设置发一个启用蓝牙功能的请求(不终止本应用程序),这时会显示一个请求用户启用蓝牙功能的对话框。

如果用户响应“Yes”,那么系统会开始启用蓝牙功能,完成启动过程(有可能失败),焦点会返回给本应用程序。如果蓝牙功能启用成功,Activity 会在 onActivityResult( )回调中接收到 RESULT_OK 结果;如果蓝牙没有被启动(或者用户响应了“No”),那么该结果编码是 RESULT_CANCELED。

2. 查找设备

使用 BluetoothAdapter 对象,能够通过设备发现或查询已配对的设备列表来找到远程的蓝牙设备。设备发现是一个扫描过程,该过程搜索本地区域内可用的蓝牙设备,然后请求一些彼此相关的一些信息(这个过程称为“发现”“查询”或“扫描”)。

但是,本地区域内的蓝牙设备只有在它们也启用了可发现功能时才会响应发现请求。如果一个设备是可发现的,那么它会通过共享某些信息(如设备名称、类别和唯一的 MAC 地址)来响应发现请求。使用这些信息,执行发现处理的设备能够有选择地初始化与被发现设备的连接。

一旦与远程的设备建立首次连接,配对请求就会自动地展现给用户。当设备完成配对时,相关设备的基本信息(如设备名称、类别和 MAC 地址)就会被保存,并能够使用蓝牙 API 来读取。使用已知的远程设备的 MAC 地址,在任何时候都能够初始化一个连接,而不需要执行发现处理(假设设备在可连接的范围内)。

配对和连接之间的差异如下。

①配对意味着两个设备对彼此存在性的感知,它们之间有一个共享的用于验证的连接密钥,用这个密钥,两个设备之间建立被加密的连接。

②连接意味着当前设备间共享一个 RFCOMM 通道,并且能够被用于设备间的数据传输。当前 Android 蓝牙 API 在 RFCOMM 连接被建立之前,要求设备之间配对(在使用蓝牙 API 初始化加密连接时,配对是自动被执行的)。

(1)查询配对设备。

在执行设备发现之前,应该先查询已配对的设备集合,来看期望的设备是否是已知的。可调用 getBondedDevices()方法来完成这件工作,这个方法会返回一个代表已配对设备的 BluetoothDevice 对象的集合,如能够查询所有的配对设备,然后使用一个 ArrayAdapter 对象把每个已配对设备的名称显示给用户。

从 BluetoothDevice 对象来初始化一个连接所需要的所有信息就是 MAC 地址。随后,该 MAC 地址能够被提取用于初始化连接。

(2)发现设备。

简单地调用 startDiscovery()方法就可以开始发现设备。该过程是异步的,并且该方法会立即返回一个布尔值来指明发现处理是否被成功的启动。通常发现过程会查询扫描大约 12 s,接下来获取扫描发现的每个设备的蓝牙名称。

为接收每个被发现设备的的信息,应用程序必须注册一个 ACTION_FOUND 类型的广播接收器。对应每个蓝牙设备,系统都会广播 ACTION_FOUND 类型的 Intent。这个 Intent 会携带 EXTRA_DEVICE 和 EXTRA_CLASS 附加字段,这个两个字段分别包含了 Bluetooth-Device 和 BluetoothClass 对象。

执行设备发现，对于蓝牙适配器来说是一个沉重的过程，它会消耗大量的资源。一旦发现要连接设备，在尝试连接之前一定要确认用 cancelDiscovery( )方法来终止发现操作。另外，如果已经存在一个设备连接，那么执行发现会明显地减少连接的可用带宽，因此在有连接的时候不应该执行发现处理。

在 Android 应用程序中具体实现搜索附近蓝牙设备时，首先需在主 Activity 的生命周期 Oncreat( )函数中创建关于"蓝牙设备搜索结束"和"蓝牙设备发现"的广播接收器的 Intent 过滤器：

```
IntentFilter discoveryFilter =
new IntentFilter(BluetoothAdapter.ACTION_DISCOVERY_FINISHED);
registerReceiver(discoveryReceiver, discoveryFilter);
IntentFilter foundFilter = new IntentFilter(BluetoothDevice.ACTION_FOUND);
registerReceiver(foundReceiver, foundFilter);
```

最后，实现蓝牙相关广播接受器的处理函数，在蓝牙设备发现的广播中将发现到的蓝牙设备添加到蓝牙设备集合中，在蓝牙搜索结束广播中取消注册蓝牙发现广播和搜索结束广播，其代码如下：

```
private List<BluetoothDevice> deviceList = new ArrayList<BluetoothDevice>();
private BroadcastReceiver foundReceiver = new BroadcastReceiver() {
public void onReceive(Context context, Intent intent) {
BluetoothDevice device =
intent.getParcelableExtra(BluetoothDevice.EXTRA_DEVICE); //获得发现结果
deviceList.add(device); //添加到设备列表
}
};
private BroadcastReceiver discoveryReceiver = new BroadcastReceiver() {
 public void onReceive(Context context, Intent intent) {
 unregisterReceiver(foundReceiver); //取消注册广播接收器
 unregisterReceiver(this);
 }
 };
```

3. 启用设备的可发现性

如果要让本地设备可以被其他设备发现，那么就要调用 ACTION_REQUEST_DISCOVERABLE 操作意图的 startActivityForResult(Intent, int)方法。这个方法会向系统设置发出一个启用可发现模式的请求(不终止应用程序)。

默认情况下,设备的可发现模式会持续 120 s。通过给 Intent 对象添加 EXTRA_DISCOVERABLE_DURATION 附加字段,可以定义不同持续时间。应用程序能够设置的最大持续时间是 3 600 s,0 意味着设备始终是可发现的。任何小于 0 或大于 3 600 s 的值都会自动被设为 120 s,如使本地蓝牙设备处于可发现状态,并使蓝牙可发现状态的持续时间设置为 100 s,实现的具体代码如下:

```
Intent i = new Intent (BluetoothAdapter. ACTION_REQUEST_DISCOVERABLE);
i. putExtra(BluetoothAdapter. EXTRA_DISCOVERABLE_DURATION, 100);
startActivity(discoverableIntent);
```

Android 3.0 以上版本在开发蓝牙程序时,Google 官方建议使用系统自带蓝牙配置程序,配对成功后再进程操作。

如果设备没有开启蓝牙功能,那么开启设备的可发现模式会自动开启蓝牙。在可发现模式下,设备会静静地把这种模式保持到指定的时长。如果想要在可发现模式被改变时获得通知,那么可以注册一个 ACTION_SCAN_MODE_CHANGED 类型的 Intent 广播。这个Intent对象中包含了 EXTRA_SCAN_MODE 和 EXTRA_PREVIOUS_SCAN_MODE 附加字段,它表示了新旧扫描模式。每个可能的值是 SCAN_MODE_CONNECTABLE_DISCOVERABLE、SCAN_MODE_CONNECTABLE 或 SCAN_MODE_NONE,它们分别指明设备是在可发现模式下还是在可发现模式下依然可接收连接,或者是在可发现模式下并不能接收连接。

如果要初始化与远程设备的连接,则不需要启用设备的可现性。只有在把应用程序作为服务端来接收输入连接时,才需要启用可发现性,因为远程设备在跟设备连接之前必须能够发现它。

下面给出显示已配对蓝牙设备的简单实例,具体实现代码如下。

(1)在应用程序的配置文件 AndroidManifest. xml 中添加蓝牙使用权限:

```
<uses-permission android:name="android. permission. BLUETOOTH_ADMIN" />
<uses-permission android:name="android. permission. BLUETOOTH" />
```

(2)在主 Activity 生命周期函数 Oncreat()中获取蓝牙适配器 BluetoothAdapter 对象,通过使用 BluetoothAdapter. getDefaultAdepter()方法获取本机 BluetoothAdapter 对象:

```
BluetoothAdapter mBluetoothAdapter;
mBluetoothAdapter = BluetoothAdapter. getDefaultAdapter();
```

(3)通过判断 BluetoothAdapter 对象是否为空来判断当前设备中是否拥有蓝牙设备:

```
if(mBluetoothAdapter = = null){
    t_One. setText("手机没有蓝牙!");}
```

(4)判断并开启当前设备中蓝牙设备,BluetoothAdapter 对象判断 isEnabled()的返回值。如果蓝牙设备未开启,创建 Intent 对象并创建其 Action 为(BluetoothAdapter. ACTION_

REQUEST_ENABLE），并使用 startActivity（Intent 对象）方法提示用户开启蓝牙设备：

```
if(! mBluetoothAdapter. isEnabled( ) ) {
        Intent enableBtIntent =
        new Intent( BluetoothAdapter. ACTION_REQUEST_ENABLE) ;
        startActivity( enableBtIntent) ;
}
```

（5）获取所有已经配对的蓝牙设备列表，使用 BluetoothAdapter 对象 mAdapter 的 getBondedDevices（）方法获取与本机蓝牙适配器已配对的全部蓝牙设备并作为 BluetoothDevice 对象放到 Set 集合中：

```
Set < BluetoothDevice >  pairedDevices  =  mAdapter. getBondedDevices( ) ;
```

（6）将发现设备集合与绑定设备集合的全部蓝牙设备模块存放到一个新的设备集合中，并创建迭代器，遍历新蓝牙设备集合，获取全部远程蓝牙设备地址或其他信息：

```
String deviceInfoList = "设备信息:" ;
if( pairedDevices. size( ) >0) {
   while( iterator. hasNext( ) ) {
   BluetoothDevice btoothDevice  =  (BluetoothDevice) iterator. next( ) ;
   deviceInfoList =  deviceInfoList  + btoothDevice. getAddress( ) + " \n" ;
   }
   t_Two. setText( deviceInfoList) ;
}
```

Bluetooth 配对过程如图 11－1 所示。

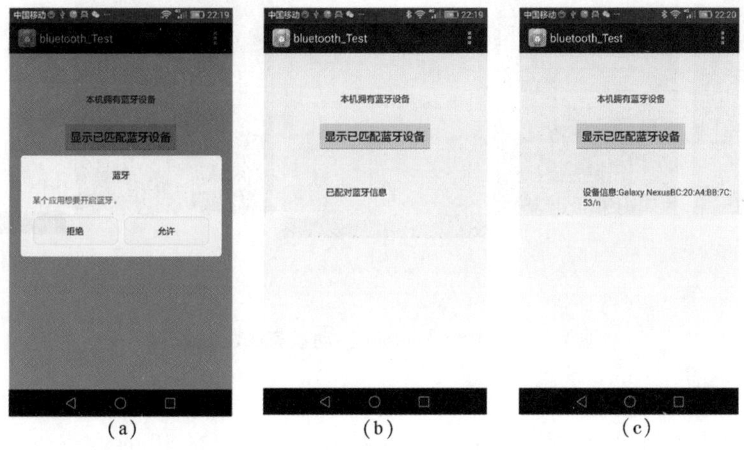

（a）　　（b）　　（c）

图 11－1　Bluetooth 配对过程

11.1.2 蓝牙的连接与数据传输

为让两个设备上的两个应用程序之间建立连接，必须同时实现服务端和客户端机制，因为一个设备必须打开服务端口，同时另一个设备必须初始化跟服务端设备的连接（使用服务端的 MAC 地址来初始化一个连接）。当服务端和客户端在相同的 RFCOMM 通道上有一个 BluetoothSocket 连接时，才能够被认为是服务端和客户端之间建立了连接。这时，每个设备能够获得输入和输出流，并且能够彼此开始传输数据。

服务端设备和客户端设备彼此获取所需的 BluetoothSocket 的方法是不同的。服务端会在接收输入连接时接收到一个 BluetoothSocket 对象，客户端会在打开与服务端的 RFCOMM 通道时接收到一个 BluetoothSocket 对象。

一种实现技术是自动的准备一个设备作为服务端，以便在每个设备都会有一个服务套接字被打开，并监听连接请求。当另一个设备初始化一个跟服务端套接字的连接时，它就会变成一个客户端。另一种实现技术是一个设备是明确的“host”连接，并且根据要求打开一个服务套接字，而其他的设备只是简单的初始化连接。

如果两个设备之前没有配对，那么 Android 框架会在连接过程期间自动显示一个配对请求通知或对话框给用户，两台 Android 移动设备的配对过程如图 11 -2 所示。因此，在试图连接设备时，应用程序不需要关心设备是否被配对。FRCOMM 的尝试性连接会一直阻塞，一直到用户成功配对或者是因用户拒绝配对或配对超时而失败。

图 11 -2 两台 Android 移动设备的配对过程

1. 服务端实现

如果想要连接两个设备，则一个必须通过持有一个打开的 BluetoothServerSocket 对象来作为服务端。服务套接字的用途是监听输入的连接请求，并且在一个连接请求被接收时提供一个 BluetoothSocket 连接对象。在从 BluetoothServerSocket 对象中获取 BluetoothSocket 时，BluetoothServerSocket 能够（并且建议）被废弃，除非想要接收更多的连接。以下是建立服务套接字和接收一个连接的基本过程。

(1)调用 listenUsingRfcommWithServiceRecord(String, UUID)方法来获得一个 BluetoothServerSocket 对象。该方法中的 String 参数是一个可识别的服务端的名称,系统会自动把它写入设备上的 Service Discovery Protocol(SDP)数据库实体(该名称是任意的,并且可以简单的使用应用程序的名称)。UUID 参数也会被包含在 SDP 实体中,并且是与客户端设备连接的基本协议。也就是说,当客户端尝试跟服务端连接时,它会携带一个它想要连接的服务端能够唯一识别的 UUID。只有在这些 UUID 完全匹配的情况下,连接才可能被接收。

(2)通过调用 accept()方法启动连接请求。这是一个阻塞调用。只有在连接被接收或发生异常的情况下,该方法才返回。只有在发送连接请求的远程设备所携带的 UUID 与监听服务套接字所注册的一个 UUID 匹配时,该连接才被接收。连接成功,accept()方法会返回一个被连接的 BluetoothSocket 对象。

(3)如果想要接收其他连接,要调用 close()方法。该方法会释放服务套接字以及它所占用的所有资源,但不会关闭被连接的已经由 accept()方法所返回的 BluetoothSocket 对象。

与 TCP/IP 不同,每个 RFCOMM 通道一次只允许连接一个客户端,因此在大多数情况下,在接收到一个连接套接字之后,立即调用 BluetoothServerSocket 对象的 close()方法是有道理的。accept()方法的调用不应该在主 Activity 的 UI 线程中执行,因为该调用是阻塞的,这会阻止应用程序的其他交互。通常在由应用程序所管理的一个新的线程中使用 BluetoothServerSocket 对象或 BluetoothSocket 对象来工作。要终止诸如 accept(1)这样的阻塞调用方法,就要从另一个线程中调用 BluetoothServerSocket 对象(或 BluetoothSocket 对象)的 close()方法,这时阻塞会立即返回。注意:在 BluetoothServerSocket 或 BluetoothSocket 对象上的所有方法都是线程安全的。

2. 客户端实现

为初始化一个与远程设备(持有打开的服务套接字的设备)的连接,首先必须获取个代表远程设备的 BluetoothDevice 对象,然后使用 BluetoothDevice 对象来获取一个 BluetoothSocket 对象,并初始化该连接。以下是一个基本的连接过程。

(1)通过调用 BluetoothDevice 的 createRfcommSocketToServiceRecord(UUID)方法,获得一个 BluetoothSocket 对象。这个方法会初始化一个连接到 BluetoothDevice 对象的 BluetoothSocket 对象。传递给这个方法的 UUID 参数必须与服务端设备打开 BluetoothServerSocket 对象时所使用的 UUID 相匹配。在应用程序中简单地使用硬编码进行比对,如果匹配,服务端和客户端代码就可以应用这个 BluetoothSocket 对象了。

(2)通过调用 connect()方法来初始化连接。在这个调用中,为找到匹配的 UUID,系统会在远程的设备上执行一个 SDP 查询。如果查询成功,并且远程设备接收了该连接请求,那么它会在连接期间共享使用 RFCOMM 通道,并且 connect()方法会返回。这个方法是一个阻塞调用。如果因为某些原因连接失败或连接超时(大约在 12 s 之后),就会抛出一个异常。

connect()方法是阻塞调用,这个连接过程始终应该在独立于主 Activity 线程之外的线程中被执行。在调用 connect()方法时,应该始终确保设备没有正在执行设备发现操作。如果是在发现操作的过程中,那么连接尝试会明显变慢,并且更像是要失败的样子。

3. 管理连接

成功连接了两个(或更多)设备时,每一个设备都有一个被连接的 BluetoothSocket 对象。这是良好的开始,因为能够在设备之间共享数据。使用 BluetoothSocket 对象来传输任意数据的过程是简单的。

(1)分别通过 getInputStream()和 getOutputStream()方法来获得通过套接字来处理传输任务的 InputStream 和 OutputStream 对象。

(2)用 read(byte[])和 write(byte[])方法来读写流中的数据。

当然,有更多实现细节要考虑。首先,对于所有数据流的读写应该使用专用的线程。这是重要的,因为 read(byte[])和 write(byte[])方法是阻塞式调用。Read(byte[])方法在从数据流中读取某些数据之前一直是阻塞的。write(byte[])方法通常是不阻塞的,但是对于流的控制,如果远程设备不是足够快的调用 read(byte[])方法,并且中间缓存被填满,那么 write(byte[])方法也会被阻塞。因此,线程中的主循环应该是专用于从 InputStream 对象中读取数据。在线程类中有一个独立的公共方法用于初始化对 OutputStream 对象的写入操作。

(3)创建一个 BluetoothProfile. ServiceListener 监听器,该监听器会在它们连接到服务器或中断与服务器的连接时通知 BluetoothProfile 的 IPC 客户端。

(4)在 onServiceConnected()事件中获得对配置代理对象的处理权。

(5)一旦获得配置代理对象,就可以用它来监视连接的状态,并执行与配置有关的其他操作。

4. 蓝牙信息传输的实例

蓝牙设备的初始化、连接、可见性设置已经完成,因此在本节实验中将其实验过程跳过。实例创建过程如下。

(1)蓝牙服务器端 Socket 创建。开启一个新的线程并通过 UUID(通用唯一标识码)接收与本地蓝牙设备器对象_bluetooth 连接的设备,创建 BluetoothServerSocket 对象,并使用 BluetoothServerSocket 的 accept()方法使服务器蓝牙 socket 对象处于等待连接请求状态:

```
BluetoothServerSocket serverSocket;
BluetoothAdapter _bluetooth = BluetoothAdapter. getDefaultAdapter();
\ _bluetooth 为本地蓝牙适配器对象
 serverSocket = _ bluetooth. listenUsingRfcommWithServiceRecord ( PROTOCOL _
SCHEME _ RFCOMM, UUID. fromString ( " 00001101 - 0000 - 1000 - 8000 -
```

```
00805F9B34FB"));
BluetoothSocket socket = _serverSocket. accept();
\socket 对象为发送连接请求的远程设备蓝牙客户端 socket
```

(2)蓝牙客户端创建。创建 BluetoothSocket 对象通过 UUID(通用唯一标识码)与远程设备对象进行连接:

```
BluetoothSocket socket = device. createRfcommSocketToServiceRecord (UUID. fromString ("00001101-0000-1000-8000-00805F9B34FB"));
\device 对象为用户在列表中所选择的蓝牙设备对象
socket. connect();
```

(3)蓝牙客户端与服务器端 Socket 通信。创建输入输出流并与客户端和服务器端 socket 对象的获取输入输出流方法进行连接,通过输出流的 write(bytes)方法发送信息,通过输入流的 read(bytes)方法接收信息,信息接收过程应放到线程中进行,并且需根据信息数据具体情况添加互斥锁:

```
private OutputStream outputStream;
private InputStream inputStream;
inputStream = socket. getInputStream();
outputStream = socket. getOutputStream();
outputStream. write(bytes);
inputStream. read(bytes);
```

11.1.3 健康设备配置

Android 3.0(API Level 14)中引入了对 Bluetooth Health Device Profile(HDP)支持,这会让程序与支持蓝牙的健康设备进行蓝牙通信的应用程序,如心率监护仪、血压测量仪、体温计、体重秤等。Bluetooth Health API 包含 BluetoothHealth、BluetoothHealthCallbackhe 和 BluetoothHealthAppConfiguration 等类。Bluetooth Health API 中的 HDP 概念见表 11-2。

表 11-2　Bluetooth Health API 中的 HDP 概念

概念	介绍
Source	HDP 中定义的一个角色,Source 是一个把医疗数据(如体重、血弹、体温等)传输给诸如 Android 手机或平板电脑等的设备
Sink	HDP 中定义的一个角色,在 HDP 中,一个 Sink 是一个接收医疗数据的小设备。在一个 Android HDP 应用程序中,Sink 用 BluetoothHealthAppConfiguration 对象来代表
Registration	给特定的健康设备注册一个 Sink
Connection	健康设备和 Android 手机或平板电脑之间打开的通信通道

以下是创建 Android HDP 应用中所涉及的基本步骤。

(1)获得 BluetoothHealth 代理对象的引用。类似于常规的耳机和 A2DP 配置设备,必须调用带有 BluetoothProfile. ServiceListener 和 HEALTH 配置类型参数的 getProfileProxy( )方法来建立与配置代理对象的连接。

(2)创建 BluetoothHealthCallback 对象,并注册一个扮演 Health Sink 角色的应用程序配(BluetoothHealthAppConfiguration)。

(3)建立跟健康设备的连接。某些设备会初始化连接,在这样的设备中进行这一步是没有必要的。

(4)当成功连接到健康设备时,就可以使用文件描述来读写健康设备。所接收到的数据需要使用健康管理器来解释,这个管理器实现了 IEEE 11073 - ×××××规范。

(5)完成以上步骤后,关闭健康通道并注销应用程序。该通道在长期被闲置时也会被关闭。

11.2 近场通信 NFC

Android 2. 3 (API Level 9) 引入了近场通信(Near Field Communication, NFC)API。NFC 是一种非接触式的技术,用于在短距离(通常小于 4 cm)内少量数据的传输。NFC 传输可以在两个支持 NFC 的设备或者一个设备和一个 NFC“标签”之间进行。NFC 标签既包括在扫描时会传输 URL 的被动标签,也包括复杂的系统,如 NFC 支付方案中使用的那些(如 Google Wallet)。

在 Android 中,NFC 消息是通过使用 NFC Data Exchange Format (NDEF)处理的。

为读取、写入或者广播 NFC 消息,应用程序需要具有 NFC 权限:

```
< uses - permission android : name = "android. permission. NFC" / >
```

11.2.1 NFC 与 NDEF

为实现标签和 NFC 设备及 NFC 设备之间的交互通信,NFC 论坛(NFC FROUM)定义了称为 NFC 数据交换格式(NDEF)的通用数据格式。NDEF 是轻量级的紧凑二进制格式,可带有 URL、vCard 和 NFC 定义的各种数据类型。NDEF 使得 NFC 的各种功能能容易地使用各种支持的标签类型传输数据,因为 NDEF 封装了标签的种类细节信息,所以应用不需要关心与何种标签在通信。

NDEF 交换的信息由一系列记录组成,每条记录包含一个有效载荷,内容可以是 URL、MIME 媒质或 NFC 定义的数据类型。使用 NFC 定义的数据类型,载荷内容必须被定义在一个 NFC 记录类型定义(RTD)文件中。

记录中数据的类型和大小由记录载荷的头部注明。头部包含:类型域,用来指定载荷的类型；载荷的长度数，单位是字节（octet）；可选的指定载荷是否带有一个 NDEF 记录。类型域的值由类型名称格式指定。NDEF 消息格式如图 11－3 所示。

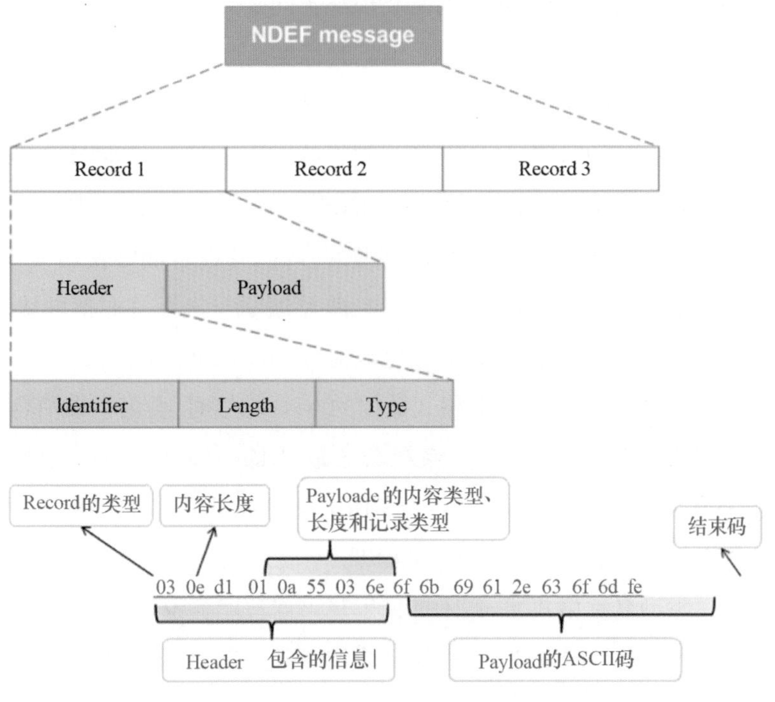

图 11－3　NDEF 消息格式

Android 对 NFC 的支持主要在 android. nfc 和 android. nfc. tech 两个包中。Android. nfc 包中主要类如下。

（1）NfcManager。可以用来管理 Android 移动设备中指出的所有 NFC Adapter，但由于大部分移动设备只支持一个 NFC Adapter，因此可以直接使用 getDefaultAapater 来获取系统支持的Adapter。

（2）NfcAdapter。表示本设备的 NFC adapter，可以定义 Intent 来请求将系统检测到 tags 的提醒发送到 Activity，并提供方法去注册前台 tag 提醒发布和前台 NDEF 推送。

（3）NdefMessage。NDEF 是 NFC 论坛定义的数据结构，用来有效存数据到 NFC tags，如文本、URL 和其他 MIME 类型。一个 NdefMessage 扮演一个容器，这个容器储存那些发送和读到的数据。一个 NdefMessage 对象包含 0 个或多个 NdefRecord，每个 NDEF record 有一个类型，如文本、URL、智慧型海报/广告或其他 MIME 数据。在 NDEFMessage 中，第一个 NfcRecord 的类型用来发送 tag 到一个 Android 移动设备上的 Activity。

（4）Tag。表示一个被动的 NFC 目标，如 tag、card、钥匙挂扣，甚至是一个电话模拟的 NFC 卡。

当一个 tag 被检测到时，一个 tag 对象将被创建并且封装到一个 Intent 中，然后 NFC 发布系统将这个 Intent 用 startActivity 发送到注册了接受这种 Intent 的 Activity 中。可以用

getTechList()方法得到这个 tag 支持的技术细节,创建一个 android. nfc. tech 提供的相应的 TagTechnology 对象。

开始编写 NFC 应用程序之前,重要的是要理解不同类型的 NFC 标签、标签调度系统是如何解析 NFC 标签的,以及在检测到 NDEF 消息时标签调度系统所做的特定的工作等。

NFC 标签涉及广泛的技术,并且有很多不同的方法向标签中写入数据。Android 支持由 NFC Forum 定义的 NDEF 标准。NDEF 数据被封装在一个消息(NdefMessage)中,该消息中包含了一条或多条记录(NdefRecord),每个 NDEF 记录必须具有记录类型的规范格式。

Android 也支持其他不包含 NDEF 数据类型的标签,开发者能够使用 android. nfc. tech 包中的类来工作。要使用其他类型标签来工作,涉及编写与该标签通信的协议栈,因此建议使用 NDEF,以便减少开发难度,并且最大化地支持 Android 设备。

当 Android 设备扫描到包含 NDEF 格式数据的 NFC 标签时,会解析该消息,并尝试分析数据的 MIME 类型或 URI 标识。首先,系统会读取消息(NdefMessage)中的第一条 NdefRecord,判断如何解释整个 NDEF 消息(一个 NDEF 消息能够有多条 NDEF 记录)。在格式良好的 NDEF 消息中,第一条 NdefRecord 包含以下字段信息。

①3 - bit TNF(类型名称格式)。指示如何解释可变长度类型字段。

②可变长度类型。说明记录的类型,如果使用 TNF_WELL_KNOWN,则使用这个字段来指定记录的类型定义(RTD)。

③可变长度 ID。唯一标识该记录。这个字段不经常使用,但是如果需要唯一地标识一个标记,那么就可以为该字段创建一个 ID。

④可变长度负载。读/写的实际的数据负载。一个 NDEF 消息能够包含多个 NDEF 记录,因此不要以为在 NDEF 消息的第一条 NDEF 记录中包含了所有的负载。

类型名称格式的映射见表 11 - 3,记录类型定义的映射见表 11 - 4。

表 11 - 3　类型名称格式的映射

类型名称格式(TNF)	映射
TNF_ABSOLUTE_URI	基于类型字段的 URI
TNF_EMPTY	退化到 ACTION_TECH_DISCOVERED 类型的 Intent 对象
TNF_EXTERNAL_TYPE	基于类型字段中 URN 的 URI URN 是缩短的格式(<domain_name>:<service_name>)被编码到 NDEF 类型中,Android 会把这个 URN 映射成以下格式的 URI:vnd. android. nfc://ext/<domain_name>:<service_name>
TNF_MIME_MEDIA	基于类型字段的 MIME 类型
TNF_UNCHANGED	退化到 ACTION_TECH_DISCOVERED 类型的 Intent 对象
TNF_UNKNOWN	退化到 ACTION_TECH_DISCOVERED 类型的 Intent 对象
TNF_WELL_KNOWN	依赖类型字段中设置的记录类型定义(RTD)的 MIME 类型或 URI

表 11－4　记录类型定义

记录类型定义(RTD)	映射
RTD_ALTERNATIVE_CARRIER	退化到 ACTION_TECH_DISCOVERED 类型的 Intent 对象
RTD_HANDOVER_CARRIER	退化到 ACTION_TECH_DISCOVERED 类型的 Intent 对象
RTD_HANDOVER_REQUEST	退化到 ACTION_TECH_DISCOVERED 类型的 Intent 对象
RTD_HANDOVER_SELECT	退化到 ACTION_TECH_DISCOVERED 类型的 Intent 对象
RTD_SMART_POSTER	基于负载解析的 URI
RTD_TEXT	text/plain 类型的 MIME
RTD_URI	基于有效负载的 URI

标签调度系统使用 TNF 和类型字段来尝试把 MIME 类型或 URI 映射到 NDEF 消息中。如果成功,它会把信息跟实际的负载一起封装到 ACTION_NEDF_DISCOVERED 类型的 Intent 中。但是,会有标签调度系统不能根据第一条 NDEF 记录来判断数据类型的情况,这样就会有 NDEF 数据不能被映射到 MIME 类型或 URI,或者是 NFC 标签没有包含 NDEF 开始数据的情况发生。在这种情况下,就会用一个标签技术信息相关的 Tag 对象和封装在 ACTION_TECH_DISCOVERED 类型 Intent 对象内部负载来代替。

表 11－3 和表 11－4 介绍标签调度系统映射如何把 TNF 和类型字段映射到 MIME 型或 URI 上,同时也介绍了哪种类型的 TNF 不能被映射到 MIME 类型或 URI 上。在这种情况下,标签调度系统会退化到 ACTION_TECH_DISCOVERED 类型的 Intent 对象。

例如,如果标签调度系统遇到一个 TNF_ABSOLUTE_URI 类型的记录,它会把这个记录的可变长度类型字段映射到一个 URI 中,标签调度系统会把这个 URI 跟其他相关的标签的信息(如数据负载)一起封装到 ACTION_NDEF_DISCOVERED 的 Intent 对象中。如果遇到 TNF_UNKNOWN 类型,它会创建一个封装了标签技术信息的 Intent 对象来代替。

当标签调度系统完成对 NFC 标签和它的标识信息封装的 Intent 对象的创建时,它会把该 Intent 对象发送给感兴趣的应用程序。如果有多个应用程序能够处理该 Intent 对象,就会显示 Activity 选择器,让用户选择 Activity。标签调度系统定义了三种 Intent 对象,下面按照由高到低的优先级列出这三种 Intent 对象。

(1)ACTION_NDEF_DISCOVERED。这种 Intent 用于启动包含 NDEF 负载和已知类型的标签的 Activity。这是最高优先级的 Intent,并且标签调度系统在任何其他 Intent 之前都会尽可能地尝试使用这种类型的 Intent 来启动 Activity。

(2)ACTION_TECH_DISCOVERED。如果没有注册处理 ACTION_NDEF_DISCOVERED 类型的 Intent 的 Activity,那么标签调度系统会尝试使用这种类型的 Intent 来启动应用程序。

如果被扫描到的标签包含了不能被映射到 MIME 类型或 URI 的 NDEF 数据，或者没有包含 NDEF 数据，但是已知的标签技术，那么也会直接启动这种类型的 Intent 对象，而不是先启动 ACTION_NDEF_DISCOVERED 类型的 Intent。

(3) ACTION_TAB_DISCOVERED。如果没有处理以上两种类型 Intent 的 Activity，就会启动这种类型的 Intent：

①用解析 NFC 标签时由标签调度系统创建的 Intent 对象（ACTION_NDEF_DISCOVERED 或 ACTION_TECH_DISCOVERED）来尝试启动 Activity；

②如果没有对应的处理 Intent 的 Activity，那么就会尝试使用下一个优先级的 Intent（ACTION_TECH_DISCOVERED 或 ACTION_TAG_DISCOVERED）来启动 Activity，直到有对应的应用程序来处理这个 Intent，或者是直到标签调度系统尝试了所有可能的 Intent；

③如果没有应用程序来处理任何类型的 Intent，那么就不做任何事情。

在可能的情况下，都会使用 NDEF 消息和 ACTION_NDEF_DISCOVERED 类型的 Intent 来工作，因为它是这三种 Intent 中最标准的。这种 Intent 与其他两种 Intent 相比，会允许使用者在更加合适的时机来启动其的应用程序，从而给用户带来更好的体验。标签调度系统工作流程图如图 11-4 所示。

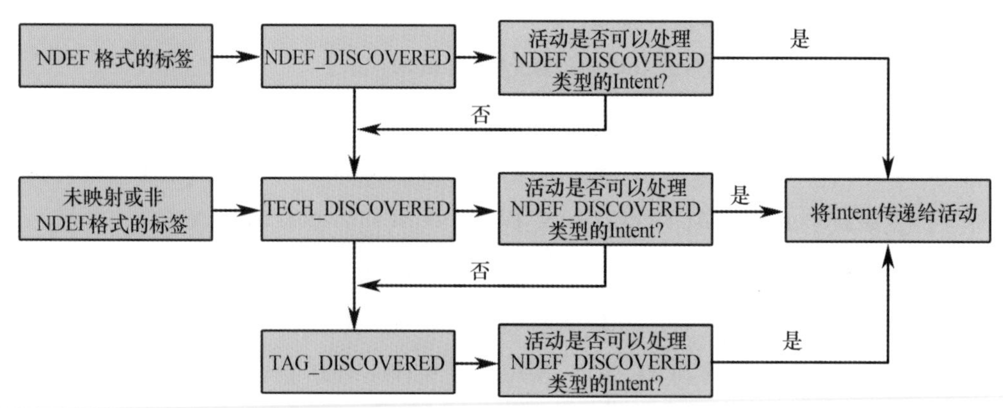

图 11-4　标签调度系统工作流程图

如果想要处理被扫描到的 NFC 标签时启动某个应用程序 A，就可以在这个应用程序 A 的 Android 清单中针对一种、两种或全部三种类型 NFC 的 Intent 来过滤。

但是，通常想要在应用程序启动时控制最常用的 ACTION_NDEF_DISCOVERED 类型的 Intent。在没有过滤 ACTION_NDEF_DISCOVERED 类型的 Intent 的应用程序，或数据负载不是 NDEF 时，才会从 ACTION_NDEF_DISCOVERED 类型的 Intent 回退到 ACTION_TECH_DISCOVERED 类型的 Intent。通常，ACTION_TAB_DISCOVERED 是最一般化的过滤分类。很多应用程序都会在过滤 ACTION_TAG_DISCOVERED 之前过滤前两种类型，这样就会降低应用程序 A 被启动的可能性。ACTION_TAG_DISCOVERED 只是在没有应用程序处理

ACTION_NDEF_DISCOVERED 或 ACTION_TECH_DISCOVERED 类型的 Intent 的情况下才使用的最后手段。

因为 NFC 标签的多样性，并且很多时候不在开发者控制之下，所以在必要时需要回退到其他两种类型的 Intent。如果开发者能够控制标签的类型和写入的数据，则建议使用 NDEF 格式。

要过滤 ACTION_NDEF_DISCOVERED 类型的 Intent，就要在清单中与需要过滤的数据一起声明该类型的 Intent 过滤器。例如，过滤 text/plain 类型的 MIME 的 ACTION_NDEF_DISCOVERED 类型过滤器的声明如下：

```
<intent-filter>
    <action android:name="android.nfc.action.NDEF_DISCOVERED"/>
    <category android:name="android.intent.category.DEFAULT"/>
    <data android:mimeType="text/plain" />
</intent-filter>
```

如果 Activity 要过滤 ACTION_TECH_DISCOVERED 类型的 Intent，则必须创建一个 XML 资源文件，该文件在 tech-list 集合中指定 Activity 所支持的技术。如果 tech-list 集合是标签所支持的技术的一个子集，那么 Activity 被认为是匹配的，通过调用 getTechList()方法来获得标签所支持的技术集合。

例如，如果扫描到的标签支持 MifareClassic、NdefFormatable 和 NfcA，那么为了与它们匹配，tech-list 集合就必须指定所有这三种技术，或者指定其中的两种或一种。

以下示例定义了所有的相关的技术，开发者可以根据需要删除其中一些设置，然后把这个文件保存到 <project-root>/res/xml 文件夹中：

```
<resources xmlns:xliff="urn:oasis:names:tc:xliff:document:1.2">
    <tech-list>
    <tech>android.nfc.tech.IsoDep</tech>
    <tech>android.nfc.tech.NfcA</tech>
    <tech>android.nfc.tech.NfcB</tech>
    <tech>android.nfc.tech.NfcF</tech>
    <tech>android.nfc.tech.NfcV</tech>
    <tech>android.nfc.tech.Ndef</tech>
    <tech>android.nfc.tech.NdefFormatable</tech>
    <tech>android.nfc.tech.MifareClassic</tech>
    <tech>android.nfc.tech.MifareUltralight</tech>
    </tech-list>
```

```
</resources>
```

也可以指定多个 tech - list 集合,每个 tech - list 集合被认为是独立的,并且如果任何一个 tech - list 集合是由 getTechList( )返回的技术的子集,那么 Activity 就被认为是匹配的。下列示例能够与支持 NfcA 和 Ndef 技术 NFC 标签或者支持 NfcB 和 Ndef 技术的标签相匹配:

```
<resources xmlns:xliff = "urn:oasis:names:tc:xliff:document:1.2" >
    <tech - list >
    <tech >android.nfc.tech.NfcA </tech >
    <tech >android.nfc.tech.Ndef </tech >
    </tech - list >
</resources >
<resources xmlns:xliff = "urn:oasis:names:tc:xliff:document:1.2" >
    <tech - list >
    <tech >android.nfc.tech.NfcB </tech >
     <tech >android.nfc.tech.Ndef </tech >
     </tech - list >
  </resources >
```

在 AndroidManifest. xml 文件中, < activity >元素内的 < meta - data >元素中指定开发者所创建的资源文件:

```
<intent - filter >
    <action android:name = "android.nfc.action.TECH_DISCOVERED"/ >
</intent - filter >
<meta - data android:name = "android.nfc.action.TECH_DISCOVERED"
    android:resource = "@xml/nfc_tech_filter" / >
```

使用下列 Intent 过滤器来过滤 ACTION_TAG_DISCOVERED 类型的 Intent:

```
<intent - filter >
    <action android:name = "android.nfc.action.TAG_DISCOVERED"/ >
</intent - filter >
```

在使用 NFC 标签和 Android 设备来进行工作的时候,使用的读写 NFC 标签上数据的主要格式是 NDEF。当设备扫描到带有 NDEF 的数据时,Android 会提供对消息解析的支持,并在可能的时候会以 NdefMessage 对象的形式来发送它。但是,有些情况下,设备扫描到的 NFC 标签没有包含 NDEF 数据,或者该 NDEF 数据没有被映射到 MIME 类型或 URI。在这些情况下,程序需要打开跟 NFC 标签的通信,并用自己的协议(原始的字节形式)来读写它。

Android 用 android. nfc. tech 包提供了对这些情况的一般性支持，Android 支持的标签技术见表 11－5。程序能够使用 getTechList( ) 方法来判断 NFC 标签所支持的技术，并且用 android. nfc. tech 提供的一个类来创建对应的 TagTechnology 对象。

表 11－5　Android 支持的标签技术

类	介绍
TagTechnology	所有的 NFC 标签技术类必须实现的接口
NfcA	提供对 NFC－A(ISO 14443－3A)属性和 I/O 操作的访问
NfcB	提供对 NFC－B(ISO 14443－3B)属性和 I/O 操作的访问
NfcF	提供对 NFC－F(ISO 6319－4)属性和 I/O 操作的访问
NfcV	提供对 NFC－V(ISO 15693)属性和 I/O 操作的访问
IsoDep	提供对 NFC－A(ISO 14443－4)属性和 I/O 操作的访问
Ndef	提供对 NDEF 格式的 NFC 标签上的 NDEF 数据和操作的访问
NdefFormatable	提供对可以被 NDEF 格式化的 NFC 标签的格式化操作
MifareClassic	如果 Android 设备支持 MIFARE，那么它提供对经典的 MIFARE 类型标签属性和 I/O 操作的访问
MifareUltralight	如果 Android 设备支持 MIFARE，那么它提供对超薄的 MIFARE 类型标签属性和 I/O 操作的访问

11.2.2　读取 NFC 标签

当一个 Android 设备用于扫描一个 NFC 标签时，其系统将使用自己的标签分派系统解码传入的有效载荷。这个标签分派系统会分析标签，将数据归类，并使用 Intent 启动一个应用程序来接收数据。

为使应用程序能够接收 NFC 数据，需要添加一个 Activity Intent Filter 来监听以下的某个 Intent 动作。

(1) NfcAdapter. ACTION_NDEF_DISCOVERED。这是优先级最高也是最具体的 NFC 消息动作，使用这个动作的 Intent 包括 MIME 类型和/或 URI 数据。最好的做法是只要有可能，就监听这个广播，因为其 extra 数据允许更加具体地定义要响应的标签。

(2) NfcAdapter. ACTION_TECH_DISCOVERED。当 NFC 技术已知，但是标签不包含数据(或者包含的数据不能被映射为 MIME 类型或 URI)时广播这个动作。

(3) NfcAdapter. ACTION_TAG_DISCOVERED。如果从未知技术收到一个标签，则使用此 Intent 动作广播该标签。

下面这段 Android 应用程序配置文件 AndroidManifest. xml 中的代码显示了如何注册一个 Activity,使其只响应对应于百度 URI 的 NFC 标签:

```
<! --监听 NFC 标签-->
<activity android:name=".SinaBlogViewer">
<intent-filter>
<action android:name="android.nfc.action.NDEF_DISCOVERED"/>
<category android:name="android.intent.category.DEFAULT"/>
<data android:scheme="http"
android:host=" https://www.baidu.com"/>
</intent-filter>
</activity>
```

NFC Intent Filter 尽可能地具体是一种很好的做法,这样可以将能够响应指定 NFC 标签的应用程序减到最少,从而提供最好、最快的用户体验。

很多时候,应用程序使用 Intent 数据/URI 和 MIME 类型就足以做出合适的响应。但是,需要时可以通过 Intent(启动 Activity 的 Intent)内的 extra 使用 NFC 消息提供的有效载荷。

NfcAdapter. EXTRA_TAG extra 包含一个代表扫描的标签的原始 Tag 对象。NfcAdapter. EXTRA_NDEF_MESSAGES extra 中包含一个 NDEF Messages 的数组。下面给出读取 NFC 标签有效载荷的代码实现:

```
private void processIntent(Intent intent) {
String action = getIntent().getAction();
if (NfcAdapter.ACTION_NDEF_DISCOVERED.equals(action)) {
Parcelable[] messages =
intent.getParcelableArrayExtra(NfcAdapter.EXTRA_NDEF_MESSAGES);
for (int i = 0; i < messages.length; i++) {
NdefMessage message = (NdefMessage)messages[i];
NdefRecord[] records = message.getRecords();
for (int j = 0; j < records.length; j++) {
 NdefRecord record = records[j];
 //处理的单独的记录
 }
 }
 }
```

```
}
```

这里给出一个读取 NFC 标签的简单实例，NFC 实例应用程序的运行界面如图 11－5 所示。下面首先给出配置文件中的关键代码：

```
<manifest… >
<uses - feature android:name = "android. hardware. nfc" android:required = "true" />
<application… >
<activity
android:name = "edu. hrbust. nfc. MainActivity"
android:label = "@ string/app_name" >
<intent - filter >
<action android:name = "android. intent. action. MAIN" />
<category android:name = "android. intent. category. LAUNCHER" />
</intent - filter >
<! - - 为 ACTION_TECH_DISCOVERED 类型的 Intent 注册过滤器 - - >
<intent - filter >
<action android:name = "android. nfc. action. TECH_DISCOVERED"/>
</intent - filter >
<meta - data
android:name = "android. nfc. action. TECH_DISCOVERED"
android:resource = "@ xml/nfc_tech_filter" />
</activity >
</application >
<uses - permission android:name = "android. permission. NFC" />
</manifest >
```

然后再给出主活动的 Java 代码：

```
public class NFCMainActivity extends Activity {
TextView IDText;
NfcAdapter nfcAdapter;
protected void onCreate(Bundle savedInstanceState) {
super. onCreate(savedInstanceState);
setContentView(R. layout. activity_main);
IDText = (TextView)findViewById(R. id. textView2);
nfcAdapter = NfcAdapter. getDefaultAdapter(this); //获取 NFC 适配器
```

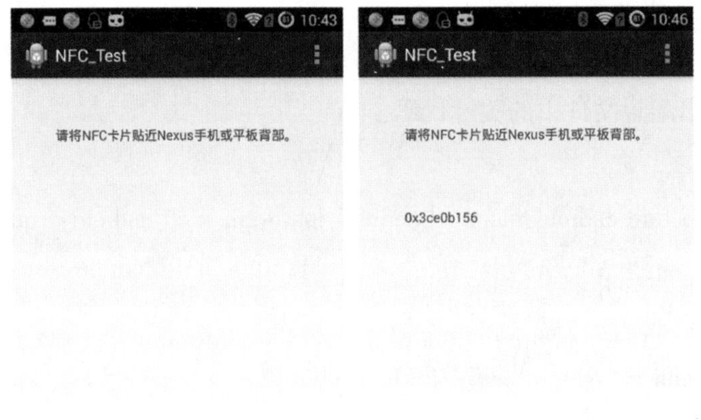

(a)　　　　(b)

图 11－5　NFC 实例应用程序的运行界面

```
if(nfcAdapter = = null){
IDText.setText("不支持 NFC");
finish();
return;
}
if(! nfcAdapter.isEnabled()){
IDText.setText("请先打开 NFC 设备");
finish();
return;
}
}
protected void onResume() {
super.onResume();
//判断当前的动作类型是否是 ACTION_TECH_DISCOVERED
if(NfcAdapter.ACTION_TECH_DISCOVERED.equals(getIntent()
.getAction())){
//通过 Intent 内的 extra 获取 NFC 标签的 ID：EXTRA_ID
byte[] bytesId =
```

```
getIntent( ). getByteArrayExtra( NfcAdapter. EXTRA_ID) ;
String Id  = bytesToHexString( bytesId) ;
IDText. setText( Id) ;
}
}
//自定义函数,将字节转换成十六进制
private String bytesToHexString( byte[ ] src) {
StringBuilder stringBuilder  =  new StringBuilder( "0x" ) ;
if ( src  = =  null | | src. length  < =  0) {
return null;
}
char[ ] buffer  =  new char[ 2 ] ;
for ( int i  =  0; i  <  src. length; i + + ) {
buffer[ 0 ]  =  Character. forDigit( ( src[ i ]  > > >  4)  & 0x0F, 16) ;
buffer[ 1 ]  =  Character. forDigit( src[ i ]  & 0x0F, 16) ;
stringBuilder. append( buffer) ;
}
return stringBuilder. toString( ) ;
}
}
```

11.2.3　使用前台分派系统

通常,Android 移动设备会在非锁屏的状态下搜索 NFC 所支持的标签,除非是在设备的设置菜单中 NFC 被禁用。当 Android 设备发现 NFC 标签时,期望使用最合适的 Activity 来处理该 Intent,而不是询问用户使用什么应用程序。因为设备可扫描到 NFC 标签的距离很短,所以强制地让用户手动的选择一个 Activity 很可能会导致设备离开 NFC 标签,从而中断该连接。

开发者应该开发自己的 Activity 来处理其所关心的 NFC 标签,从而阻止 NFC 应用选择器的操作。因此,Android 提供了特殊的标签调度系统,来分析扫描到的 NFC 标签,通过解析数据,在被扫描到的数据中尝试找到感兴趣的应用程序,具体做法如下:

①解析 NFC 标签并搞清楚标签中标识数据负载的 MIME 类型或 URI;

②把 MIME 类型或 URI 以及数据负载封装到一个 Intent 中;

③基于 Intent 来启动 Activity。

在默认情况下,标签分派系统会根据标准的 Intent 解析过程确定哪个应用程序应该收到特定的标签。在 Intent 解析过程中,位于前台的 Activity 并不比其他应用程序的优先级高。因此,如果几个应用程序都被注册为接收扫描类型的标签,用户就需要选择使用哪个应用程序,即使此时应用程序位于前台。

通过使用前台分派系统,可以指定特定的一个具有高优先级的 Activity,使得当它位于前台时成为默认接收标签的应用程序。使用 NFC Adapter 的 enable/disableForegroundDispatch 方法可以切换前台分派系统。只有当一个 Activity 位于前台时才能使用前台分派系统,所以应该分别在 onResume 和 onPause 处理程序内启动和禁用该系统。下面给出使用前台分派系统的实现代码:

```
public void onPause() {
super.onPause();
nfcAdapter.disableForegroundDispatch(this);
}
public void onResume() {
super.onResume();
nfcAdapter.enableForegroundDispatch(
this,
//用于打包 Tag Intent 的 Intent
 nfcPendingIntent,
 //用于声明想要拦截的 Intent 的 Intent Filter 数组
 intentFiltersArray,
 //想要处理的标签技术的数组
 techListsArray);
 String action = getIntent().getAction();
 if (NfcAdapter.ACTION_NDEF_DISCOVERED.equals(action)) {
 processIntent(getIntent());//读取 NFC 标签的有效载荷
 }
 }
```

Intent Filter 数组应该声明想要拦截的 URI 或 MIME 类型。如果收到的任何标签的类型与这些条件不匹配,那么将会被使用标准的标签分派系统处理。为确保良好的用户体验,只指定应用程序处理的标签内容是很重要的。

通过显式指定想要处理的技术(通常是添加 NfcF 类),可以进一步细化收到的标签。NFC Adapter 最好会填充 Pending Intent,以便把收到的标签直接传输给程序。

下面给出为使用前台分派系统所需要的参数，即 Pending Intent、MIME 类型数组和技术数组：

```
PendingIntent nfcPendingIntent;
IntentFilter[ ] intentFiltersArray;
String[ ][ ] techListsArray;
NfcAdapter nfcAdapter;
public void onCreate( Bundle savedInstanceState) {
super. onCreate( savedInstanceState) ;
setContentView( R. layout. main) ;
nfcAdapter = NfcAdapter. getDefaultAdapter( this) ;
//创建 Pending Intent.
int requestCode = 0;
int flags = 0;
Intent nfcIntent = new Intent( this, getClass( ) ) ;
nfcIntent. addFlags( Intent. FLAG_ACTIVITY_SINGLE_TOP) ;
nfcPendingIntent = PendingIntent. getActivity( this, requestCode, nfcIntent, flags) ;
//创建局限为 URI 或 MIME 类型的 Intent Filter,以从中拦截 TAG 扫描
IntentFilter tagIntentFilter =
new IntentFilter( NfcAdapter. ACTION_NDEF_DISCOVERED) ;
tagIntentFilter. addDataScheme( "http" ) ;
tagIntentFilter. addDataAuthority( "blog. sina. com. cn/hrbustmachao" , null) ;
intentFiltersArray = new IntentFilter[ ] { tagIntentFilter } ;
//创建要处理的技术数组
techListsArray = new String[ ][ ] {
new String[ ] {
NfcF. class. getName( )
}
} ;
}
```

11.2.4　Android Beam 简介

Android 4.0（API Level 14）中引入的 Android Beam 提供了一个简单的 API。应用程序可以使用该 API 在使用 NFC 的两个设备之间传输数据，只要将这两个设备背靠背放在一起

即可。例如,原生的联系人、浏览器和 YouTube 应用程序就使用 Android Beam 来与其他设备共享当前查看的联系人、网页和视频。

为使用 Android Beam 传输消息,应用程序必须位于前台,而且接收数据的设备不能处于锁住状态。通过将两个支持 NFC 的 Android 设备放在一起,可以启动 Android Beam。用户会看到一个"touch to beam"(触摸以传输)UI,此时可以选择把前台应用程序"beam"(传输)到另外一个设备。

当设备被放到一起时,Android Beam 会使用 NFC 在设备之间推送 NDEF 消息。通过在应用程序内启用 Android Beam,可以定义所传输的消息的有效载荷。如果没有自定义消息,应用程序的默认动作会在目标设备上启动它。如果目标设备上没有安装应用程序,那么 Google Play 就会启动,并显示应用程序的详细信息页面。

为定义应用程序传输的消息,需要在 manifest 文件中请求 NFC 权限:

```
<uses-permission android:name="android.permission.NFC" />
```

定义自己的有效载荷的过程如下:

①创建一个包含 NdefRecord 的 NdefMessage 对象,NdefRecord 中包含了消息的有效载荷;

②将 NdefMessage 作为 Android Beam 的有效载荷分配给 NFC Adapter;

③配置应用程序来监听传入的 Android Beam 消息。

(1)创建 Android Beam 消息。

要创建一个新的 Ndef Message,需要创建一个 NdefMessage 对象,并在其中创建至少一个 NdefRecord,用于包含想要传递给目标设备上的应用程序的有效载荷。

创建新的 Ndef Record 时,必须指定它表示的记录类型、一个 MIME 类型、一个 ID 和有效载荷。有几种公共的 Ndef Record 类型,可以用在 Android Beam 中类传递数据。要注意,它们总是应该作为第一条记录添加到要传输的消息中。

使用 NdefRecord. TNF_MIME_MEDIA 类型可以传输绝对 URI:

```
NdefRecord uriRecord = new NdefRecord(
NdefRecord.TNF_ABSOLUTE_URI,
"http://blog.sina.com.cn/hrbustmachao".getBytes(Charset.forName("US-ASCII")),
new byte[0], new byte[0]);
```

这是使用 Android Beam 传输的最常见的 Ndef Record,收到的 Intent 与任何启动 Activity 的 Intent 具有一样的形式,用来确定特定的 Activity 应该接收哪些 NFC 消息的 Intent Filter 可以使用 scheme、host 和 path Prefix 属性。

如果需要传输的消息所包含的信息不容易被解释为 URI,NdefRecord. TNF_MIME_

MEDIA 类型支持创建一个应用程序特定的 MIME 类型，并包含相关的有效载荷：

```
byte[] mimeByte =
"application/edu.hrbust.nfcbeam".getBytes(Charset.forName("US-ASCII"));
byte[] tagId = new byte[0];
byte[] payload = "Not a URI".getBytes(Charset.forName("US-ASCII"));
NdefRecord uriRecord = new NdefRecord(
NdefRecord.TNF_MIME_MEDIA, mimeByte, tagId, payload);
```

包含 Android Application Record(ARR)形式的 Ndef Record 是一种很好的做法，这可以保证应用程序会在目标设备上启动。如果目标设备上没有安装应用程序，则会启动 Google Play Store，让用户可以安装它。要创建一个 AAR Ndef Record，需要使用 Ndef Record 类的 createApplicationRecord 静态方法，并制定应用程序的包名。下面给出创建一条 Android Beam NDEF 消息，并在其中添加 AAR 的代码：

```
String payload = "Two to beam across";
String mimeType = "application/edu.hrbust.nfcbeam";
byte [] mimeBytes = mimeType.getBytes(Charset.forName("US-ASCII"));
NdefMessage nfcMessage = new NdefMessage(new NdefRecord[] {
//创建 NFC 有效载荷
new NdefRecord(
NdefRecord.TNF_MIME_MEDIA, mimeBytes,
new byte[0], payload.getBytes()),
//添加 AAR (Android Application Record)
 NdefRecord.createApplicationRecord("edu.hrbust.nfcbeam")
 });
```

(2)分配 Android Beam 有效载荷。

使用 NFC Adapter 可以指定 Android Beam 的有效载荷。通过使用 NfcAdapter 类的 getDefaultAdapter 静态方法，可以访问默认的 NFC Adapter：

```
NfcAdapter nfcAdapter = NfcAdapter.getDefaultAdapter(this);
```

有两种方法可以把创建的 NDEF Message 指定为应用程序的 Android Beam 有效载荷。最简单的方法是使用 setNdefPushMessage 方法来分配，当 Android Beam 启动时，总是应该从当前 Activity 发送的消息。通常，这种分配只需要在 Activity 的 onResume 方法中完成一次：

```
nfcAdapter.setNdefPushMessage(nfcMessage, this);
```

更好的方法是使用 setNdefPushMessageCallback 方法。该处理程序在消息被传输之前立即触发，允许根据应用程序当前的上下文，如正在看哪个视频、浏览哪个网页或者哪个地图

坐标居中,动态设置有效载荷的内容。下面给出动态设置 Android Beam 消息的代码:

```
nfcAdapter.setNdefPushMessageCallback(new
CreateNdefMessageCallback()
{
public NdefMessage createNdefMessage(NfcEvent event) {
String payload = "Beam me up, Android! \n\n" +
"Beam Time: " + System.currentTimeMillis();
NdefMessage message = createMessage(payload);
return message;
}
}, this);
```

如果使用回调处理程序同时设置了静态消息和动态消息,那么只有动态消息会被传输。

(3)接收 Android Beam 消息。

Android Beam 消息的接收方式与本章前面介绍的 NFC 标签十分类似。为接收在前一节打包的有效载荷,首先要在 Activity 中添加一个新的 Intent Filter,即 Android Beam 的 Intent Filter,具体代码如下:

```
<intent-filter>
<action android:name="android.nfc.action.NDEF_DISCOVERED"/>
<category android:name="android.intent.category.DEFAULT"/>
<data android:mimeType="application/edu.hrbust.nfcbeam"/>
</intent-filter>
```

Android Beam 启动后,接收设备上的 Activity 就会被启动。如果接收设备上没有安装应用程序,那么 Google Play Store 将会启动,以允许用户下载应用程序。

传输的数据会使用一个具有 NfcAdapter.ACTION_NDEF_DISCOVERED 动作的 Intent 传输给 Activity,其有效载荷可作为一个 NdfMessage 数组用于存储对应的 NfcAdapter.EXTRA_NDEF_MESSAGES extra。提取 Android Beam 有效载荷的实现代码如下:

```
Parcelable[] messages = intent.getParcelableArrayExtra(
NfcAdapter.EXTRA_NDEF_MESSAGES);
NdefMessage message = (NdefMessage)messages[0];
NdefRecord record = message.getRecords()[0];
String payload = new String(record.getPayload());
```

通常,有效载荷字符串是一个 URI 的形式,可以像对待 Intent 内封装的数据一样提取和处理它,以显示合适的视频、网页或地图坐标。

第 12 章　Android 传感器开发

Android 系统中提供了对传感器的支持。传感器在 Android 的应用中起到了非常重要的作用，有时可以实现一些意想不到的功能，如指南针、计步器等。本章将介绍一些传感器的开发及应用。

12.1　Sensor 开发基础

12.1.1　Sensor 简介

Android 系统的一大亮点就是对传感器的应用，它提供了十余种传感器。Android 中支持的 Sensor 类型见表 12－1。

表 12－1　Android 中支持的 Sensor 类型

感应检测	说明
TYPE_ACCELEROMETER	加速度传感器
TYPE_AMBIENT_TEMPERATURE	温度传感器
TYPE_GRAVITY	重力传感器
TYPE_GYROSCOPE	回转仪传感器
TYPE_LIGHT	光传感器
TYPE_LINEAR_ACCELERATION	线性加速度传感器
TYPE_MAGNETIC_FIELD	磁场传感器
TYPE_PRESSURE	压力传感器
TYPE_PROXIMITY	接近传感器
TYPE_RELATIVE_HUMIDITY	相对湿度传感器
TYPE_ROTATION_VECTOR	旋转矢量传感器

下面给出常用的传感器介绍。

方向传感器（Orientation）简称 O－sensor，主要感应方位的变化，现在已经被 SensorManager. getOrientation()取代，可以通过磁力计 MagneticField 和加速度传感器 Accelerometer 来

获得方位信息。该传感器同样捕获三个参数，分别代表手机沿传感器坐标系的 X 轴、Y 轴和 Z 轴转过的角度。

磁力传感器(MagneticField)简称 M－sensor，主要读取的是磁场的变化，通过该传感器便可开发出指南针、罗盘等磁场应用。磁场传感器读取的数据同样是空间坐标系三个方向的磁场值，其数据单位为 μT，即微特斯拉。

加速度传感器(Accelerometer)简称 G－sensor，主要用于感应设备的运动。该传感器捕获三个参数，分别表示空间坐标系中 X 轴、Y 轴、Z 轴方向上的加速度减去重力加速度在相应轴上的分量，其单位均为 m/s^2。

重力传感器(Gravity)简称 GV－sensor，主要用于输出重力数据。在地球上，重力数值为 9.8，单位是 m/s^2。坐标系统与加速度传感器坐标系相同。当设备复位时，重力传感器的输出与加速度传感器相同。

光传感器(Light)主要用来检测设备周围光线强度。光强单位是勒克斯(lx)，其物理意义是照射到单位面积上的光通量。

12.1.2　Sensor 开发过程

在开发传感器应用之前，首先要了解传感器的开发过程。要测试感应检测 Sensor 的功能，必须在装有 Android 系统的真机设备上进行。为方便对 Sensor 的访问，Android 提供了用于访问硬件的 API：android.hardware 包，该包提供了用于访问 Sensor 的类和接口。在 Android 应用程序中使用 Sensor 要依赖于 android.hardware.SensorEventListener 接口，通过该接口可以监听 Sensor 的各种事件。下面给出 Android 中 Sensor 应用程序的开发步骤。

(1)调用 Context.getSystemService(SENSOR_SERVICE)方法获取传感器管理服务，其代码格式如下：

```
SensorManager manager = (SensorManager)getSystemService(SENSOR_SERVICE);
```

(2)调用 SensorManager 的 getDefaultSensor(int type)方法获取指定类型的传感器，其代码格式如下：

```
SensorManager.getDefaultSensor(int type);
```

(3)在 Activity 的 onResume( )中调用 SensorManager 的 registerListener(SensorEventListener listener, Sensor sensor, int rate)方法注册监听，其代码格式如下：

```
SensorManager.registerListener( // 注册监听器
SensorEventListener listener,   // 监听传感器事件
Sensor sensor,                  // 传感器对象
int rate)                       // 延迟时间精密度
```

参数 rate 可以取值如下。

①Sensor. manager. SENSOR_DELAY_FASTEST。延迟 0 ms。

②Sensor. manager. SENSOR_DELAY_GAME。延迟 20 ms,适合游戏的频率。

③Sensor. manager. SENSOR_DELAY_UI。延迟 60 ms,适合普通界面的频率。

④Sensor. manager. SENSOR_DELAY_NORMAL。延迟 200 ms,正常频率。

(4)实现 SensorEventListener 接口中下列两个方法,监听并取得传感器 Sensor 的状态,其代码格式如下:

```
public interface SensorEventListener{
//传感器精度发生改变时调用
public abstract void onAccuracyChanged(Sensor sensor, int accuracy);
//传感器采样值发生变化时调用
public abstract void onSensorChanged(SensorEvent event);
}
```

SensorEventListener 接口包含了 onAccuracyChanged 和 onSensorChanged 两个方法。前者在一般场合中较少使用到,常用到的是 onSensorChanged 方法,它只有一个 SensorEvent 类型的参数 event,SensorEvent 类代表了一次传感器的响应事件,当系统从传感器获取到信息的变更时,会捕获该信息并向上层返回一个 SensorEvent 类型的对象,这个对象包含了传感器类型(public Sensor sensor)、传感器的时间戳(public long timestamp)、传感器数值的精度(public int accuracy)以及传感器的具体数值(public final float[ ] values)。

其中,values 值非常重要,其数据类型是 float[ ],它代表了从各种传感器采集回的数值信息。该 float 型的数组最多包含三个成员,根据传感器的不同,values 中每个成员代表的含义也不同。例如,通常温度传感器仅传回一个用于表示温度的数值,而加速度传感器则需要传回一个包含 X、Y、Z 三个轴上的加速度数值,同样的一个数据“10”,如果是从温度传感器传回则可能代表 10 ℃,而如果从亮度传感器传回则可能代表数值为 10 lx。

12.1.3　Sensor 坐标系

在 Android 中开发 Sensor 应用程序时,可以通过 Sensor 类型和 values 数组的值来正确地处理并使用传感器传回的值。为正确理解传感器所传回的数值,首先需要了解 Android 所定义的两个坐标系:世界坐标系(world coordinate - system)和旋转坐标系(rotation coordinate - system)。

世界坐标系(world coordinate - system)定义了从一个特定的 Android 设备上看待外部世界的方式,主要是以设备的屏幕为基准而定义,并且该坐标系依赖的是屏幕的默认方向,不因屏幕显示方向的改变而改变。坐标系以屏幕的中心为圆点,X 轴方向是沿着屏幕的水平方向从左向右。手机默认的正方状态,一般来说是如图 12 - 1 所示的世界坐标系默认长边在左右两侧，并且听筒在上方的情况，如果是特殊的设备，则 X 轴和 Y 轴可能会互换。

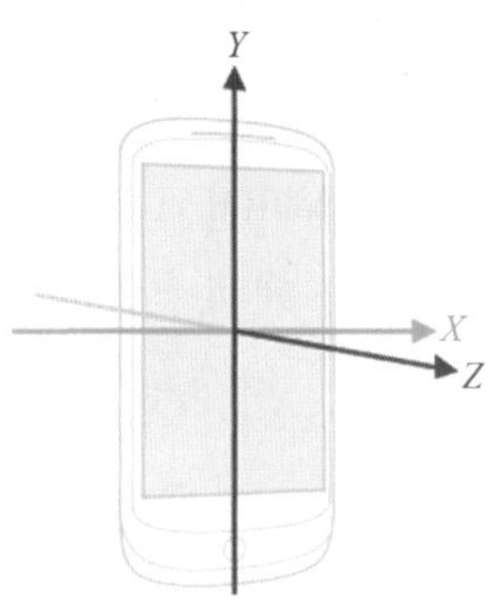

图 12－1　世界坐标系

Y 轴方向与屏幕的侧边平行，是从屏幕的正中心开始沿着平行屏幕侧边的方向指向屏幕的顶端。Z 轴的方向比较直观，即将手机屏幕朝上平放在桌面上时屏幕所朝的方向。有了约定好的世界坐标系，重力传感器、加速度传感器等所传回的数据和解析数据的方法就能够按照这种约定来确定联系了。

旋转坐标系如图 12－2 所示，球体可以理解为地球，这个坐标系是专用于旋转矢量传感器（Rotation Vector Sensor）的，可以理解为一个“反向的”世界坐标系。旋转矢量传感器用于描述设备所朝向的方向的传感器，而 Android 为描述这个方向定义了一个坐标系，这个坐标系也由 X 轴、Y 轴、Z 轴构成，特别之处是旋转矢量传感器所传回的数值是屏幕从标准位置（屏幕水平朝上且正北）开始，分别以这三个坐标轴为轴所旋转的角度。使用旋转矢量传感器的典型实例即“电子罗盘”。在这个坐标系中，X 轴即是 Y 轴与 Z 轴的向量积 $Y \cdot Z$，它的方位是与地球球面相切并且指向地理的西方；Y 轴为设备当前所在的位置与地面相切并且指向地磁北极的方向；Z 轴为设备所在位置指向地心的方向，垂直于地面。

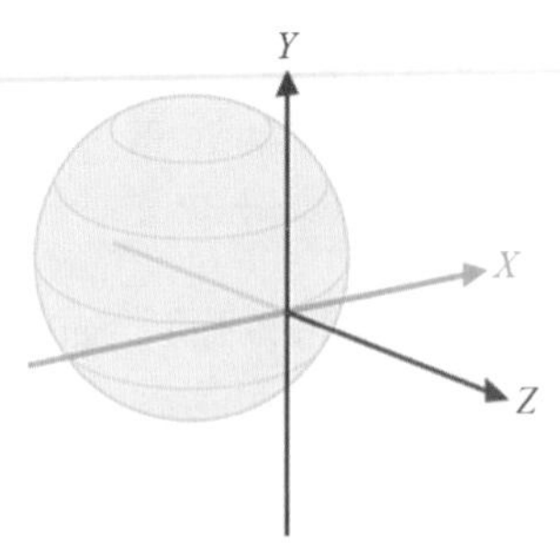

图 12－2　旋转坐标系

由于这个坐标系是专用于确定设备方向的，因此这里进一步给出访问旋转矢量传感器所传回的 values[ ]数组中各个数值所表示的含义，作为对 values[ ]值的一种示例说明。当旋转矢量传感器感应到方位变化时会返回一个包含变化结果数值的数组，即 values[ ]，数组长度为 3。

①values[0]。方位角,即手机绕 Z 轴所旋转的角度。

②values[1]。倾斜角,即手机绕 X 轴所旋转的角度。

③values[2]。翻滚角,即手机绕 Y 轴所旋转的角度。

以上所指明的角度都是逆时针方向的。

12.2　Sensor 应用实例

由于 Android 模拟器不支持 Sensor 感应功能,因此本节实例均需要在真机上运行。

12.2.1　获取 Sensor 清单

如图 12－3 所示为获取 Sensor 清单应用实例的程序运行界面,下面首先给出实例的主活动的代码:

```
public class MainActivity extends ActionBarActivity {
private SensorManager sensorManager; // 传感器的管理类
private TextView sensorList;
private TextView label;
private List<Sensor> list = null;
protected void onCreate(Bundle savedInstanceState) {
super.onCreate(savedInstanceState);
setContentView(R.layout.activity_main);
sensorManager =
(SensorManager) getSystemService (Context.SENSOR_SERVICE);
list =sensorManager.getSensorList(Sensor.TYPE_ALL);
sensorList = (TextView)findViewById(R.id.sensorList);
label = (TextView)findViewById(R.id.label);
for(Sensor sensor:list) {
sensorList.append(sensor.getName() + "\n");
}
Build build = new Build();
label.setText( build.MODEL + "包含的传感器清单:");
}
}
```

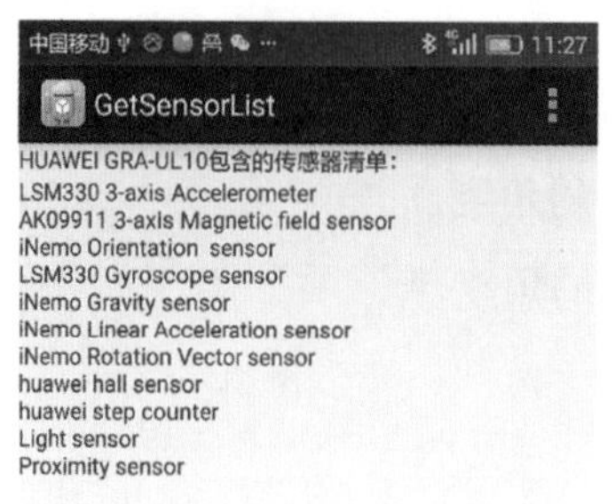

(a)

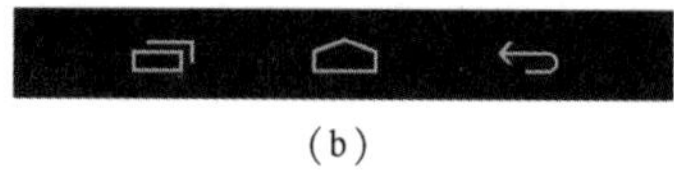

(b)

图 12－3　获取 Android 应用实例的程序运行界面清单

感应检测 Sensor 的硬件组件由不同的厂商提供。不同的 Sensor 设备组件所检测的事件也不同，可以使用 Sensor 类的 get × × ×( ) 方法检测设备所支持的 Sensor 的相关信息。除上面实例中用到的 public String getName( ) 获取传感器名称的方法外，还可以使用表 12－2 中的 Sensor 的相关信息获取方法。

表 12－2　Sensor 的相关信息获取方法

方法名称	方法说明
public float getMaximumRange( )	获取 Sensor 最大值
public int getMinDelay( )	获取 Sensor 的最小延迟
public float getPower( )	获取 Sensor 使用时所耗功率
public float getResolution( )	获取 Sensor 的精度
public int getType( )	获取 Sensor 类型
public String getVendor( )	获取 Sensor 供应商信息
public int getVersion( )	获取 Sensor 版本号信息

12.2.2　指南针应用实例

综上所述，方向传感器是基于软件的，并且它的数据是通过加速度传感器和磁场传感器共同获得的。如图 12－4 所示为指南针应用实例的程序运行界面，下面首先给出实例的布局文件代码：

```
<LinearLayout... >
<TextView
android:text = "指南针"
android:textSize = "24sp"
android:layout_width = "fill_parent"
android:layout_height = "wrap_content"/>
<ImageView
android:id = "@ +id/compass"
android:layout_width = "320dp"
android:layout_height = "320dp"
android:src = "@drawable/compass"/>
</LinearLayout>
```

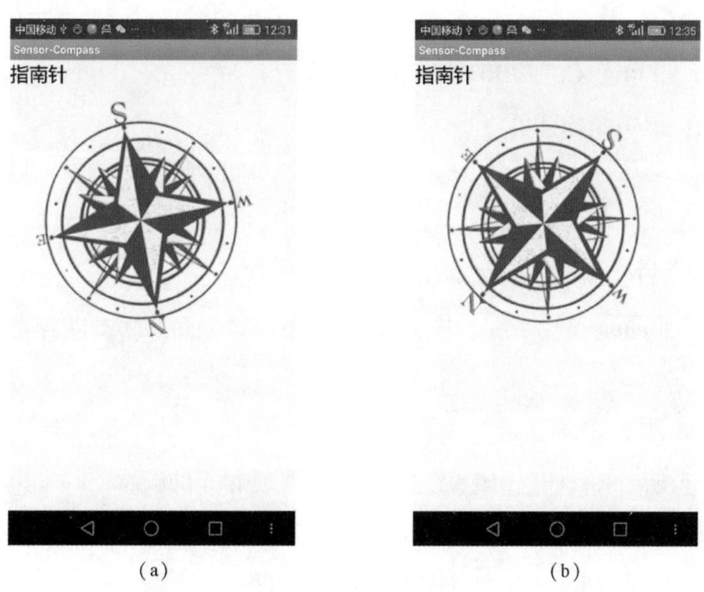

(a)　　　　(b)

图 12－4　指南针应用实例的程序运行界面

主活动的代码如下：

```
public class CompassActivity extends Activity implements SensorEventListener {
private SensorManager mSensorManager;
private Sensor mOrientation; //方向传感器
private Sensor accelerometer; //加速度传感器
```

```
private Sensor magnetic; //地磁场传感器
private float[ ] accelerometerValues = new float[3];
private float[ ] magneticFieldValues = new float[3];
private ImageView compass;
private float currentDegree = 0;
public void onCreate(Bundle savedInstanceState) {
super. onCreate (savedInstanceState);
setContentView (R. layout. main);
compass = (ImageView)findViewById(R. id. compass);
//获取传感器管理服务
mSensorManager =
(SensorManager) getSystemService (SENSOR_SERVICE);
//初始化加速度传感器
accelerometer =
mSensorManager. getDefaultSensor
(Sensor. TYPE_ACCELEROMETER);
//初始化地磁场传感器
magnetic =
mSensorManager. getDefaultSensor
(Sensor. TYPE_MAGNETIC_FIELD);
calculateOrientation(); //自定义函数,用来计算方向并转动图片
}
protected void onResume() {
//注册监听,SENSOR_DELAY_UI 为适合普通界面的频率
mSensorManager. registerListener (this, accelerometer,
SensorManager. SENSOR_DELAY_UI);
mSensorManager. registerListener (this, magnetic,
SensorManager. SENSOR_DELAY_UI);
}
protected void onPause() {
super. onPause ();
mSensorManager. unregisterListener (this);
}
```

```
public void onAccuracyChanged(Sensor sensor, int accuracy) {
}
public void onSensorChanged(SensorEvent event) {
if (event. sensor. getType( ) = = Sensor. TYPE_ACCELEROMETER) {
accelerometerValues = event. values;
}
if (event. sensor. getType( ) = = Sensor. TYPE_MAGNETIC_FIELD) {
magneticFieldValues = event. values;
}
calculateOrientation( );//自定义函数,用来计算方向并转动图片
}
//以指南针图像中心为轴旋转,从起始度数 currentDegree 旋转至
//targetDegree
private void rotateCompass(float currentDegree, float targetDegree) {
RotateAnimation ra; //旋转变化动画类
ra = new RotateAnimation(currentDegree,
targetDegree, Animation. RELATIVE_TO_SELF, 0. 5f,
Animation. RELATIVE_TO_SELF, 0. 5f);
ra. setDuration(200); //在 200 毫秒之内完成旋转动作
compass. startAnimation(ra); //开始旋转图像
}
//自定义函数,用来计算方向并转动图片
private void calculateOrientation( ) {
float[ ] values = new float[3];
float[ ] R = new float[9];
SensorManager. getRotationMatrix(R, null, accelerometerValues,
magneticFieldValues);
SensorManager. getOrientation(R, values);
values[0] = (float) Math. toDegrees(values[0]);
//处理传感器传回的数值并反映到图像的旋转上,
//需要注意的是由于指南针图像的旋转是与手机(传感器)相反的,
//因此需要旋转的角度为负的角度( - event. values[0])
float targetDegree = - values[0];
```

```
rotateCompass(currentDegree, targetDegree);
currentDegree = targetDegree;
}
}
```

在上述代码中,RotateAnimation 类是 Android 系统中的旋转变化动画类,用于控制 View 对象的旋转动作,该类继承于 Animation 类。该类中最常用的方法便是 RotateAnimation 构造方法:public RotateAnimation (float fromDegrees, float toDegrees, int pivotXType, float pivotXValue, int pivotYType, float pivotYValue)。其中,参数 fromDegrees 表示旋转的开始角度;参数 toDegrees 表示旋转的结束角度;参数 pivotXType 表示 *X* 轴的伸缩模式,可以取值为 ABSOLUTE、RELATIVE_TO_SELF、RELATIVE_TO_PARENT;参数 pivotXValue 表示 *X* 轴的伸缩值参数;pivotYType 表示 *Y* 轴的伸缩模式,可以取值为 ABSOLUTE、RELATIVE_TO_SELF、RELATIVE_TO_PARENT;参数 pivotYValue 表示 *Y* 坐标的伸缩值。

12.2.3 计步器应用实例

如图 12-5 所示为计步器应用实例的程序运行界面,下面直接给出活动的关键代码:

```
public class PedometerActivity extends Activity implements SensorEventListener {
private SensorManager mSensorManager;
private Sensor mAccelerometer; //加速度传感器
private TextView pedometerStatus, stepcount, debug;
private static final float GRAVITY = 9.80665f;
private static final float GRAVITY_RANGE = 0.01f;
public void onCreate(Bundle savedInstanceState) {
super.onCreate(savedInstanceState);
setContentView(R.layout.main);
stepcount = (TextView)findViewById(R.id.stepcount);
debug = (TextView)findViewById(R.id.debug);
mSensorManager =
(SensorManager)getSystemService(SENSOR_SERVICE);
mAccelerometer =
mSensorManager.getDefaultSensor(Sensor.TYPE_ACCELEROMETER);
}
protected void onResume() {
super.onResume();
```

```
mSensorManager.registerListener(this, mAccelerometer,
SensorManager.SENSOR_DELAY_UI);
}
protected void onPause() {
super.onPause();
mSensorManager.unregisterListener(this);
}
public void onAccuracyChanged(Sensor sensor, int accuracy) {
}
public void onSensorChanged(SensorEvent event) {
switch (event.sensor.getType()){
case Sensor.TYPE_ACCELEROMETER:{
debug.setText("values[0]-->" +event.values[0] +"\nvalues[1]-->" +
event.values[1] + "\nvalues[2]-->" + event.values[2]);
if (justFinishedOneStep(event.values[2])){
stepcount.setText((Integer.parseInt(stepcount.getText().toString(
)) +1) + "");
}
break;
}
default:
break;
}
}
//存储一步的过程中传感器传回值的数组便于分析
private ArrayList<Float> dataOfOneStep = new ArrayList<Float>();
private boolean justFinishedOneStep(float newData){
boolean finishedOneStep = false;
dataOfOneStep.add(newData);//将新数据加入到用于存储数据的列表中
dataOfOneStep = eliminateRedundancies(dataOfOneStep);//消除冗余数据
//分析是否完成了一步动作
finishedOneStep = analysisStepData(dataOfOneStep);
if(finishedOneStep){ //若分析结果为完成了一步动作
```

```
dataOfOneStep. clear( ); //则清空数组
return true; //并返回真
} else { //若分析结果为尚未完成一步动作
if( dataOfOneStep. size( ) > = 100){//防止占资源过大
dataOfOneStep. clear( );
}
return false; //则返回假
}
}
/ * 分析数据子程序
 * /
private boolean analysisStepData( ArrayList < Float > stepData) {
boolean answerOfAnalysis = false;
boolean dataHasBiggerValue = false;
boolean dataHasSmallerValue = false;
for( int i = 1; i < stepData. size( ) - 1; i + + ) {
//是否存在一个极大值
if( stepData. get( i). floatValue( ) > GRAVITY + GRAVITY_RANGE) {
if( ( stepData. get( i). floatValue( ) > stepData. get( i + 1 ). floatValue( ) ) &&
( stepData. get( i). floatValue( ) > stepData. get( i - 1 ). floatValue( ) ) ) {
dataHasBiggerValue = true;
}
}
//是否存在一个极小值
if( stepData. get( i). floatValue( ) < GRAVITY - GRAVITY_RANGE) {
if( ( stepData. get( i). floatValue( ) < stepData. get( i + 1 ). floatValue( ) ) &&
( stepData. get( i). floatValue( ) < stepData. get( i - 1 ). floatValue( ) ) ) {
dataHasSmallerValue = true;
}
}
}
answerOfAnalysis = dataHasBiggerValue && dataHasSmallerValue;
return answerOfAnalysis;
```

```
}
/* 消除 ArrayList 中的冗余数据,节省空间,降低干扰
 * 函数的输入参数 rawData:原始数据
 * 函数的返回值:处理后的数据 */
private ArrayList<Float> eliminateRedundancies(ArrayList<Float> rawData){
for(int i =0; i<rawData.size() -1 ;i++){
if((rawData.get(i) < GRAVITY + GRAVITY_RANGE) &&
(rawData.get(i) > GRAVITY - GRAVITY_RANGE)&&
(rawData.get(i+1) < GRAVITY + GRAVITY_RANGE) &&
(rawData.get(i+1) > GRAVITY - GRAVITY_RANGE)){
rawData.remove(i);
}else{
break;
}
}
 return rawData;
 }
 }
```

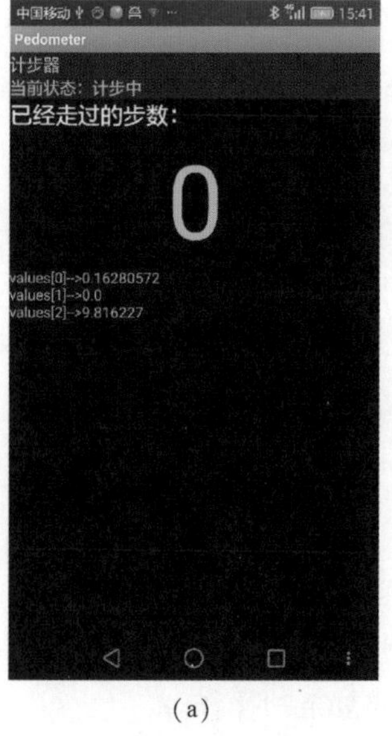

(a)

(b)

图 12-5 计步器应用实例的程序运行界面

第 13 章　综合开发案例——理财日记本

13.1　系统分析

13.1.1　需求分析

为方便地理清人们的财务状况，开发了一款基于 Android 系统的理财日记本软件。通过此软件，用户可以清楚地记录自己的收入、支出等信息。了解生活中的财务情况，面向日常记账用户。该软件是用户的专业理财管家。

13.1.2　可行性分析

根据《计算机软件文档编制规范》(GB/T 8567—2006)中可行性分析的要求，制定可行性研究报告如下。

1. 引言

(1)编写目的 。

说明该软件开发项目的实现在技术、经济和社会条件方面的可行性；评述为了合理地达到开发目标而可能选择的各种方案；说明并论证所选定的方案。

(2)背景 。

为更好地记录用户每月的收入及支出详细情况，现委托其他公司开发一个个人理财相关的软件，项目名称为“理财日记本”。

2. 可行性研究的前提

(1)要求 。

①系统的功能符合本人的实际情况。

②方便地对收入及支出进行增、删、改、查等操作。

③系统的功能操作要方便、易懂，不要有多余或复杂的操作。

④保证软件的安全性。

(2)目标。

详细记录个人用户的收入及支出的财务情况。

3. 投资及效益分析

(1)支出。

根据预算,公司计划投入 3 个人,为此需要支付 1.5 万元的工资及各种福利待遇,项目的安装、调试以及用户培训等费用支出需要 5 000 元,项目后期维护阶段预计需要投入 5 000 元的资金,累计项目投入需要 2.5 万元资金。

(2)收益。

客户提供项目开发资金 5 万元,对于项目后期进行的改动,采取协商的原则,根据改动规模额外提供资金。因此,从投资与收益的效益比上,公司大致可以获得 2.5 万元的利润。

项目完成后,会给公司提供资源储备,包括技术、经验的积累。

4. 结论

根据上面的分析,在技术上,不会存在问题,因此项目延期的可能性很小;在效益上,公司投入 3 个人,一个月的时间获利 2.5 万元,比较可观。另外,公司还可以储备项目开发的经验和资源。因此,认为该项目可以开发。

13.1.3　编写项目计划书

根据《计算机软件文档编制规范》(GB/T 8567—2006)中的项目开发计划要求,结合单位实际情况,设计项目计划书如下。

1. 引言

(1)编写目的。

为使项目按照合理的顺序开展,并保证按时、高质量地完成,现拟订项目计划书,将项目开发生命周期中的任务范围、团队组织结构、团队成员的工作任务、团队内外沟通协作方式、开发进度、检查项目工作等内容描述出来,作为项目相关人员之间的共识、约定以及项目生命周期内的所有项目活动的行动基础。

(2)背景。

理财日记本的项目性质为个人记账类型,它可以方便地记录用户的收入、支出等信息,项目周期为一个月。项目背景规划见表 13 -1。

表 13 -1　项目背景规划

项目名称	签定项目单位	项目负责人	参与开发部门
理财日记本	甲方	甲方	设计部门 开发部门 测试部门
	乙方	乙方	

2. 概述

(1)项目目标。

项目应当符合 SMART 原则,把项目要完成的工作用清晰的语言描述出来。理财日记本的主要目标是为用户提供一套能够方便地管理个人收入及支出信息的软件。

(2)应交付成果。

项目开发完成后,交付的内容如下。

①以光盘的形式提供理财日记本的源程序、APK 安装文件和系统使用说明书。

②系统发布后,进行无偿维护和服务 6 个月,超过 6 个月进行系统有偿维护与服务。

(3)项目开发环境。

开发本项目所用的操作系统可以是 Windows 或者 Linux 操作系统,开发工具为 Android Studio,数据库采用 Android 自带的 SQLite3。

(4)项目验收方与依据。

项目验收分为内部验收和外部验收两种方式。项目开发完成后,首先进行内部验收,由测试人员根据用户需求和项目目标进行验收。项目在通过内部验收后,交给客户进行外部验收,验收的主要依据为需求规格说明书。

3. 项目团队组织

本公司针对该项目组建了一个由软件工程师、界面设计师和测试人员构成的开发团队,为明确项目团队中每个人的任务分工,现制定人员分工表,见表 13-2。

表 13-2　人员分工表

姓名	技术水平	所属部门	角色	工作描述
王某	软件工程师	项目开发部	软件工程师	负责需求分析、软件设计与编码
刘某	美工设计师	设计部	界面设计师	负责软件的界面设计
李某	系统测试工程师	软件测试部	测试人员	对软件进行测试、编写软件测试文档

13.2　系统设计

13.2.1　系统目标

根据个人对理财日记本软件的要求制定目标如下。

①操作简单方便，界面简洁美观。

②方便地对收入及支出进行增、删、改、查等操作。

③通过便签，方便地记录用户的计划。

④能够通过设置密码保证程序的安全性。

⑤系统运行稳定，安全可靠。

13.2.2　系统功能结构

理财日记本的功能结构图如图 13－1 所示。

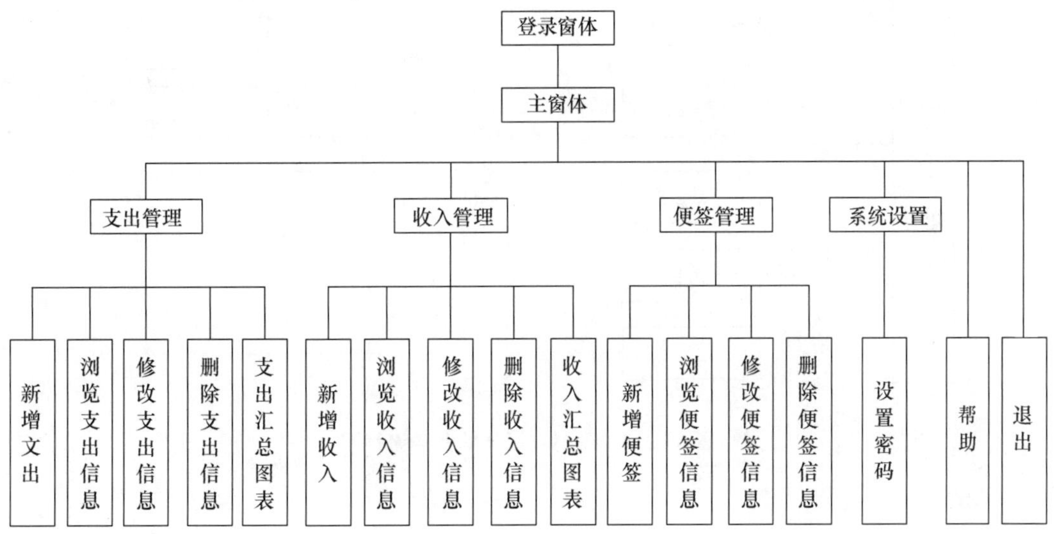

图 13－1　理财日记本的功能结构图

13.2.3　系统业务流程图

理财日记本的业务流程图如图 13－2 所示。

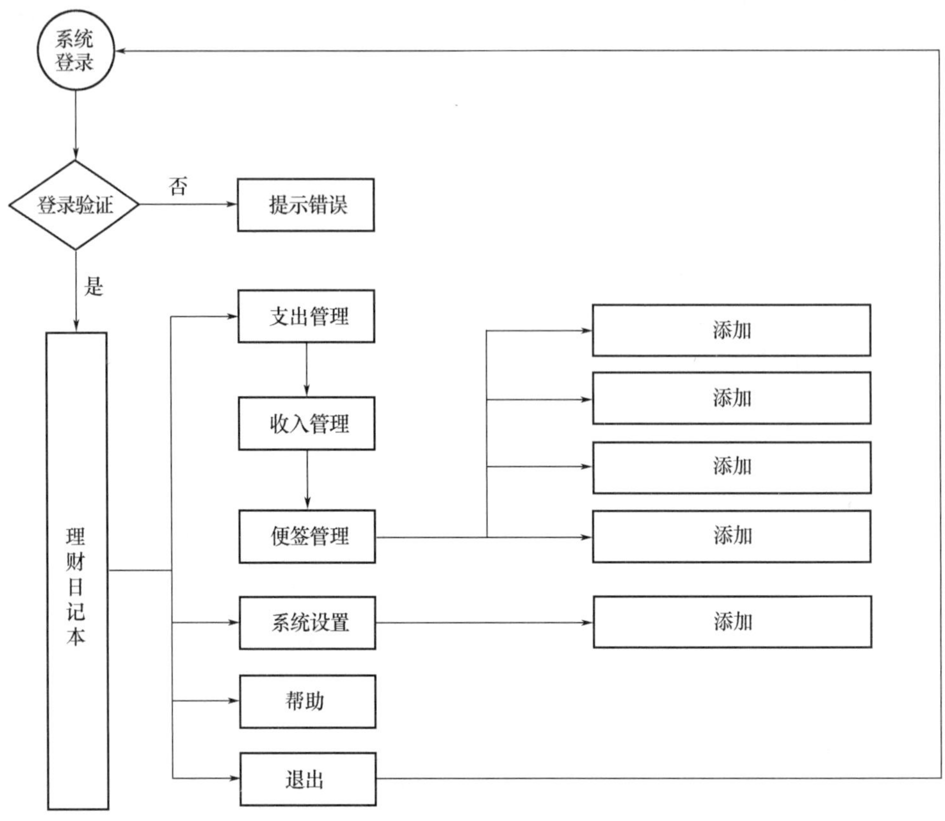

图 13－2　理财日记本的业务流程图

13.2.4　系统编码规范

开发应用程序常常需要团队合作完成,每个人负责不同的业务模块。为使程序的结构与代码风格统一标准化、增加代码的可读性,需要在编码之前制定一套统一的编码规范。下面介绍理财日记本系统开发中的编码规范。

1. 数据库命名规范

(1)数据库。

数据库以数据库相关英文单词或缩写进行命名,数据库命名见表 13－3。

表 13－3　数据库命名

数据库名称	描述
account. db	理财日记本数据库

(2)数据表。

数据表以字母 tb 开头(小写),后面加数据表相关英文单词或缩写。数据表命名见表 13 - 4。

表 13 - 4　数据表命名

数据表名称	描述
tb_outaccount	支出信息表

(3)字段。

字段一律采用英文单词或词组(可利用翻译软件)命名,如果找不到专业的英文单词或词组,可以用相同意义的英文单词或词组代替。字段命名见表 13 - 5。

表 13 - 5　字段命名

字段名称	描述
_id	编号
money	金额

2. 程序代码命名规范

(1)数据类型简写规则。

程序中定义常量、变量或方法等内容时,常常需要制定类型。下面介绍一种常见的数据类型简写规则,见表 13 - 6。

表 13 - 6　数据类型简写规则

数据类型	简写
整型	int
字符串	str
布尔型	bl
单精度浮点型	flt
双精度浮点型	dbl

(2)组件命名规则。

所有的组件对象名称都为组件名称的拼音简写,出现冲突时可采用不同的简写规则。

组件命名规则见表 13－7。

表 13－7　组件命名规则

控件	缩写形式
EditText	txt
Button	btn
Spinner	sp
ListView	lv
……	……

13.3　系统开发及运行环境

本系统的软件开发环境及运行环境具体如下。

操作系统:Windows 7。

JDK 环境:Java SE Development KET(JDK) version 7。

开发工具:Android Studio 2.3。

开发语言:Java、XML。数据库管理软件:SQLite 3。

运行平台:Windows、Linux 各版本。

分辨率:最佳效果 1 440 px×900 px。

13.4　数据库与数据表设计

开发应用程序时,对数据库的操作是必不可少的,数据库设计是根据程序的需求及其实现功能所制定的,数据库设计的合理性将直接影响程序的开发过程。

13.4.1　数据库分析

理财日记本是一款运行在 Android 系统上的程序,在 Android 系统中集成了一种轻量型的数据库,即 SQLite,该数据库是使用 C 语言编写的开源嵌入式数据库,支持的数据库大小为 2 TB,使用该数据库,用户可以像使用 SQL Server 数据库或者 Oracle 数据库那样来存储、管理和维护数据。本系统采用了 SQLite 数据库,并且命名为 account. db。该数据库中用到了四个数据表,分别是 tb_flag、tb_inaccount、tb_outaccount 和 tb_pwd。

13.4.2　创建数据库

理财日记本系统在创建数据库时，是通过使用 SQLiteOpenHelper 类的构造函数来实现的，实现代码如下：

```
private static final int VERSION = 1;                    //定义数据库版本号
Private static String DBNAME = "account. db"             //定义数据库名
public DBOpenHelper( Context context) {                  //定义构造函数
  super( context , jDBNAME , null , VERSION); //重写基类的构造函数，以创建数据库
}
```

13.4.3　创建数据表

在创建数据表前，首先要根据项目实际要求规划相关的数据表结构，然后在数据库中创建相应的数据表。

1. tb_pwd(密码信息表)

tb_pwd 表用于保存理财日记本的密码信息，tb_pwd 表的结构见表 13 –8。

表 13 –8　tb_pwd 表的结构

字段名	数据类型	主键否	描述
password	varchar(20)	否	用户密码

2. tb_outaccount(支出信息表)

tb_outaccount 表用于保存用户的支出信息，tb_outaccount 表的结构见表 13 –9。

表 13 –9　tb_outaccount 表的结构

字段名	数据类型	主键否	描述
_id	integer	是	编号
money	decimal	否	支出金额
time	varchar(10)	否	支出时间
type	varchar(10)	否	支出类别
address	varchar(100)	否	支出地点
mark	varchar(200)	否	备注

3. tb_inaccount(收入信息表)

tb_inaccount 表用于保存用户的收入信息,tb_inaccount 表的结构见表 13 - 10。

表 13 - 10　tb_inaccount 表的结构

字段名	数据类型	主键否	描述
_id	integer	是	编号
money	decimal	否	收入金额
time	varchar(10)	否	收入时间
type	varchar(10)	否	收入类别
handler	varchar(100)	否	付款方
mark	varchar(200)	否	备注

3. tb_flag(便签信息表)

tb_flag 表用于保存理财日记本的便签信息,tb_flag 表的结构见表 13 - 11。

表 13 - 11　tb_flag 表的结构

字段名	数据类型	主键否	描述
_id	integer	是	编号
flag	varchar(200)	否	便签内容

13.5　创 建 项 目

理财日记本系统的项目名称为 AccountMS,该系统是使用 Android Studio 2.3 开发的一个项目,即本章的例 13 - 1 所创建的项目。

13.6　系统文件夹组织结构

在编写项目代码之前,需要制定好项目的系统文件夹组织结构,如不同的 Java 包存放不同的窗体、公共类、数据模型、工具类或者图片资源等,这样不仅可以保证团队开发的一致性,也可以规范系统的整体架构。创建完系统中可能用到的文件夹或者 Java 包后,在开发时,只需将创建的类文件或者资源文件保存到相应的文件夹中即可。理财日记本系统的文件夹组织结构如图 13 - 3 所示。

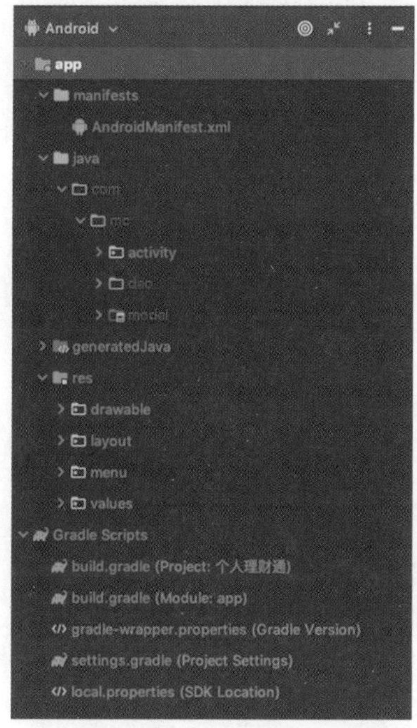

图 13－3　文件夹组织结构

13.7　公共类设计

公共类是代码重用的一种形式，它将各个功能模块经常调用的方法提取到公共的 Java 类中。例如，访问数据库的 Dao 类容纳了所有访问数据库的方法，并同时管理着数据库的连接、关闭等内容。使用公共类，不仅实现了项目代码的重用，还提供了程序的性能和代码的可读性。本节将介绍理财日记本中的公共类设计。

13.7.1　数据模型公共类

在 com. mc. model 包中存放的是数据模型公共类，它们对应着数据库中不同的数据表，这些模型将被访问数据库的 Dao 类和程序中各个模块甚至各个组件使用。数据模型是对数据表中所有字段的封装，它主要用于存储数据，并通过相应的 get( ) 方法和 set( ) 方法实现不同属性的访问原则。现在以收入信息表为例，介绍它所对应的数据模型类的实现代码，主要代码如下：

```
package com. mc. model;
public class Tb_inaccount// 收入信息实体类
{
    private int _id;// 存储收入编号
```

```
    private double money;// 存储收入金额
    private String time;// 存储收入时间
    private String type;// 存储收入类别
    private String handler;// 存储收入付款方
    private String mark;// 存储收入备注
    public Tb_inaccount(){// 默认构造函数
        super();
    }
    // 定义有参构造函数,用来初始化收入信息实体类中的各个字段
    public Tb_inaccount(int id, double money, String time, String type,
            String handler, String mark) {
        super();
        this._id = id;// 为收入编号赋值
        this.money = money;// 为收入金额赋值
        this.time = time;// 为收入时间赋值
        this.type = type;// 为收入类别赋值
        this.handler = handler;// 为收入付款方赋值
        this.mark = mark;// 为收入备注赋值
    }
    public int getid(){// 设置收入编号的可读属性
        return _id;
    }
    public void setid(int id){// 设置收入编号的可写属性
        this._id = id;
    }
    public double getMoney(){// 设置收入金额的可读属性
        return money;
    }
    public void setMoney(double money){// 设置收入金额的可写属性
        this.money = money;
    }
    public String getTime(){// 设置收入时间的可读属性
        return time;
```

```
    }
    public void setTime(String time){// 设置收入时间的可写属性
        this.time = time;
    }
    public String getType(){// 设置收入类别的可读属性
        return type;
    }
    public void setType(String type){// 设置收入类别的可写属性
        this.type = type;
    }
    public String getHandler(){// 设置收入付款方的可读属性
        return handler;
    }
    public void setHandler(String handler){// 设置收入付款方的可写属性
        this.handler = handler;
    }
    public String getMark(){// 设置收入备注的可读属性
        return mark;
    }
    public void setMark(String mark){// 设置收入备注的可写属性
        this.mark = mark;
    }
}
```

其他数据模型类的定义与收入数据模型类的定义方法类似,其属性内容就是数据表中相应的字段。

com.mc.model 包中包含的数据模型类见表 13－12。

表 13－12　数据模型类

类名	说明
tb_flag	便签信息数据表模型类
tb_inaccount	收入信息数据表模型类
tb_outaccount	支出信息数据表模型类
tb_pwd	密码信息数据表模型类

13.7.2 Dao 公共类

Dao 的全称是 Data Access Object,即数据访问对象。本系统中创建了 com. mc. dao 包,该包中包含了 DBOpenHelper、FlagDAO、IncaaountDAO、OutaccountDAO 和 PwdDAO 等五个数据访问类。其中,DBOpenHelper 类用来实现创建数据库、数据表等功能;FlagDAO 类用来对便签信息进行管理;IncaaountDAO 类用来对收入信息进行管理;OutaccountDAO 类用来对支出信息进行管理;PwdDAO 类用来对密码信息进行管理。下面主要对 DBOpenHelper 类和 IncaaountDAO 类进行详细讲解。

1. DBOpenHelper. Java 类

DBOpenHelper 类主要用来实现创建数据库和数据表的功能,该类继承自 SQLiteOpen-Helper 类。在该类中,首先需要在构造函数中创建数据库,然后在重写的 onCreate( )方法中使用 SQLiteDatabase 对象的 execSQL( )方法分别创建 tb_outaccount、tb_inaccount、tb_pwd 和 tb_flag 等四个数据表。DBOpenHelper 类实现代码如下:

```
package com. mc. dao;
import android. content. Context;
import android. database. sqlite. SQLiteDatabase;
import android. database. sqlite. SQLiteOpenHelper;
public class DBOpenHelper extends SQLiteOpenHelper {
    private static final int VERSION  = 1;// 定义数据库版本号
    private static final String DBNAME  = "account. db";// 定义数据库名
    public DBOpenHelper( Context context) {// 定义构造函数
        super( context, DBNAME, null, VERSION);// 重写基类的构造函数
    }
    @ Override
  public void onCreate( SQLiteDatabase db) {// 创建数据库
        db. execSQL( "create table tb_outaccount ( _id integer primary key,
        money decimal,time varchar(10) ,"
                + "type varchar(10) ,address varchar(100),
                mark varchar(200))");// 创建支出信息表
        db. execSQL( "create table tb_inaccount ( _id integer primary key,
        money decimal,time varchar(10) ,"
                + "type varchar(10) ,handler varchar(100),
                mark varchar(200))");// 创建收入信息表
```

```
        db.execSQL("create table tb_pwd (password varchar(20))");// 创建密码表
        db.execSQL("create table tb_flag (_id integer primary key,
        flag varchar(200))");// 创建便签信息表
    }
    @Override
    public void onUpgrade(SQLiteDatabase db, int oldVersion, int newVersion){

    }
}
```

2. InaccountDAO.java 类

InaccountDAO 类主要用来对收入信息进行管理,包括收入信息的添加、修改、删除、查询及获取最大编号、总记录数等功能。下面对该类中的方法进行详细讲解。

(1)InaccountDAO 类的构造函数。

在 InaccountDAO 类中定义两个对象,分别是 DBOpenHelper 对象和 SQLiteDatabase 对象,然后创建该类的构造函数,在构造函数中初始化 DBOpenHelper。主要代码如下:

```
private DBOpenHelper helper;// 创建 DBOpenHelper 对象
    private SQLiteDatabase db;// 创建 SQLiteDatabase 对象
    public InaccountDAO(Context context){// 定义构造函数
        helper = new DBOpenHelper(context);// 初始化 DBOpenHelper 对象
        db = helper.getWritableDatabase();// 初始化 SQLiteDatabase 对象
    }
```

(2)add(Tb_inaccount tb_inaccount)方法。

```
public void add(Tb_inaccount tb_inaccount) {
    // 执行添加收入信息操作
    db.execSQL(
            "insert into tb_inaccount (_id,money,time,type,handler,mark) "
            + "values (?,?,?,?,?,?)",
            new Object[] { tb_inaccount.getid(), tb_inaccount.getMoney(),
                    tb_inaccount.getTime(), tb_inaccount.getType(),
                    tb_inaccount.getHandler(), tb_inaccount.getMark() });
}
```

(3)ubdate(Tb_inaccount tb_inaccount)方法。

该方法的主要功能是根据指定的编号修改收入信息。其中,tb_inaccount 参数表示收入数据表对象。主要代码如下：

```
public void update(Tb_inaccount tb_inaccount) {
//      db = helper.getWritableDatabase();// 初始化 SQLiteDatabase 对象
        // 执行修改收入信息操作
        db.execSQL(
                "update tb_inaccount set money = ?,time = ?,type = ?,handler = ?,"
                + "mark = ? where _id = ?",
                new Object[] { tb_inaccount.getMoney(), tb_inaccount.getTime(),
                        tb_inaccount.getType(), tb_inaccount.getHandler(),
                        tb_inaccount.getMark(), tb_inaccount.getid() });
}
```

(4)find(int id)方法。

该方法的主要功能是根据指定的编号查找收入信息。其中,id 参数表示要查找的收入编号,返回值为 Tb_inaccount 对象。主要代码如下：

```
public Tb_inaccount find(int id) {
//      db = helper.getWritableDatabase();// 初始化 SQLiteDatabase 对象
        Cursor cursor = db
                .rawQuery("select _id,money,time,type,handler,mark from tb_inaccount"
                        + " where _id = ?",
                        new String[] { String.valueOf(id) });// 根据编号查找收入信息,并存储到 Cursor 类中
        if (cursor.moveToNext()){// 遍历查找到的收入信息
          // 将遍历到的收入信息存储到 Tb_inaccount 类中
          return new Tb_inaccount(cursor.getInt(cursor.getColumnIndex("_id")),
                    cursor.getDouble(cursor.getColumnIndex("money")),
                    cursor.getString(cursor.getColumnIndex("time")),
                    cursor.getString(cursor.getColumnIndex("type")),
                    cursor.getString(cursor.getColumnIndex("handler")),
                    cursor.getString(cursor.getColumnIndex("mark")));
        }
        cursor.close();// 关闭游标
        return null;// 如果没有信息,则返回 null
```

```
}
```

(5)delete(Interger... ids)方法。

该方法的主要功能是根据指定的一系列编号删除收入信息。其中,ids 参数表示要删除的收入编号的集合。主要代码如下:

```
public void detele(Integer... ids) {
    if (ids.length > 0){// 判断是否存在要删除的 id
        StringBuffer sb = new StringBuffer();// 创建 StringBuffer 对象
        for (int i = 0; i < ids.length; i++){// 遍历要删除的 id 集合
            sb.append('? ').append(',');// 将删除条件添加到 StringBuffer 对象中
            }
            sb.deleteCharAt(sb.length() - 1);// 去掉最后一个","字符
            //db = helper.getWritableDatabase();// 初始化 SQLiteDatabase 对象
            // 执行删除收入信息操作
            db.execSQL("delete from tb_inaccount where _id in (" + sb + ")",
                    (Object[]) ids);
    }
}
```

(6)getScrollData(int start, int count)方法。

该方法的主要功能是从收入数据表的指定索引处,获取指定数量的收入数据。其中,start 参数表示要从此处开始获取数据的索引;count 参数表示要获取的数量;返回值为 List<Tb_inaccount>对象。主要代码如下:

```
public List<Tb_inaccount> getScrollData(int start, int count) {
    List<Tb_inaccount> tb_inaccount = new ArrayList<Tb_inaccount>(); // 创建集合对象
    //db = helper.getWritableDatabase();// 初始化 SQLiteDatabase 对象
    // 获取所有收入信息
    Cursor cursor = db.rawQuery("select * from tb_inaccount limit ?,?",
            new String[] { String.valueOf(start), String.valueOf(count) });
    while (cursor.moveToNext()){// 遍历所有的收入信息
        // 将遍历到的收入信息添加到集合中
        tb_inaccount.add(new Tb_inaccount(cursor.getInt(cursor
                .getColumnIndex("_id")), cursor.getDouble(cursor
                .getColumnIndex("money")), cursor.getString(cursor
```

```
                        . getColumnIndex("time")), cursor. getString(cursor
                        . getColumnIndex("type")), cursor. getString(cursor
                        . getColumnIndex("handler")), cursor. getString(cursor
                        . getColumnIndex("mark"))));
        }
        cursor. close();// 关闭游标
        return tb_inaccount;// 返回集合
    }
```

(7)getCount()方法。

该方法的主要功能是获取收入数据报表中的总记录数,返回值为获取到的总记录数。主要代码如下:

```
public long getCount() {
    db = helper. getWritableDatabase();// 初始化 SQLiteDatabase 对象
    Cursor cursor = db
            . rawQuery("select count(_id) from tb_inaccount", null);
    // 获取收入信息的记录数
    if (cursor. moveToNext()){// 判断 Cursor 中是否有数据
        return cursor. getLong(0);// 返回总记录数
    }
    cursor. close();// 关闭游标
    return 0;// 如果没有数据,则返回 0
}
```

(8)getMaxId()方法。

该方法的主要功能是获取收入数据表中的最大编号,返回值为获取到的最大编号。主要代码如下:

```
public int getMaxId() {
    db = helper. getWritableDatabase();// 初始化 SQLiteDatabase 对象
    Cursor cursor = db. rawQuery("select max(_id) from tb_inaccount", null);
    // 获取收入信息表中的最大编号
    while (cursor. moveToLast()) {// 访问 Cursor 中的最后一条数据
        return cursor. getInt(0);// 获取访问到的数据,即最大编号
    }
    cursor. close();// 关闭游标
```

```
        return 0;// 如果没有数据,则返回 0
    }
```

13.8　登录模块设计

登录模块主要是通过输入正确的密码进入理财日记本的主窗体,它可以提高程序的安全性,保护数据资料不外泄。登录模块运行结果如图 13-4 所示。

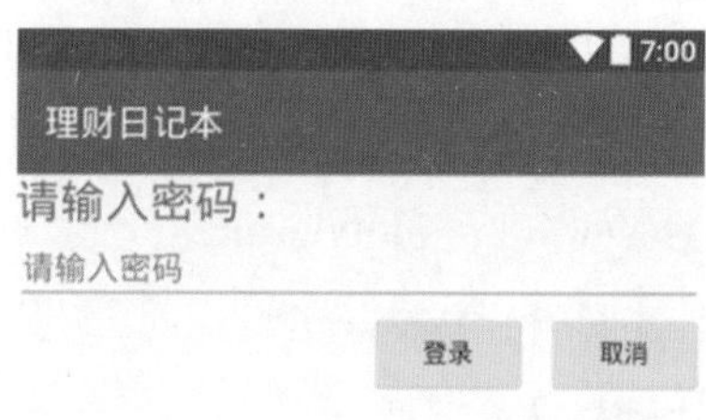

图 13-4　登录模块运行结果

13.8.1　设计登录布局文件

在 res/layout 目录下新建一个 login.xml 用来作为登录窗体的布局文件。该布局文件中添加一个 TextView 组件、一个 EditText 组件和两个 Button 组件,实现代码如下:

```
<RelativeLayout xmlns:android="http://schemas.android.com/apk/res/android"
  android:layout_width="match_parent"
  android:layout_height="match_parent"
    android:paddingBottom="@dimen/activity_vertical_margin"
    android:paddingLeft="@dimen/activity_horizontal_margin"
    android:paddingRight="@dimen/activity_horizontal_margin"
    android:paddingTop="@dimen/activity_vertical_margin"
  >
  <TextView android:id="@+id/tvLogin"
```

```
        android:layout_width = "wrap_content"
        android:layout_height = "wrap_content"
        android:text = "请输入密码:"
        android:textSize = "25sp"
        android:textColor = "#8C6931"
    />
    <EditText android:id = "@+id/txtLogin"
        android:layout_width = "match_parent"
        android:layout_height = "wrap_content"
        android:layout_below = "@id/tvLogin"
        android:inputType = "textPassword"
        android:hint = "请输入密码"
    />
    <Button android:id = "@+id/btnClose"
        android:layout_width = "wrap_content"
        android:layout_height = "wrap_content"
        android:layout_below = "@id/txtLogin"
        android:layout_alignParentRight = "true"
        android:layout_marginLeft = "10dp"
        android:text = "取消"
    />
    <Button android:id = "@+id/btnLogin"
        android:layout_width = "wrap_content"
        android:layout_height = "wrap_content"
        android:layout_below = "@id/txtLogin"
        android:layout_toLeftOf = "@id/btnClose"
        android:text = "登录"
    />
</RelativeLayout>
```

13.8.2 登录功能的实现

在 com.mc.activity 包中创建一个 Login.java 文件，该文件的布局文件设置为 login.xml。当用户在“请输入密码”文本框中输入密码时，单击“登录”按钮，为“登录”按钮设置监听事件，在监听事件中判断数据库中是否设置了密码并且输入的密码为空,或者输入的密码是否与数据库中的密码一致。如果条件满足,则登录主 Activity;否则,弹出信息提示框。代码如下:

```
txtlogin = (EditText) findViewById(R.id.txtLogin);// 获取密码文本框
btnclose = (Button) findViewById(R.id.btnClose);// 获取取消按钮
btnlogin.setOnClickListener(new OnClickListener() {// 为登录按钮设置监听事件
  @Override
  public void onClick(View arg0) {
    // TODO Auto-generated method stub
    Intent intent = new Intent(Login.this, MainActivity.class);// 创建 Intent 对象
    PwdDAO pwdDAO = new PwdDAO(Login.this);// 创建 PwdDAO 对象
    // 判断是否有密码及是否输入了密码
    if (pwdDAO.getCount() == 0 || pwdDAO.find().getPassword().isEmpty()) {
      if(txtlogin.getText().toString().isEmpty()){
        startActivity(intent);// 启动主 Activity
      }else{
        Toast.makeText(Login.this, "请不要输入任何密码登录系统!",
              Toast.LENGTH_SHORT).show();
      }
    } else {
        // 判断输入的密码是否与数据库中的密码一致
        if (pwdDAO.find().getPassword()
              .equals(txtlogin.getText().toString())) {
          startActivity(intent);// 启动主 Activity
        } else {
          // 弹出信息提示
          Toast.makeText(Login.this, "请输入正确的密码!",
                Toast.LENGTH_SHORT).show();
        }
    }
    txtlogin.setText("");// 清空密码文本框
  }
});
```

说明:本系统中,在 com.mc.antivity 包中创建的.java 类文件都是基于 Activity 类的,下

面遇到时,将不再说明。

13.8.3 退出登录窗口

单击“取消”按钮,为“取消”按钮设置监听事件,在监听事件中调用 finish( )方法实现退出当前程序的功能。代码如下:

```
btnlogin = (Button) findViewById(R.id.btnLogin);// 获取登录按钮
btnclose.setOnClickListener(new OnClickListener() {// 为取消按钮设置监听事件
    @Override
    public void onClick(View arg0) {
        // TODO Auto-generated method stub
        finish();// 退出当前程序
    }
});
```

13.9 系统主窗体设计

主窗体是程序操作过程中必不可少的,它是与用户交互中的重要环节。通过主窗体,用户可以调用系统相关的各子模块,快速掌握本系统中所实现的各个功能。理财日记本系统中,当登录窗体,验证成功后,用户将进入主窗体。主窗体中以图标和文本相结合的方式显示各功能按钮,单击这些功能按钮时,可以打开相应功能的 Activity。主窗体运行结果如图 13-5 所示。

图 13-5 主窗体运行结果

13.9.1 设计系统主窗体布局文件

在 res/layout 目录下新建一个 main. xml 用来作为主窗体的布局文件。该布局文件中，添加一个 GridView 组件，用来显示功能图标及文本，实现代码如下：

```
<GridView xmlns:android = "http://schemas.android.com/apk/res/android"
    android:id = "@ +id/gvInfo"
    android:layout_width = "fill_parnt"
    android:layout_height = "fill_parent"
    android:columnWidth = "90dp"
    android:numColumns = "auto_fit"
    android:verticalSpacing = "10dp"
    android:horizontalSpacing = "10dp"
    android:stretchMode = "spacingWidthUniform"
     android:gravity = "center"
/>
```

在 res/latout 目录下再新建一个 gvitem. xml 用来为 main. xml 布局文件中的 GridView 组件提供资源。该文件中，添加一个 ImageView 组件和一个 TextView 组件，实现代码如下：

```
<LinearLayout xmlns:android = "http://schemas.android.com/apk/res/android"
    android:id = "@ +id/item"
    android:orientation = "vertical"
    android:layout_width = "wrap_content"
    android:layout_height = "wrap_content"
    android:layout_marginTop = "5dp"
  >
  <ImageView android:id = "@ +id/ItemImage"
      android:layout_width = "75dp"
       android:layout_height = "75dp"
       android:layout_gravity = "center"
       android:scaleType = "fitXY"
       android:padding = "4dp"
  />
   <TextView android:id = "@ +id/ItemTitle"
       android:layout_width = "wrap_content"
```

```
            android:layout_height = "wrap_content"
            android:layout_gravity = "center"
            android:gravity = "center_horizontal"
    />
</LinearLayout>
```

13.9.2 显示各功能窗口

在 com. mc. activity 包中创建一个 MainActivity. java 文件,该文件的布局文件设置为 main. xml。在 MainActivity. java 文件中,首先创建一个 GridView 组件对象,然后分别定义一个 String 类型的数组和一个 int 类型的数组,它们分别用来存储系统狗能的文本及对应的图标,代码如下:

```
GridView gvInfo;// 创建 GridView 对象
// 定义字符串数组,存储系统功能
String[] titles = new String[] { "新增支出", "新增收入", "我的支出", "我的收入",
            "数据管理","系统设置", "收支便签", "帮助","退出" };
// 定义 int 数组,存储功能对应的图标
int[] images = new int[] { R.drawable.addoutaccount,
            R.drawable.addinaccount, R.drawable.outaccountinfo,
            R.drawable.inaccountinfo, R.drawable.showinfo, R.drawable.sysset,
            R.drawable.accountflag, R.drawable.help,R.drawable.exit };
```

当用户在主窗体中单击各功能按钮时,使用相应功能所对应的 Activity 初始化 Intent 对象,然后使用 start Activity()方法启动相应的 Activity,如果用户单击的是“退出”功能按钮,则调用 finish()方法关闭当前 Activity。代码如下:

```
@Override
public void onCreate(Bundle savedInstanceState) {
      super.onCreate(savedInstanceState);
      setContentView(R.layout.main);
      gvInfo = (GridView) findViewById(R.id.gvInfo);
      pictureAdapter adapter = new pictureAdapter(titles, images, this);
      gvInfo.setAdapter(adapter);// 为 GridView 设置数据源
      gvInfo.setOnItemClickListener(new OnItemClickListener() {
            @Override
```

```
            public void onItemClick(AdapterView<?> arg0, View arg1, int arg2,
                    long arg3) {
                Intent intent = null;// 创建 Intent 对象
                switch (arg2) {
                case 0:
                        intent = new Intent(MainActivity.this, AddOutaccount.class);// 使用 AddOutaccount 窗口初始化 Intent
                        startActivity(intent);// 打开 AddOutaccount
                        break;
                case 1:
                intent = new Intent(MainActivity.this, AddInaccount.class);
                // 使用 AddInaccount 窗口初始化 Intent
                        startActivity(intent);// 打开 AddInaccount
                        break;
                case 2:
                        intent = new Intent(MainActivity.this, Outaccountinfo.class);
                        // 使用 Outaccountinfo 窗口初始化 Intent
                        startActivity(intent);// 打开 Outaccountinfo
                        break;
                case 3:
                intent = new Intent(MainActivity.this, Inaccountinfo.class);
                // 使用 Inaccountinfo 窗口初始化 Intent
                        startActivity(intent);// 打开 Inaccountinfo
                        break;
                case 4:
                        intent = new Intent(MainActivity.this, Showinfo.class);
                        // 使用 Showinfo 窗口初始化 Intent
                        startActivity(intent);// 打开 Showinfo
                        break;
                case 5:
                        intent = new Intent(MainActivity.this, Sysset.class);
                        // 使用 Sysset 窗口初始化 Intent
```

```
                    startActivity(intent);// 打开 Sysset
                    break;
                case 6:
                    intent = new Intent(MainActivity.this, Accountflag.class);
                    // 使用 Accountflag 窗口初始化 Intent
                    startActivity(intent);// 打开 Accountflag
                    break;
                case 7:
                    intent = new Intent(MainActivity.this, Help.class);
                    // 使用 Help 窗口初始化 Intent
                    startActivity(intent);// 打开 Help
                    break;
                case 8:
                    finish();// 关闭当前 Activity
                }
            }
        });
    }
```

13.9.3 定义文本及图片组件

在 MainActivity 中定义一个内部类 ViewHolder,用来定义文本组件及图片组件对象,代码如下:

```
class ViewHolder{// 创建 ViewHolder 类
    public TextView title;// 创建 TextView 对象
    public ImageView image;// 创建 ImageView 对象
}
```

13.9.4 定义功能图标及说明文字

在 MainActivity 中定义一个内部类 Picture,用来定义功能图标及说明文字的实体,代码如下:

```
class Picture{// 创建 Picture 类
    private String title;// 定义字符串,表示图像标题
    private int imageId;// 定义 int 变量,表示图像的二进制值
```

```
    public Picture() {// 默认构造函数
        super();
    }
    public Picture(String title, int imageId) {// 定义有参构造函数
        super();
        this.title = title;// 为图像标题赋值
        this.imageId = imageId;// 为图像的二进制值赋值
    }
    public String getTitle() {// 定义图像标题的可读属性
        return title;
    }
    public void setTitle(String title) {// 定义图像标题的可写属性
        this.title = title;
    }
    public int getImageId() {// 定义图像二进制值的可读属性
        return imageId;
    }
    public void setimageId(int imageId) {// 定义图像二进制值的可写属性
        this.imageId = imageId;
    }
}
```

13.9.5　设置功能图标及说明文字

在 MainActivity 中定义一个内部类 pictureAdapter，该类继承自 BaseAdapter 类，该类用来分别为 ViewHolder 类中的 TextView 组件和 ImageView 组件设置功能的说明性文字及图标，代码如下：

```
class pictureAdapter extends BaseAdapter{// 创建基于 BaseAdapter 的子类
    private LayoutInflater inflater;// 创建 LayoutInflater 对象
    private List<Picture> pictures;// 创建 List 泛型集合
    // 为类创建构造函数
    public pictureAdapter(String[] titles, int[] images, Context context) {
        super();
        pictures = new ArrayList<Picture>();// 初始化泛型集合对象
```

```
        inflater = LayoutInflater.from(context);// 初始化 LayoutInflater 对象
        for (int i = 0; i < images.length; i++){// 遍历图像数组
            Picture picture = new Picture(titles[i], images[i]);
            pictures.add(picture);// 将 Picture 对象添加到泛型集合中
        }
    }
    @Override
    public int getCount() {// 获取泛型集合的长度
        if (null != pictures) {// 如果泛型集合不为空
            return pictures.size();// 返回泛型长度
        } else {
            return 0;// 返回 0
        }
    }
    @Override
    public Object getItem(int arg0) {
        return pictures.get(arg0);// 获取泛型集合指定索引处的项
    }
    @Override
    public long getItemId(int arg0) {
        return arg0;// 返回泛型集合的索引
    }
    @Override
    public View getView(int arg0, View arg1, ViewGroup arg2) {
        ViewHolder viewHolder;// 创建 ViewHolder 对象
        if (arg1 == null){// 判断图像标识是否为空
            arg1 = inflater.inflate(R.layout.gvitem, null);// 设置图像标识
            viewHolder = new ViewHolder();// 初始化 ViewHolder 对象
            viewHolder.title = (TextView) arg1.findViewById(R.id.ItemTitle);
            viewHolder.image =
            (ImageView) arg1.findViewById(R.id.ItemImage);
            // 设置图像的二进制值
            arg1.setTag(viewHolder);// 设置提示
```

```
        } else {
            viewHolder = (ViewHolder) arg1.getTag();//设置提示
        }
    viewHolder.title.setText(pictures.get(arg0).getTitle());//设置图像标题
    viewHolder.image.setImageResource(pictures.get(arg0).getImageId());
    // 设置图像的二进制值
    return arg1;// 返回图像标识
    }
}
```

13.10　收入管理模块设计

收入管理模块主要包括四部分,分别是“新增收入”“收入信息浏览”“修改/删除收入信息”和“收入信息汇总图表”。其中,“新增收入”用来添加收入信息;“收入信息浏览”用来显示所有的收入信息;“修改/删除收入信息”用来根据编号修改或者删除收入信息;“收入信息汇总图表”用来统计收入信息并以图表形式显示。本节将从这四个方面对收入管理模块进行详细介绍。

首先来看“新增收入”模块,“新增收入”窗口运行结果如图 13－6 所示。

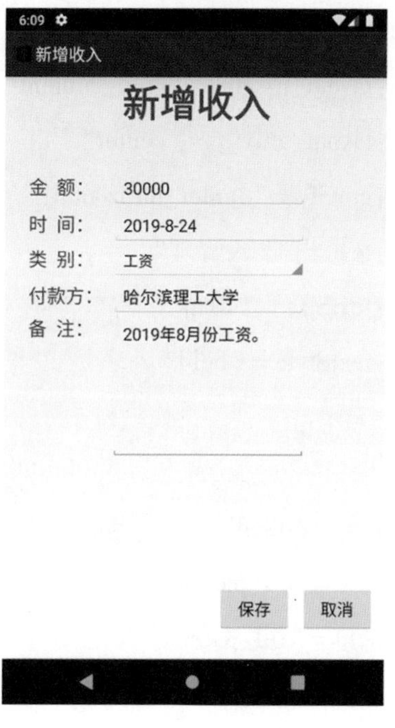

图 13－6　“新增收入”窗口运行结果

13.10.1 设计新增收入布局文件

在 res/layout 目录下新建一个 addinaccount. xml 用来作为新增收入窗体的布局文件。该布局文件使用 LineraLayout 结合 RelativeLayout 进行布局,在该布局文件中添加五个 TextView 组件、四个 EditText 组件、一个 Spinner 组件和两个 Button 组件,实现代码如下:

```
<LinearLayout xmlns:android = "http://schemas.android.com/apk/res/android"
    android:id = "@ +id/initem"
    android:orientation = "vertical"
    android:layout_width = "fill_parent"
    android:layout_height = "fill_parent"
    >
    <LinearLayout
        android:orientation = "vertical"
        android:layout_width = "fill_parent"
         android:layout_height = "fill_parent"
         android:layout_weight = "3"
         >
         <TextView
             android:layout_width = "wrap_content"
             android:layout_height = "wrap_content"
             android:layout_gravity = "center"
             android:gravity = "center_horizontal"
             android:text = "收入管理"
             android:textSize = "40sp"
             android:textStyle = "bold" />
    </LinearLayout>
    <LinearLayout
      android:orientation = "vertical"
      android:layout_width = "fill_parent"
      android:layout_height = "fill_parent"
      android:layout_weight = "1"
      >
      <RelativeLayout android:layout_width = "fill_parent"
```

```
    android:layout_height = "fill_parent"
    android:padding = "10dp"
     >
     <TextView android:layout_width = "90dp"
    android:id = "@ +id/tvInMoney"
    android:textSize = "20sp"
    android:text = "金　额:"
    android:layout_height = "wrap_content"
    android:layout_alignBaseline = "@ +id/txtInMoney"
    android:layout_alignBottom = "@ +id/txtInMoney"
    android:layout_alignParentLeft = "true"
    android:layout_marginLeft = "16dp" >
     </TextView >
     <EditText
    android:id = "@ +id/txtInMoney"
    android:layout_width = "210dp"
    android:layout_height = "wrap_content"
android:layout_toRightOf = "@id/tvInMoney"
    android:inputType = "number"
    android:numeric = "integer"
    android:maxLength = "9"
    android:hint = "0.00"
/ >
     <TextView android:layout_width = "90dp"
    android:id = "@ +id/tvInTime"
    android:textSize = "20sp"
    android:text = "时　间:"
    android:layout_height = "wrap_content"
    android:layout_alignBaseline = "@ +id/txtInTime"
    android:layout_alignBottom = "@ +id/txtInTime"
    android:layout_toLeftOf = "@ +id/txtInMoney" >
     </TextView >
     <EditText
```

```
android:id = "@+id/txtInTime"
android:layout_width = "210dp"
android:layout_height = "wrap_content"
android:layout_toRightOf = "@id/tvInTime"
android:layout_below = "@id/txtInMoney"
android:inputType = "datetime"
android:hint = "2015-01-01"
/>
<TextView android:layout_width = "90dp"
android:id = "@+id/tvInType"
android:textSize = "20sp"
android:text = "类    别:"
android:layout_height = "wrap_content"
android:layout_alignBaseline = "@+id/spInType"
android:layout_alignBottom = "@+id/spInType"
android:layout_alignLeft = "@+id/tvInTime" >
</TextView>
<Spinner android:id = "@+id/spInType"
android:layout_width = "210dp"
android:layout_height = "wrap_content"
android:layout_toRightOf = "@id/tvInType"
android:layout_below = "@id/txtInTime"
android:entries = "@array/intype"
/>
<TextView android:layout_width = "90dp"
android:id = "@+id/tvInHandler"
android:textSize = "20sp"
android:text = "付款方:"
android:layout_height = "wrap_content"
android:layout_alignBaseline = "@+id/txtInHandler"
android:layout_alignBottom = "@+id/txtInHandler"
android:layout_toLeftOf = "@+id/spInType" >
</TextView>
```

```
        <EditText
        android:id="@+id/txtInHandler"
        android:layout_width="210dp"
        android:layout_height="wrap_content"
        android:layout_toRightOf="@id/tvInHandler"
         android:layout_below="@id/spInType"
         android:singleLine="false"
         />
          <TextView android:layout_width="90dp"
         android:id="@+id/tvInMark"
         android:textSize="20sp"
         android:text="备  注:"
         android:layout_height="wrap_content"
         android:layout_alignTop="@+id/txtInMark"
         android:layout_toLeftOf="@+id/txtInHandler">
          </TextView>
          <EditText
         android:id="@+id/txtInMark"
         android:layout_width="210dp"
         android:layout_height="150dp"
         android:layout_toRightOf="@id/tvInMark"
         android:layout_below="@id/txtInHandler"
         android:gravity="top"
         android:singleLine="false"
         />
          </RelativeLayout>
</LinearLayout>
<LinearLayout
  android:orientation="vertical"
  android:layout_width="fill_parent"
  android:layout_height="fill_parent"
  android:layout_weight="3"
   >
```

```
        <RelativeLayout android:layout_width = "fill_parent"
            android:layout_height = "fill_parent"
            android:padding = "10dp"
             >
        <Button
            android:id = "@ +id/btnInCancel"
            android:layout_width = "80dp"
            android:layout_height = "wrap_content"
            android:layout_alignParentRight = "true"
            android:layout_marginLeft = "10dp"
            android:text = "取消"
            />
          <Button
                android:id = "@ +id/btnInSave"
                android:layout_width = "80dp"
                android:layout_height = "wrap_content"
                android:layout_toLeftOf = "@id/btnInCancel"
                android:text = "修改" />
          </RelativeLayout>
      </LinearLayout>
</LinearLayout>
```

13.10.2 设置收入时间

在 com. mc. activity 包中创建一个 AddInaccount. java 文件,该文件的布局文件设置为 addinaccount. xml。在 AddInaccount. java 文件中,首先创建类中需要用到的全局对象及变量,代码如下:

```
protected static final int DATE_DIALOG_ID = 0;// 创建日期对话框常量
  EditText txtInMoney, txtInTime, txtInHandler, txtInMark;// 创建四个 EditText 对象
  Spinner spInType;// 创建 Spinner 对象
  Button btnInSaveButton;// 创建 Button 对象"保存"
  Button btnInCancelButton;// 创建 Button 对象"取消"
  private int mYear;// 年
  private int mMonth;// 月
```

```
private int mDay;// 日
```

在重写的 onCreate( )方法中初始化创建的 EditText 对象、Spinner 对象和 Button 对象,代码如下:

```
txtInMoney = (EditText) findViewById(R.id.txtInMoney);// 获取金额文本框
txtInTime = (EditText) findViewById(R.id.txtInTime);// 获取时间文本框
txtInHandler = (EditText) findViewById(R.id.txtInHandler);// 获取付款方文本框
txtInMark = (EditText) findViewById(R.id.txtInMark);// 获取备注文本框
spInType = (Spinner) findViewById(R.id.spInType);// 获取类别下拉列表
btnInSaveButton = (Button) findViewById(R.id.btnInSave);// 获取保存按钮
btnInCancelButton = (Button) findViewById(R.id.btnInCancel);// 获取取消按钮
```

单击"时间"文本框,为该文本框设置监听事件,在监听事件中使用 showDialog( )方法弹出时间选择对话框,并且在 Activity 创建时默认显示当前的系统时间,代码如下:

```
txtInTime.setOnClickListener(new OnClickListener() {// 为时间文本框设置单击监听事件
        @Override
        public void onClick(View arg0) {
            // TODO Auto-generated method stub
            showDialog(DATE_DIALOG_ID);// 显示日期选择对话框
        }
    });
```

上面的代码中用到了 updateDisplay( )方法,该方法用来显示设置的时间,代码如下:

```
private void updateDisplay() {
    // 显示设置的时间
    txtInTime.setText(new StringBuilder().append(mYear).append("-")
            .append(mMonth + 1).append("-").append(mDay));
}
```

在为"时间"文本框设置监听事件时,弹出了时间选择对话框,该对话框的弹出需要重写 onCreateDialog( )方法,该方法用来根据指定的标识弹出时间选择对话框,代码如下:

```
@Override
protected Dialog onCreateDialog(int id) {// 重写 onCreateDialog 方法
    switch (id) {
    case DATE_DIALOG_ID:// 弹出日期选择对话框
        return new DatePickerDialog(this, mDateSetListener, mYear, mMonth,
```

```
                    mDay);
        }
        return null;
}
```

上面的代码中用到了 mDateSetListener 对象，该对象是 OnDateSetListener 类的一个对象，用来显示用户设置的时间，代码如下：

```
private DatePickerDialog. OnDateSetListener mDateSetListener =
        new DatePickerDialog. OnDateSetListener() {
    public void onDateSet(DatePicker view, int year, int monthOfYear,
            int dayOfMonth) {
        mYear = year;// 为年份赋值
        mMonth = monthOfYear;// 为月份赋值
        mDay = dayOfMonth;// 为天赋值
        updateDisplay();// 显示设置的日期
    }
};
```

13.10.3 添加收入信息

填写完信息后，单击“保存”按钮，为该按钮设置监听事件。在监听事件中，使用 InaccountDAO 对象的 add()方法将用户的输入保存到收入信息表中，代码如下：

```
btnInSaveButton. setOnClickListener(new OnClickListener() {// 为保存按钮设置监听事件
            @ Override
            public void onClick(View arg0) {
                // TODO Auto - generated method stub
                String strInMoney = txtInMoney. getText(). toString();// 获取金额文本框的值
                if (! strInMoney. isEmpty()) {// 判断金额不为空
                    // 创建 InaccountDAO 对象
                    InaccountDAO inaccountDAO = new InaccountDAO(
                            AddInaccount. this);
                    // 创建 Tb_inaccount 对象
                    Tb_inaccount tb_inaccount = new Tb_inaccount(
```

```
                                inaccountDAO. getMaxId( ) + 1, Double
                                        . parseDouble( strInMoney), txtInTime
                                        . getText( ). toString( ), spInType
                                        . getSelectedItem( ). toString( ),
                                txtInHandler. getText( ). toString( ),
                                txtInMark. getText( ). toString( ) );
                        inaccountDAO. add( tb_inaccount);// 添加收入信息
                        // 弹出信息提示
                        Toast. makeText ( AddInaccount. this, "【新增收入】数据添加成功!",
                                Toast. LENGTH_SHORT). show( );
                    } else {
                        Toast. makeText( AddInaccount. this, "请输入收入金额!",
                                Toast. LENGTH_SHORT). show( );
                    }
                }
            });
```

13.10.4　重置新增收入窗口中的各个控件

单击“取消”按钮,重置新增收入窗口中的各个控件,代码如下:

```
btnInCancelButton. setOnClickListener( new OnClickListener( ) {// 为取消按钮设置监听事件
            @ Override
            public void onClick( View arg0) {
                // TODO Auto - generated method stub
                txtInMoney. setText( "" );// 设置金额文本框为空
                txtInMoney. setHint( "0. 00" );// 为金额文本框设置提示
                txtInTime. setText( "" );// 设置时间文本框为空
                txtInTime. setHint( "2015 - 01 - 01" );// 为时间文本框设置提示
                txtInHandler. setText( "" );// 设置付款方文本框为空
                txtInMark. setText( "" );// 设置备注文本框为空
                spInType. setSelection(0);// 设置类别下拉列表默认选择第一项
            }
```

```
        });
```

13.10.5 设计收入信息浏览布局文件

收入信息浏览窗口运行效果如图 13－7 所示。

图 13－7 收入信息浏览窗口运行效果

在 res/layout 目录下新建一个 inaccountinfo. xml 用来作为收入信息浏览窗体的布局文件。该布局文件使用 LinearLayout 结合 RelativeLayout 进行布局，在该布局文件中添加一个 TextView 组件和一个 ListView 组件，代码如下：

```
<LinearLayout xmlns:android="http://schemas.android.com/apk/res/android"
    android:id="@+id/iteminfo" android:orientation="vertical"
    android:layout_width="wrap_content"
    android:layout_height="wrap_content"
    android:layout_marginTop="5dp"
    android:weightSum="1" >
  <LinearLayout android:id="@+id/linearLayout1"
          android:layout_height="wrap_content"
          android:layout_width="match_parent"
          android:orientation="vertical"
          android:layout_weight="0.06" >
      <RelativeLayout android:layout_height="wrap_content"
```

```
                android:layout_width = "match_parent" >
            <TextView android:text = "我的收入"
                android:layout_width = "fill_parent"
                android:layout_height = "wrap_content"
                android:gravity = "center"
                android:textSize = "20sp"
                android:textColor = "#8C6931"
            />
        </RelativeLayout >
            </LinearLayout >
            <LinearLayout android:id = "@ +id/linearLayout2"
                android:layout_height = "wrap_content"
                android:layout_width = "match_parent"
                android:orientation = "vertical"
                android:layout_weight = "0.94" >
            <ListView android:id = "@ +id/lvinaccountinfo"
                android:layout_width = "match_parent"
                android:layout_height = "match_parent"
                android:scrollbarAlwaysDrawVerticalTrack = "true"
                />
      </LinearLayout >
   </LinearLayout >
```

13.10.6　显示所有的收入信息

在 com. mc. activity 包中创建一个 Inaccountinfo. java 文件,该文件的布局文件设置为 inaccountinfao. xml。在 Inaccountinfo. java 文件中,首先创建类中需要用到的全局对象及变量,代码如下:

```
public static final String FLAG  =  "id";// 定义一个常量,用来作为请求码
ListView lvinfo;// 创建 ListView 对象
String strType  =  "";// 创建字符串,记录管理类型
```

在重写的 onCreate()方法中,初始化创建的 ListView 对象,并显示所有的收入信息,代码如下:

```
lvinfo  =  (ListView) findViewById(R. id. lvinaccountinfo);// 获取布局文件中的 List-
```

```
View 组件
ShowInfo(R.id.btnininfo);// 调用自定义方法显示收入信息
```

上面的代码中用到了 ShowInfo()方法，该方法用来根据参数中传入的管理类型 id，显示相应的信息，代码如下：

```
private void ShowInfo(int intType) {// 用来根据传入的管理类型，显示相应的信息
        String[] strInfos = null;// 定义字符串数组，用来存储收入信息
        ArrayAdapter<String> arrayAdapter = null;// 创建 ArrayAdapter 对象
        strType = "btnininfo";// 为 strType 变量赋值
        InaccountDAO inaccountinfo = new InaccountDAO(Inaccountinfo.this);// 创建 InaccountDAO 对象
        // 获取所有收入信息，并存储到 List 泛型集合中
        List<Tb_inaccount> listinfos = inaccountinfo.getScrollData(0,
                (int) inaccountinfo.getCount());
        strInfos = new String[listinfos.size()];// 设置字符串数组的长度
         int m = 0;// 定义一个开始标识
         for (Tb_inaccount tb_inaccount : listinfos) {// 遍历 List 泛型集合
             // 将收入相关信息组合成一个字符串，存储到字符串数组的相应位置
             strInfos[m] = tb_inaccount.getid() + "|" + tb_inaccount.getType()
                     + " " + String.valueOf(tb_inaccount.getMoney()) + "元"
                     + tb_inaccount.getTime();
             m++;// 标识加 1
         }
         // 使用字符串数组初始化 ArrayAdapter 对象
         arrayAdapter = new ArrayAdapter<String>(this,
                 android.R.layout.simple_list_item_1, strInfos);
         lvinfo.setAdapter(arrayAdapter);// 为 ListView 列表设置数据源
    }
```

13.10.7 单机制定选项时打开详细信息

当用户单击 ListView 列表中的某条收入记录时，为其设置监听事件。在监听事件中，根据用户单击的收入信息的编号，打开相应的 Acvitity，代码如下：

```
lvinfo.setOnItemClickListener(new OnItemClickListener(){// 为 ListView 添加项单击事件
```

```
// 重写 onItemClick 方法
@ Override
public void onItemClick( AdapterView < ?  > parent, View view,
        int position, long id) {
    String strInfo = String. valueOf(((TextView) view). getText());// 记录收入信息
    String strid = strInfo. substring(0, strInfo. indexOf('|'));// 从收入信息中截取收入编号
    Intent intent = new Intent( Inaccountinfo. this, InfoManage. class);// 创建 Intent 对象
    intent. putExtra( FLAG, new String[] { strid, strType });// 设置传递数据
    startActivity( intent);// 执行 Intent 操作
  }
});
```

13.10.8　设计修改/删除收入布局文件

修改/删除收入信息窗口运行结果如图 13 –8 所示。在 res/layout 目录下新建一个 infomanage. xml 用来作为修改、删除收入信息和支出信息窗体的布局文件。该布局文件使用 LinearLayout 结合 Relativelayout 进行布局,在该布局文件中添加五个 TextView 组件、四个 EditText 组件、一个 Spinner 组件和两个 Button 组件,实现代码如下:

图 13 –8　修改/删除收入信息窗口运行结果

```
<LinearLayout xmlns:android = "http://schemas.android.com/apk/res/android"
    android:id = "@ +id/inoutitem"
    android:orientation = "vertical"
    android:layout_width = "fill_parent"
    android:layout_height = "fill_parent"
    >
    <LinearLayout
      android:orientation = "vertical"
      android:layout_width = "fill_parent"
        android:layout_height = "fill_parent"
    android:layout_weight = "3"
        >
        <TextView android:id = "@ +id/inouttitle"
            android:layout_width = "wrap_content"
            android:layout_gravity = "center"
            android:gravity = "center_horizontal"
            android:text = "支出管理"
            android:textSize = "40sp"
            android:textStyle = "bold"
            android:layout_height = "wrap_content"/>
    </LinearLayout>
    <LinearLayout
      android:orientation = "vertical"
      android:layout_width = "fill_parent"
      android:layout_height = "fill_parent"
      android:layout_weight = "1"
        >
        <RelativeLayout android:layout_width = "fill_parent"
            android:layout_height = "fill_parent"
            android:padding = "10dp"
            >
            <TextView android:layout_width = "90dp"
            android:id = "@ +id/tvInOutMoney"
```

```
android:textSize = "20sp"
android:text = "金　额:"
android:layout_height = "wrap_content"
android:layout_alignBaseline = "@ + id/txtInOutMoney"
android:layout_alignBottom = "@ + id/txtInOutMoney"
android:layout_alignParentLeft = "true"
android:layout_marginLeft = "16dp" >
</TextView >
< EditText
android:id = "@ + id/txtInOutMoney"
android:layout_width = "210dp"
android:layout_height = "wrap_content"
android:layout_toRightOf = "@ id/tvInOutMoney"
android:inputType = "number"
android:numeric = "integer"
android:maxLength = "9"
/ >
< TextView android:layout_width = "90dp"
android:id = "@ + id/tvInOutTime"
android:textSize = "20sp"
android:text = "时　间:"
android:layout_height = "wrap_content"
android:layout_alignBaseline = "@ + id/txtInOutTime"
android:layout_alignBottom = "@ + id/txtInOutTime"
android:layout_toLeftOf = "@ + id/txtInOutMoney" >
</TextView >
< EditText
android:id = "@ + id/txtInOutTime"
android:layout_width = "210dp"
android:layout_height = "wrap_content"
android:layout_toRightOf = "@ id/tvInOutTime"
android:layout_below = "@ id/txtInOutMoney"
android:inputType = "datetime"
```

```
/>
<TextView android:layout_width="90dp"
android:id="@+id/tvInOutType"
android:textSize="20sp"
android:text="类  别:"
android:layout_height="wrap_content"
android:layout_alignBaseline="@+id/spInOutType"
android:layout_alignBottom="@+id/spInOutType"
android:layout_alignLeft="@+id/tvInOutTime" >
</TextView>
<Spinner android:id="@+id/spInOutType"
android:layout_width="210dp"
android:layout_height="wrap_content"
android:layout_toRightOf="@id/tvInOutType"
android:layout_below="@id/txtInOutTime"
android:entries="@array/intype"
/>
<TextView android:layout_width="90dp"
android:id="@+id/tvInOut"
android:textSize="20sp"
android:text="付款方:"
android:layout_height="wrap_content"
android:layout_alignBaseline="@+id/txtInOut"
android:layout_alignBottom="@+id/txtInOut"
android:layout_toLeftOf="@+id/spInOutType" >
</TextView>
<EditText
android:id="@+id/txtInOut"
android:layout_width="210dp"
android:layout_height="wrap_content"
android:layout_toRightOf="@id/tvInOut"
android:layout_below="@id/spInOutType"
android:singleLine="false"
```

```
            />
             <TextView android:layout_width = "90dp"
            android:id = "@ +id/tvInOutMark"
            android:textSize = "20sp"
            android:text = "备　注:"
            android:layout_height = "wrap_content"
            android:layout_alignTop = "@ +id/txtInOutMark"
            android:layout_toLeftOf = "@ +id/txtInOut" >
             </TextView >
             < EditText
            android:id = "@ +id/txtInOutMark"
            android:layout_width = "210dp"
            android:layout_height = "150dp"
            android:layout_toRightOf = "@ id/tvInOutMark"
            android:layout_below = "@ id/txtInOut"
            android:gravity = "top"
            android:singleLine = "false"
            />
             </RelativeLayout >
        </LinearLayout >
        < LinearLayout
          android:orientation = "vertical"
          android:layout_width = "fill_parent"
          android:layout_height = "fill_parent"
          android:layout_weight = "3"
           >
          < RelativeLayout android:layout_width = "fill_parent"
            android:layout_height = "fill_parent"
            android:padding = "10dp"
             >
        < Button
            android:id = "@ +id/btnInOutDelete"
            android:layout_width = "80dp"
```

```
            android:layout_height = "wrap_content"
            android:layout_alignParentRight = "true"
            android:layout_marginLeft = "10dp"
            android:text = "删除"
            / >
             < Button
            android:id = "@ + id/btnInOutEdit"
            android:layout_width = "80dp"
            android:layout_height = "wrap_content"
            android:layout_toLeftOf = "@ id/btnInOutDelete"
            android:text = "修改"
            / >
        </RelativeLayout >
    </LinearLayout >
</LinearLayout >
```

13.10.9　显示指定编号的收入信息

在 com. mc. activity 包中创建 InfoManage. java 文件,该文件的布局文件设置为 infomanage. xml。在 InfoManage. java 文件中,首先创建类中需要用到的全局对象及变量,代码如下:

```
protected static final int DATE_DIALOG_ID  = 0;// 创建日期对话框常量
TextView tvtitle, textView;// 创建两个 TextView 对象
EditText txtMoney, txtTime, txtHA, txtMark;// 创建四个 EditText 对象
Spinner spType;// 创建 Spinner 对象
Button btnEdit, btnDel;// 创建两个 Button 对象
String[ ] strInfos;// 定义字符串数组
String strid, strType;// 定义两个字符串变量,分别用来记录信息编号和管理类型
private int mYear;// 年
private int mMonth;// 月
private int mDay;// 日
```

修改/删除收入信息和支出信息的功能都是在 InfoManage. java 文件中实现的,所以在第 13. 10. 10 节和第 13. 10. 11 节中讲解修改、删除收入信息时,可能会涉及支出信息的修改与删除。

在重写的 onCreate( )方法中初始化创建的 EditText 对象、Spinner 对象和 Button 对象,代

码如下:

```
tvtitle = (TextView) findViewById(R.id.inouttitle);// 获取标题标签对象
textView = (TextView) findViewById(R.id.tvInOut);// 获取地点/付款方标签对象
txtMoney = (EditText) findViewById(R.id.txtInOutMoney);// 获取金额文本框
txtTime = (EditText) findViewById(R.id.txtInOutTime);// 获取时间文本框
spType = (Spinner) findViewById(R.id.spInOutType);// 获取类别下拉列表
txtHA = (EditText) findViewById(R.id.txtInOut);// 获取地点/付款方文本框
txtMark = (EditText) findViewById(R.id.txtInOutMark);// 获取备注文本框
btnEdit = (Button) findViewById(R.id.btnInOutEdit);// 获取修改按钮
btnDel = (Button) findViewById(R.id.btnInOutDelete);// 获取删除按钮
```

在重写的 onCreate()方法中,初始化和组件对象后,使用字符串记录传入的 id 和类型,并根据类型判断显示收入信息还是支出信息,代码如下:

```
Intent intent = getIntent();// 创建 Intent 对象
Bundle bundle = intent.getExtras();// 获取传入的数据,并使用 Bundle 记录
strInfos = bundle.getStringArray(Showinfo.FLAG);// 获取 Bundle 中记录的信息
strid = strInfos[0];// 记录 id
strType = strInfos[1];// 记录类型
if (strType.equals("btnoutinfo")){// 如果类型是 btnoutinfo
  tvtitle.setText("支出管理");// 设置标题为"支出管理"
  textView.setText("地　点:");// 设置"地点/付款方"标签文本为"地点:"
  // 根据编号查找支出信息,并存储到 Tb_outaccount 对象中
   Tb_outaccount tb_outaccount = OutaccountDAO.find(Integer
                .parseInt(strid));
   txtMoney.setText(String.valueOf(tb_outaccount.getMoney()));// 显示金额
   txtTime.setText(tb_outaccount.getTime());// 显示时间
   /********************修改下拉列表项*************************/
   //创建一个适配器
   ArrayAdapter<CharSequence> adapter = ArrayAdapter.createFromResource(
          this, R.array.outtype,android.R.layout.simple_dropdown_item_1line);
   spType.setAdapter(adapter);    // 将适配器与选择列表框关联
/***************************************************************/
```

```
    spType. setPrompt( tb_outaccount. getType( ) ) ;// 显示类别
    txtHA. setText( tb_outaccount. getAddress( ) ) ;// 显示地点
    txtMark. setText( tb_outaccount. getMark( ) ) ;// 显示备注
} else if ( strType. equals( "btnininfo" ) ) {// 如果类型是 btnininfo
    tvtitle. setText( "收入管理" ) ;// 设置标题为"收入管理"
    textView. setText( "付款方:" ) ;// 设置"地点/付款方"标签文本为"付款方:"
    // 根据编号查找收入信息,并存储到 Tb_outaccount 对象中
    Tb_inaccount tb_inaccount = inaccountDAO. find( Integer
            . parseInt( strid) ) ;
    txtMoney. setText( String. valueOf( tb_inaccount. getMoney( ) ) ) ;// 显示金额
    txtTime. setText( tb_inaccount. getTime( ) ) ;// 显示时间
    spType. setPrompt( tb_inaccount. getType( ) ) ;// 显示类别
    txtHA. setText( tb_inaccount. getHandler( ) ) ;// 显示付款方
    txtMark. setText( tb_inaccount. getMark( ) ) ;// 显示备注
}
txtTime. setOnClickListener( new OnClickListener( ) {// 为时间文本框设置单击监听事件
@ Override
public void onClick( View arg0) {
    // TODO Auto - generated method stub
    showDialog( DATE_DIALOG_ID) ;// 显示日期选择对话框
}
```

13.10.10 修改收入信息

当用户修改完显示的收入或者支出信息后,单击“修改”按钮。如果显示的是支出信息,则调用 OutaccountDAO 对象的 update( )方法修改支出信息;如果显示的是收入信息,则调用 InaccountDAO 对象的 update( )方法修改收入信息。代码如下:

```
btnEdit. setOnClickListener( new OnClickListener( ) {
// 为修改按钮设置监听事件
    @ Override
    public void onClick( View arg0) {
        if ( strType. equals( "btnoutinfo" ) ) {// 判断类型如果是 btnoutinfo
            Tb_outaccount tb_outaccount = new Tb_outaccount( ) ;// 创建 Tb_outac-
```

```
count 对象
            tb_outaccount.setid(Integer.parseInt(strid));// 设置编号
            tb_outaccount.setMoney(Double.parseDouble(txtMoney
                    .getText().toString()));// 设置金额
            tb_outaccount.setTime(txtTime.getText().toString());// 设置时间
            tb_outaccount.setType(spType.getSelectedItem().toString());// 设置类别
            tb_outaccount.setAddress(txtHA.getText().toString());// 设置地点
            tb_outaccount.setMark(txtMark.getText().toString());// 设置备注
            OutaccountDAO.update(tb_outaccount);// 更新支出信息
        } else if (strType.equals("btnininfo")){// 判断类型如果是 btnininfo
            Tb_inaccount tb_inaccount = new Tb_inaccount();// 创建 Tb_inaccount 对象
            tb_inaccount.setid(Integer.parseInt(strid));// 设置编号
    tb_inaccount.setMoney(Double.parseDouble(txtMoney.getText()
                    .toString()));// 设置金额
            tb_inaccount.setTime(txtTime.getText().toString());// 设置时间
            tb_inaccount.setType(spType.getSelectedItem().toString());// 设置类别
            tb_inaccount.setHandler(txtHA.getText().toString());// 设置付款方
            tb_inaccount.setMark(txtMark.getText().toString());// 设置备注
            inaccountDAO.update(tb_inaccount);// 更新收入信息
            }
            // 弹出信息提示
            Toast.makeText(InfoManage.this, "【数据】修改成功!",
            Toast.LENGTH_SHORT)
                    .show();
        }
    });
```

13.10.11　删除收入信息

单击“删除”按钮。如果显示的是支出信息,则调用 OutaccountDAO 对象的 delete( )方法删除支出信息;如果显示的是收入信息,则调用 InaccountDAO 对象的 delete( )方法删除收入信息。代码如下：

```
btnDel.setOnClickListener(new OnClickListener() {
// 为删除按钮设置监听事件
    @Override
    public void onClick(View arg0) {
        // TODO Auto-generated method stub
        if (strType.equals("btnoutinfo")){// 判断类型如果是 btnoutinfo
            outaccountDAO.detele(Integer.parseInt(strid));// 根据编号删除支出信息
        } else if (strType.equals("btnininfo")){// 判断类型如果是 btnininfo
          inaccountDAO.detele(Integer.parseInt(strid));// 根据编号删除收入信息
        }
        Toast.makeText(InfoManage.this, "【数据】删除成功!",
        Toast.LENGTH_SHORT)   .show();
    }
});
```

13.10.12 收入信息汇总图表

在系统主窗口中选择“数据管理”，进入数据管理页面。在该页面中单击“收入汇总”按钮，将显示收入统计图表，如图 13-9 所示。

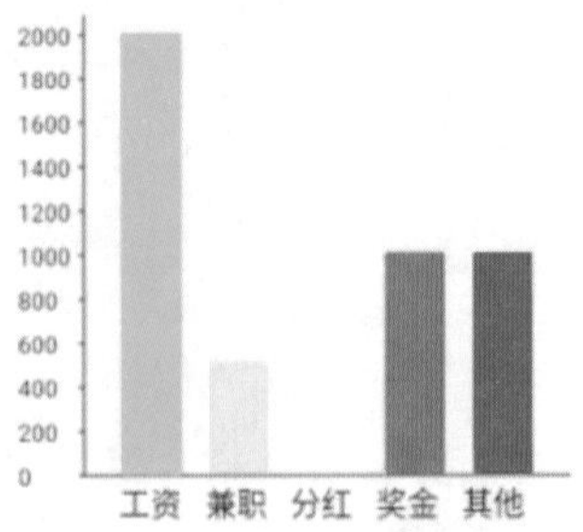

图 13-9 收入统计图表

在 InaccountDAO 类中,创建 getTotal( )方法,用于统计收入汇总信息,并保存到 Map 对象中返回,关键代码如下:

```
public Map < String,Float > getTotal( ) {
//      db = helper. getWritableDatabase( );// 初始化 SQLiteDatabase 对象
        // 获取所有收入汇总信息
        Cursor cursor = db. rawQuery( "select type,sum( money) "
                + "from tb_inaccount group by type",null);
        int count =0;
        count = cursor. getCount( );
        Map < String,Float > map = new HashMap < String,Float > ( );   //创建一个 Map 对象
        cursor. moveToFirst( );  //移动第一条记录
        for( int i =0;i < count;i + + ){// 遍历所有的收入汇总信息
            map. put( cursor. getString(0),cursor. getFloat(1));
            System. out. println( "收入:" +cursor. getString(0));
            cursor. moveToNext( );//移到下条记录
        }
        cursor. close( );// 关闭游标
        return map;// 返回 Map 对象
    }
```

在 res/layout 目录下创建布局文件 chart. xml,在该布局文件中添加一个帧布局管理器,用于显示自定义的绘图类。具体代码如下:

```
<FrameLayout xmlns:android = "http://schemas. android. com/apk/res/android"
    android:layout_width = "match_parent"
    android:layout_height = "match_parent"
    android:id = "@ +id/canvas"
    >
</FrameLayout >
```

在 com. mc. activity 包中创建一个名为 TotalChart 的 Activity,在该文件中定义所需的成员变量和常量。具体代码如下:

```
private float[ ] money =new float[ ]{600,1000,600,300,1500};   //各项金额的默认值
private int[ ] color = new int[ ]{Color. GREEN,Color. YELLOW,
Color. RED,Color. MAGENTA,Color. BLUE};   //各项颜色
```

```
private final int WIDTH = 30;  //柱型的宽度
private final int OFFSET = 15;  //间距
private int x =70;  //起点 x
private int y =329;  //终点 y
private int height =220;  //高度
String[] type = null;  //金额的类型
private String passType = "";  //记录是收入信息还是支出信息
```

在 TotalChart 中编写自定义方法 maxMoney(),用于计算支出金额数组中的最大值。具体代码如下:

```
//计算最大金额
float maxMoney(float[] money){
    float max = money[0];  //将第一个数组元素赋值给变量 max
    for(int i =0;i < money.length -1;i++){
        if(max < money[i +1]){
            max = money[i +1];  //更新 max
        }
    }
    return max;
}
```

在 TotalChart 中编写自定义方法 getMoney()获取收入或者支出汇总信息,并保存到 float 型数组中。具体代码如下。

```
//获取收支数据
float[] getMoney(String flagType){
    Map mapMoney = null;
    System.out.println(flagType);
    if("ininfo".equals(flagType)){
        InaccountDAO inaccountinfo = new InaccountDAO(TotalChart.this);
        mapMoney = inaccountinfo.getTotal();  //获取收入汇总信息
    }else if("outinfo".equals(flagType)){
        OutaccountDAO outaccountinfo = new OutaccountDAO(TotalChart.this);
        mapMoney = outaccountinfo.getTotal();  //获取支出汇总信息
    }
    int size = type.length;
```

```
    float[ ] money1 = new float[ size ];
    for( int i = 0; i < size; i + + ) {
        money1[ i ] = ( mapMoney. get( type[ i ] ) !  = null?
        ( ( Float) mapMoney. get( type[ i ] ) ) :0) ;
    }
    return money1;
}
```

在 TotalChart 中创建一个名称为 MyView 的内部类,该类继承自 android. view. View 类,并添加构造方法和重写 onDraw(Canvas canvas) 方法,然后在 onDraw( ) 方法中绘制统计图表。关键代码如下:

```
public class MyView extends View{
    public MyView( Context context) {
        super( context) ;
    }
    @ Override
    protected void onDraw( Canvas canvas) {
        super. onDraw( canvas) ;
        canvas. drawColor( Color. WHITE) ;  //指定画布的背景色为白色
        Paint paint = new Paint( ) ;  //创建采用默认设置的画笔
        paint. setAntiAlias( true) ;  //使用抗锯齿功能
        /* * * * * * * * * * * *绘制坐标轴* * * * * * * * * * * * * * * * * * * * * * */
        paint. setStrokeWidth( 1 ) ;  //设置笔触的宽度
        paint. setColor( Color. BLACK) ;  //设置笔触的颜色
        canvas. drawLine( 50, 330, 300, 330, paint) ;  //横
        canvas. drawLine( 50, 100, 50, 330, paint) ;  //竖
        /* * * * * * * * * * * * * * * * * * * * * * * * * * * * * * * * * * * * * * * * * * */
        /* * * * * * * * * * * *绘制柱型* * * * * * * * * * * * * * * * * * * * * * */
        paint. setStyle( Style. FILL) ;  //设置填充样式为填充
        int left = 0;  //每个柱型的起点 X 坐标
        money = getMoney( passType) ;  //新获取的金额* * * * * * * * * * *
```

```
* * * * * * * *
        float max = maxMoney(money);
        for(int i =0;i < money.length;i + + ){
            paint.setColor(color[i]);  //设置笔触的颜色
            left = x + i * (OFFSET + WIDTH);  //计算每个柱型起点 X 坐标
            canvas.drawRect(left, y - height/max * money[i], left + WIDTH, y, paint);
        }
        /* * * * * * * * * * * * * * * * * * * * * * * * * * * * * * * * * * * * * * * * * * */
        /* * * * * * * * * * * *绘制纵轴的刻度* * * * * * * * * * * * * * * * * * * * * * */
        paint.setColor(Color.BLACK);  //设置笔触的颜色
        int tempY =0;
        for(int i =0;i <11;i + + ){
            tempY = y - height + height/10 * i +1;
            canvas.drawLine(47,tempY , 50, tempY, paint);
            paint.setTextSize(12);  //设置字体大小
            canvas.drawText(String.valueOf((int)(max/10 * (10 - i))),
            15,tempY +5, paint);
            //绘制纵轴题注
        }
        /* * * * * * * * * * * * * * * * * * * * * * * * * * * * * * * * * * * * * * * * * * */
        /* * * * * * * * * * * *绘制说明文字* * * * * * * * * * * * * * * * * * * * * * * * */
        paint.setColor(Color.BLACK);  //设置笔触的颜色
        paint.setTextSize(21);  //设置字体大小
        /* * * * * * * * * * * * * * * * * * * * * *绘制标题* * * * * * * * * * * * * * * * * * * * * * * * * * * * * * * * * * * * * * */
        if("outinfo".equals(passType)){
            canvas.drawText("个人理财通的支出统计图", 40,55, paint);  //绘制标题
```

```
        } else if("ininfo".equals(passType)){
            canvas.drawText("个人理财通的收入统计图", 40,55, paint);   //绘制标题
        }
/*****************************************************************/
        paint.setTextSize(16);  //设置字体大小
        String str_type = "";
        for(int i =0;i < type.length;i++){
            str_type += type[i] + "    ";
        }
        canvas.drawText(str_type, 68,350, paint);  //绘制横轴题注
    }
}
```

在 TotalChart 中重写 onCreate()方法,并且在该方法中首先获取 Intent 对象来获取传递的数据包,用于确定是统计支出数据还是收入数据,然后获取布局文件中添加的帧布局管理器,并将 MyView 试图添加到该帧布局管理器中。关键代码如下:

```
@Override
protected void onCreate(Bundle savedInstanceState) {
    super.onCreate(savedInstanceState);
    setContentView(R.layout.chart);  //设置使用的布局文件
    Intent intent = getIntent();  //获取 Intent 对象
    Bundle bundle = intent.getExtras();  //获取传递的数据包
    passType = bundle.getString("passType");
    Resources res = getResources();  //获取 Resources 对象
    if("outinfo".equals(passType)){
        type = res.getStringArray(R.array.outtype);  //获取支出类型数组
    } else if("ininfo".equals(passType)){
        type = res.getStringArray(R.array.intype);  //获取收入类型数组
    }
    FrameLayout ll = (FrameLayout)findViewById(R.id.canvas);
    //获取布局文件中添加的帧布局管理器
    ll.addView(new MyView(this));
```

```
    //将自定义的 MyView 视图添加到帧布局管理器中
}
```

13.11　便签管理模块设计

便签管理模块主要包括三部分,分别是“新增便签”“便签信息浏览”和“修改/删除便签信息”。其中,“新增便签”用来添加便签信息;“便签信息浏览”用来显示所有的便签信息;“修改/删除便签信息”用来根据编号修改或者删除便签信息。本节将从这三个方面对便签管理模块进行详细介绍。

13.11.1　设计新增便签布局文件

“新增便签”窗口运行结果如图 13 – 10 所示。

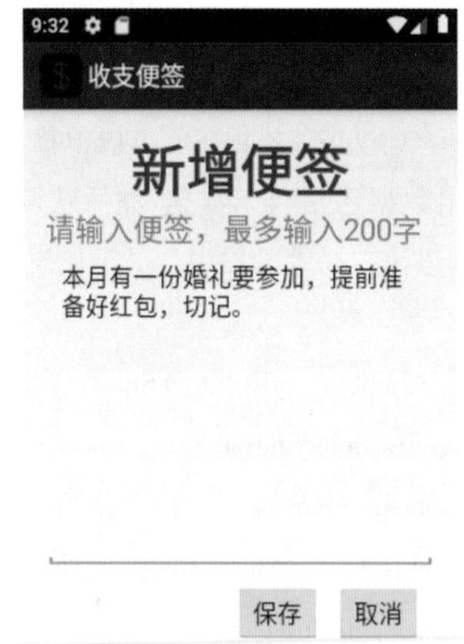

图 13 – 10　“新增便签”窗口运行结果

在 res/layout 目录下新建一个 accountflag. xml 用来作为新增便签窗体的布局文件,该布局文件使用 LinearLayout 结合 RelativeLayout 进行布局,在该布局文件中添加两个 TextView 组件、一个 EditText 组件和两个 Button 组件。实现代码如下:

```
<LinearLayout xmlns:android = "http://schemas. android. com/apk/res/android"
  android:id = "@ +id/itemflag"
  android:layout_width = "fill_parent"
  android:layout_height = "fill_parent"
```

```
android:orientation = "vertical"
android:paddingBottom = "@dimen/activity_vertical_margin"
android:paddingLeft = "@dimen/activity_horizontal_margin"
android:paddingRight = "@dimen/activity_horizontal_margin"
android:paddingTop = "@dimen/activity_vertical_margin" >
<!-- 显示标题文本框 -->
  <TextView
      android:layout_width = "wrap_content"
      android:layout_height = "wrap_content"
      android:layout_gravity = "center"
      android:gravity = "center_horizontal"
      android:text = "新增便签"
      android:textSize = "40sp"
      android:textStyle = "bold" />
<!-- 显示提示文字文本框 -->
  <TextView
      android:id = "@+id/tvFlag"
      android:layout_width = "350dp"
      android:layout_height = "wrap_content"
      android:text = "请输入便签,最多输入 200 字"
      android:textColor = "#8C6931"
      android:textSize = "22sp" />
<!-- 输入便签内容编辑框 -->
  <EditText
      android:id = "@+id/txtFlag"
      android:layout_width = "match_parent"
      android:layout_height = "wrap_content"
      android:layout_below = "@id/tvFlag"
      android:gravity = "top"
      android:lines = "10" />
  <RelativeLayout
      android:layout_width = "fill_parent"
      android:layout_height = "fill_parent"
```

```
            android:padding = "10dp"  >
        <Button
                android:id = "@ +id/btnflagCancel"
                android:layout_width = "wrap_content"
                android:layout_height = "wrap_content"
                android:layout_alignParentRight = "true"
                android:layout_marginLeft = "10dp"
                android:text = "取消" />
        <Button
                android:id = "@ +id/btnflagSave"
                android:layout_width = "wrap_content"
                android:layout_height = "wrap_content"
                android:layout_toLeftOf = "@ id/btnflagCancel"
                android:text = "保存" />
    </RelativeLayout>
</LinearLayout>
```

13.11.2 添加便签信息

在 com.mc.activity 包中创建一个 Accountflag.java 文件，该文件的布局文件设置为 accountflag.xml。在 Accountflag.java 文件中首先创建类中需要用到的全局对象及变量。代码如下：

```
public class Accountflag extends Activity {
    EditText txtFlag;// 创建 EditText 组件对象
    Button btnflagSaveButton;// 创建 Button 组件对象
    Button btnflagCancelButton;// 创建 Button 组件对象
```

在重写的 onCreate()方法中初始化创建的 EditText 对象和 Button 对象。代码如下：

```
    txtFlag = (EditText) findViewById(R.id.txtFlag);// 获取便签文本框
    btnflagSaveButton = (Button) findViewById(R.id.btnflagSave);// 获取保存按钮
    btnflagCancelButton = (Button) findViewById(R.id.btnflagCancel);// 获取取消按钮
```

填写完信息后，单击“保存”按钮，为该按钮设置监听事件。在监听事件中使用 FlagDAO 对象的 add()方法将用户的输入保存到便签信息表中。代码如下：

```
    btnflagSaveButton.setOnClickListener(new OnClickListener() {// 为保存按钮设置
```

```
监听事件
        @ Override
        public void onClick(View arg0) {
        // TODO Auto - generated method stub
            String strFlag = txtFlag.getText().toString();// 获取便签文本框的值
            if (! strFlag.isEmpty()) {// 判断获取的值不为空
              FlagDAO flagDAO = new FlagDAO(Accountflag.this);// 创建 FlagDAO 对象
              Tb_flag tb_flag = new Tb_flag(
              flagDAO.getMaxId() + 1, strFlag);// 创建 Tb_flag 对象
              flagDAO.add(tb_flag);// 添加便签信息
              // 弹出信息提示
              Toast.makeText(Accountflag.this, "【新增便签】数据添加成功!",
                    Toast.LENGTH_SHORT).show();
            } else {
              Toast.makeText(Accountflag.this, "请输入便签!",
                    Toast.LENGTH_SHORT).show();
            }
          }
  });
```

13.11.3　清空便签文本框

单击“取消”按钮,清空便签文本框中的内容。代码如下:

```
btnflagCancelButton.setOnClickListener(new OnClickListener() {// 为取消按钮设置监听事件
        @ Override
        public void onClick(View arg0) {
            // TODO Auto - generated method stub
            txtFlag.setText("");// 清空便签文本框
        }
  });
```

13.11.4　设计便签信息浏览布局文件

“便签信息浏览”窗口运行结果如图 13 - 11 所示。

图 13－11　“便签信息浏览”窗口运行结果

便签信息浏览功能是数据管理窗体中实现的，该窗体的布局文件是 showinfo. xml，对应的 java 文件是 Showinfo. java，所以下面讲解时会通过对 showinfo. xml 布局文件 Showinfo. java 文件的讲解来介绍便签信息浏览功能的实现过程。

在 res/layout 目录下新建一个 showinfo. xml 用来作为数据管理窗体的布局文件，该布局文件中可以调用支出/收入汇总图标和显示便签信息。showinfo. xml 布局文件使用 LinearLayout 结合 RelativeLayout 进行布局，在该布局文件中添加三个 Button 组件和一个 ListView 组件。代码如下：

```
<LinearLayout xmlns:android = "http://schemas.android.com/apk/res/android"
  android:id = "@ +id/iteminfo"  android:orientation = "vertical"
  android:layout_width = "wrap_content"  android:layout_height = "wrap_content"
  android:layout_marginTop = "5dp"
  android:weightSum = "1" >
   <LinearLayout android:id = "@ +id/linearLayout1"
          android:layout_height = "wrap_content"
          android:layout_width = "match_parent"
          android:orientation = "vertical"
          android:layout_weight = "0.06" >
      <RelativeLayout android:layout_height = "wrap_content"
          android:layout_width = "match_parent" >
```

```
    <Button android:text = "支出汇总"
        android:id = "@ +id/btnoutinfo"
        android:layout_width = "wrap_content"
        android:layout_height = "wrap_content"
        android:textSize = "20dp"
        android:textColor = "#8C6931"
   />
    <Button android:text = "收入汇总"
        android:id = "@ +id/btnininfo"
        android:layout_width = "wrap_content"
        android:layout_height = "wrap_content"
        android:layout_toRightOf = "@ id/btnoutinfo"
        android:textSize = "20dp"
        android:textColor = "#8C6931"
        />
    <Button android:text = "便签信息"
        android:id = "@ +id/btnflaginfo"
        android:layout_width = "wrap_content"
        android:layout_height = "wrap_content"
        android:layout_toRightOf = "@ id/btnininfo"
        android:textSize = "20dp"
        android:textColor = "#8C6931"
        />
    </RelativeLayout >
    </LinearLayout >
    <LinearLayout android:id = "@ +id/linearLayout2"
        android:layout_height = "wrap_content"
        android:layout_width = "match_parent"
        android:orientation = "vertical"
        android:layout_weight = "0.94" >
    <ListView android:id = "@ +id/lvinfo"
        android:layout_width = "match_parent"
        android:layout_height = "match_parent"
```

```
            android:scrollbarAlwaysDrawVerticalTrack = "true"
            />
    </LinearLayout>
</LinearLayout>
```

13.11.5 显示所有的便签信息

在 com. mc. activity 包中创建一个 Showinfo. java 文件，该文件的布局文件设置为 showinfo. xml。单击“便签信息”按钮，为该按钮设置监听事件。在监听事件中调用 ShowInfo( ) 方法显示便签信息。代码如下：

```
btnflaginfo.setOnClickListener(new OnClickListener() {// 为便签信息按钮设置监听事件
        @Override
        public void onClick(View arg0) {
            // TODO Auto-generated method stub
            ShowInfo(R.id.btnflaginfo);// 显示便签信息
        }
    });
```

上面的代码中用到了 ShowInfo( ) 方法，该方法为自定义的无返回值类型方法，主要用来根据传入的管理类型显示相应的信息。该方法中有一个 int 类型的参数，用来表示传入的管理类型。该参数的取值主要有 R. id. btnoutinfo、R. id. btnininfo 和 R. id. btnflaginfo 等三个值，分别用来显示支出信息、收入信息和便签信息。ShowInfo( ) 方法的代码如下：

```
private void ShowInfo(int intType) {// 用来根据传入的管理类型，显示相应的信息
    String[] strInfos = null;// 定义字符串数组，用来存储收入信息
    ArrayAdapter<String> arrayAdapter = null;// 创建 ArrayAdapter 对象
    Intent intent = null;// 创建 Intent 对象
    switch (intType) {// 以 intType 为条件进行判断
    case R.id.btnoutinfo:// 如果是支出按钮 btnoutinfo
        strType = "outinfo";// 为 strType 变量赋值
        intent = new Intent(Showinfo.this, TotalChart.class);
        // 使用 TotalChart 窗口初始化 Intent 对象
        intent.putExtra("passType", strType);// 设置要传递的数据
        startActivity(intent);// 执行 Intent，打开相应的 Activity
        break;
```

```
    case R. id. btnininfo:// 如果是收入按钮 btnininfo
        strType = "ininfo";// 为 strType 变量赋值
        intent = new Intent(Showinfo. this, TotalChart. class);
        // 使用 TotalChart 窗口初始化 Intent 对象
        intent. putExtra("passType", strType);// 设置要传递的数据
        startActivity(intent);// 执行 Intent,打开相应的 Activity
        break;
    case R. id. btnflaginfo:// 如果是 btnflaginfo 按钮
        strType = "btnflaginfo";// 为 strType 变量赋值
        FlagDAO flaginfo = new FlagDAO(Showinfo. this);// 创建 FlagDAO 对象
        // 获取所有便签信息,并存储到 List 泛型集合中
        List<Tb_flag> listFlags = flaginfo. getScrollData(0,
                (int) flaginfo. getCount());
        strInfos = new String[listFlags. size()];// 设置字符串数组的长度
        int n = 0;// 定义一个开始标识
        for (Tb_flag tb_flag : listFlags) {// 遍历 List 泛型集合
            // 将便签相关信息组合成一个字符串,存储到字符串数组的相应位置
            strInfos[n] = tb_flag. getid() + "|" + tb_flag. getFlag();
            if (strInfos[n]. length() > 15)// 判断便签信息的长度是否大于 15
                strInfos[n] = strInfos[n]. substring(0, 15) + "……";
                // 将位置大于 15 之后的字符串用……代替
            n++;// 标识加 1
            }
            // 使用字符串数组初始化 ArrayAdapter 对象
            arrayAdapter = new ArrayAdapter<String>(this,
                    android. R. layout. simple_list_item_1, strInfos);
            lvinfo. setAdapter(arrayAdapter);// 为 ListView 列表设置数据源
            break;
        }
    }
```

13.11.6　单击指定项时打开详细信息

当用户单击 ListView 列表中的某条便签记录时,为其设置监听事件。在监听事件中根

据用户单机的便签信息的编号，打开相应的 Activity。代码如下：

```
lvinfo. setOnItemClickListener( new OnItemClickListener( ) {@ Override
        public void onItemClick( AdapterView < ?  > parent, View view,
                int position, long id) {
            String strInfo  =  String. valueOf( ( (TextView) view). getText( ) );
            // 记录单击的项信息
            String strid  =  strInfo. substring(0, strInfo. indexOf('|') );
            // 从项信息中截取编号
            Intent intent  =  null;// 创建 Intent 对象
            if (strType  = =  "btnflaginfo") {// 判断如果是便签信息
                intent  =  new Intent( Showinfo. this, FlagManage. class) ;
                // 使用 FlagManage 窗口初始化 Intent 对象
                intent. putExtra( FLAG, strid) ;// 设置要传递的数据
                startActivity( intent) ;// 执行 Intent,打开相应的 Activity
            }
        }
});
```

13.11.7 设计修改/删除便签布局文件

“修改/删除便签信息”窗口运行结果如图 13 – 12 所示。

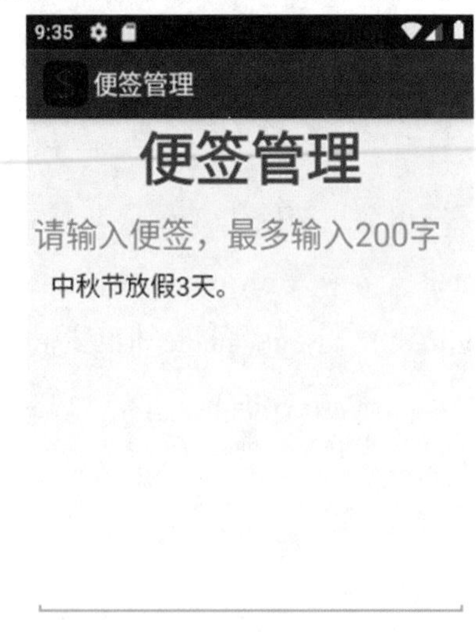

图 13 – 12 “修改/删除便签信息”窗口运行结果

在 res/layout 目录下新建一个 flagmanage.xml，用来作为修改、删除便签信息窗体的布局文件，该布局文件使用 LinearLayout 结合 RelativeLayout 进行布局。在该布局文件中添加两个 TextView 组件、一个 EditText 组件和两个 Button 组件。实现代码如下：

```
<LinearLayout xmlns:android="http://schemas.android.com/apk/res/android"
    android:id="@+id/flagmanage"
    android:orientation="vertical"
    android:layout_width="fill_parent"
    android:layout_height="fill_parent"
    <LinearLayout
      android:orientation="vertical"
      android:layout_width="fill_parent"
      android:layout_height="fill_parent"
       android:layout_weight="3"
        <TextView
            android:layout_width="wrap_content"
            android:layout_gravity="center"
            android:gravity="center_horizontal"
            android:text="便签管理"
            android:textSize="40sp"
            android:textStyle="bold"
            android:layout_height="wrap_content"/>
    </LinearLayout>
    <LinearLayout
        android:orientation="vertical"
        android:layout_width="fill_parent"
      android:layout_height="fill_parent"
      android:layout_weight="1"
      <RelativeLayout android:layout_width="fill_parent"
            android:layout_height="fill_parent"
            android:padding="5dp"
            <TextView android:layout_width="350dp"
            android:id="@+id/tvFlagManage"
            android:textSize="23sp"
```

```
                    android:text = "请输入便签,最多输入 200 字"
                    android:textColor = "#8C6931"
                    android:layout_alignParentRight = "true"
                    android:layout_height = "wrap_content"
                    />
                     <EditText
                    android:id = "@ +id/txtFlagManage"
                    android:layout_width = "350dp"
                    android:layout_height = "400dp"
                    android:layout_below = "@id/tvFlagManage"
                    android:gravity = "top"
                    android:singleLine = "false"
                    />
                     </RelativeLayout>
          </LinearLayout>
          <LinearLayout
               android:orientation = "vertical"
               android:layout_width = "fill_parent"
               android:layout_height = "fill_parent"
               android:layout_weight = "3"
               <RelativeLayout android:layout_width = "fill_parent"
                    android:layout_height = "fill_parent"
                    android:padding = "10dp"
                     >
               <Button
                    android:id = "@ +id/btnFlagManageDelete"
                    android:layout_width = "80dp"
                    android:layout_height = "wrap_content"
                    android:layout_alignParentRight = "true"
                    android:layout_marginLeft = "10dp"
                    android:text = "删除"
                    />
                     <Button
```

```
                android:id = "@ +id/btnFlagManageEdit"
                android:layout_width = "80dp"
                android:layout_height = "wrap_content"
                android:layout_toLeftOf = "@ id/btnFlagManageDelete"
                android:text = "修改"
                android:maxLength = "200"
                />
          </RelativeLayout>
        </LinearLayout>
      </LinearLayout>
```

13.11.8　显示指定编号的便签信息

在 com. mc. activity 包中创建一个 FlagManage. java 文件,该文件的布局文件设置为 flagmanage. xml。在 FlagManage. java 文件中首先创建类中需要用到的全局对象及变量。代码如下:

```
EditText txtFlag;// 创建 EditText 对象
Button btnEdit, btnDel;// 创建两个 Button 对象
String strid;// 创建字符串,表示便签的 id
```

在重写的 onCreate()方法中初始化创建的 EditText 对象和 Button 对象。代码如下:

```
txtFlag = (EditText) findViewById(R. id. txtFlagManage);// 获取便签文本框
btnEdit = (Button) findViewById(R. id. btnFlagManageEdit);// 获取修改按钮
btnDel = (Button) findViewById(R. id. btnFlagManageDelete);// 获取删除按钮
```

在重写的 onCreate()方法中初始化个组件对象后,使用字符串记录传入的 id,并根据该 id 显示便签信息。代码如下:

```
Intent intent = getIntent();// 创建 Intent 对象
Bundle bundle = intent. getExtras();// 获取便签 id
strid = bundle. getString(Showinfo. FLAG);// 将便签 id 转换为字符串
final FlagDAO flagDAO = new FlagDAO(FlagManage. this);// 创建 FlagDAO 对象
txtFlag. setText(flagDAO. find(Integer. parseInt(strid)). getFlag());// 根据便签 id 查找便签信息,并显示在文本框中
```

13.11.9　修改便签信息

当用户修改完显示的便签信息后,单击“修改”按钮,调用 FlagDAO 对象的 update()方

法修改便签信息。代码如下:

```
btnEdit. setOnClickListener( new OnClickListener( ) {// 为修改按钮设置监听事件
    @ Override
    public void onClick( View arg0) {
        // TODO Auto - generated method stub
        Tb_flag tb_flag = new Tb_flag( ) ;// 创建 Tb_flag 对象
        tb_flag. setid( Integer. parseInt( strid) ) ;// 设置便签 id
        tb_flag. setFlag( txtFlag. getText( ) . toString( ) ) ;// 设置便签值
        flagDAO. update( tb_flag) ;// 修改便签信息
        // 弹出信息提示
        Toast. makeText( FlagManage. this, "【便签数据】修改成功!",
                Toast. LENGTH_SHORT) . show( ) ;
    }
} ) ;
```

13.11.10 删除便签信息

单击"删除"按钮,调用 FlagDAO 对象的 delete( )方法删除便签信息,并弹出信息提示。代码如下:

```
btnDel. setOnClickListener( new OnClickListener( ) {// 为删除按钮设置监听事件
    @ Override
    public void onClick( View arg0) {
        // TODO Auto - generated method stub
        flagDAO. detele( Integer. parseInt( strid) ) ;// 根据指定的 id 删除便签信息
        Toast. makeText( FlagManage. this, "【便签数据】删除成功!",
                Toast. LENGTH_SHORT) . show( ) ;
    }
} ) ;
```

13.12 系统设置模块设计

系统设置模块主要对理财日记本中的登录密码进行设置。"系统设置"窗口运行结果如图 13 - 13所示。

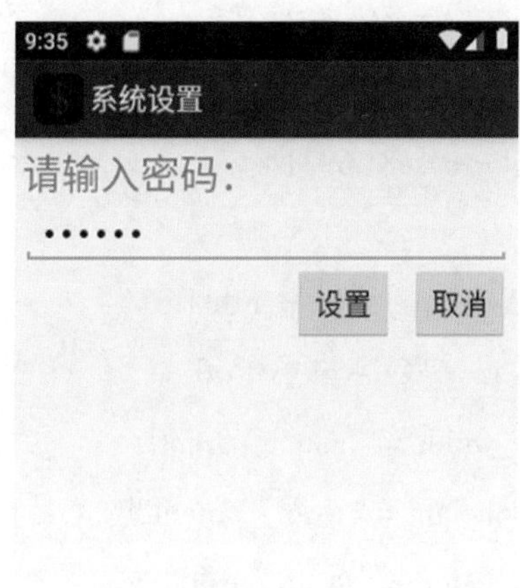

图 13－13　“系统设置”窗口运行结果

13.12.1　设计系统设置布局文件

在 res/layout 目录下新建一个 sysset.xml 用来作为系统设置窗体的布局文件。该布局文件中,将布局方式修改为 RelativeLayout,然后添加一个 TextView 组件、一个 EditText 组件和两个 Button 组件。实现代码如下:

```
<RelativeLayout xmlns:android="http://schemas.android.com/apk/res/android"
    xmlns:tools="http://schemas.android.com/tools"
    android:layout_width="match_parent"
    android:layout_height="match_parent"
    android:padding="5dp"
    android:paddingBottom="@dimen/activity_vertical_margin"
    android:paddingLeft="@dimen/activity_horizontal_margin"
    android:paddingRight="@dimen/activity_horizontal_margin"
    android:paddingTop="@dimen/activity_vertical_margin"
     tools:context="com.mc.MainActivity" >
   <!-- 添加“请输入密码”文本框 TextView -->
   <TextView android:id="@+id/tvPwd"
       android:layout_width="wrap_content"
       android:layout_height="wrap_content"
```

```
        android:text = "请输入密码:"
        android:textSize = "25sp"
        android:textColor = "#8C6931"
    />
    <!-- 添加输入密码的编辑框 EditText -->
    <EditText android:id = "@+id/txtPwd"
        android:layout_width = "match_parent"
        android:layout_height = "wrap_content"
        android:layout_below = "@id/tvPwd"
        android:inputType = "textPassword"
        android:hint = "请输入密码"
    />
    <!-- 添加"取消"按钮 -->
    <Button android:id = "@+id/btnsetCancel"
        android:layout_width = "wrap_content"
        android:layout_height = "wrap_content"
        android:layout_below = "@id/txtPwd"
        android:layout_alignParentRight = "true"
        android:layout_marginLeft = "10dp"
        android:text = "取消"
    />
    <!-- 添加"设置"按钮 -->
    <Button android:id = "@+id/btnSet"
        android:layout_width = "wrap_content"
        android:layout_height = "wrap_content"
        android:layout_below = "@id/txtPwd"
        android:layout_toLeftOf = "@id/btnsetCancel"
        android:text = "设置"
    />
</RelativeLayout>
```

13.12.2 设置登录密码

在 com. mc. activity 包中创建一个 Sysset. java 文件,该文件的布局文件设置为 sysset.

xml。在 Sysset.java 文件中首先创建一个 EditText 对象和两个 Button 对象。代码如下：

```
EditText txtpwd;// 创建 EditText 对象
Button btnSet, btnsetCancel;// 创建两个 Button 对象
```

在重写的 onCreate()方法中初始化创建的 EditText 对象和 Button 对象。代码如下：

```
txtpwd = (EditText) findViewById(R.id.txtPwd);// 获取密码文本框
btnSet = (Button) findViewById(R.id.btnSet);// 获取设置按钮
btnsetCancel = (Button) findViewById(R.id.btnsetCancel);// 获取取消按钮
```

当用户单击“设置”按钮时，为“设置”按钮添加监听事件。在监听事件中，首先创建 PwdDAO 类的对象和 Tb_pwd 类的对象，然后判断数据库中是否已经已设置密码。如果没有，则添加用户密码；否则，则修改用户密码，最后弹出提示信息。代码如下：

```
btnSet.setOnClickListener(new OnClickListener() {// 为设置按钮添加监听事件
  @Override
  public void onClick(View arg0) {
    // TODO Auto-generated method stub
    PwdDAO pwdDAO = new PwdDAO(Sysset.this);// 创建 PwdDAO 对象
    Tb_pwd tb_pwd = new Tb_pwd(txtpwd.getText().toString());// 根据输入的密码创建 Tb_pwd 对象
    if (pwdDAO.getCount() == 0) {// 判断数据库中是否已经设置了密码
      pwdDAO.add(tb_pwd);// 添加用户密码
    } else {
       pwdDAO.update(tb_pwd);// 修改用户密码
     }
     // 弹出信息提示
     Toast.makeText(Sysset.this, "【密码】设置成功！", Toast.LENGTH_SHORT)
                 .show();
  }
});
```

13.12.3 充值密码文本框

单击“取消”按钮，清空密码文本框，并为其设置初始显示。代码如下：

```
btnsetCancel.setOnClickListener(new OnClickListener() {
      @Override
      public void onClick(View arg0) {
```

```
            // TODO Auto - generated method stub
            txtpwd. setText("");// 清空密码文本框
            txtpwd. setHint("请输入密码");// 为密码文本框设置提示
        }
});
```

本章重点讲解了理财日记本系统中关键模块的开发过程。学习本章,应该能够熟悉软件的开发流程,并重点掌握如何在 Android 项目中对多个不同的数据表进行添加、修改、删除以及查询等操作。另外,还应该掌握如何使用多种布局管理器对 Acdroid 程序的界面进行布局。

第 14 章　游戏案例——围住神经猫游戏

14.1　功 能 概 述

围住神经猫是一款入门级小游戏开发的经典样例，无论是在网页上还是在手机上都可能玩过这样的游戏。本章设计中将实现一个这样的游戏，这个游戏的玩法非常简单，就是将图中的猫围住，不让它从旁边到边界处，则游戏胜利。在游戏开始时，会有几个随机分布的点亮了的格子，代表初始障碍物，而玩家需要做的就是不断点击圆点，将猫围起来，在猫移动时可以随着移动变化动作。游戏最终的目的就是用最少的步数让神经猫无路可走。围住神经猫主界面如图 14－1 所示。

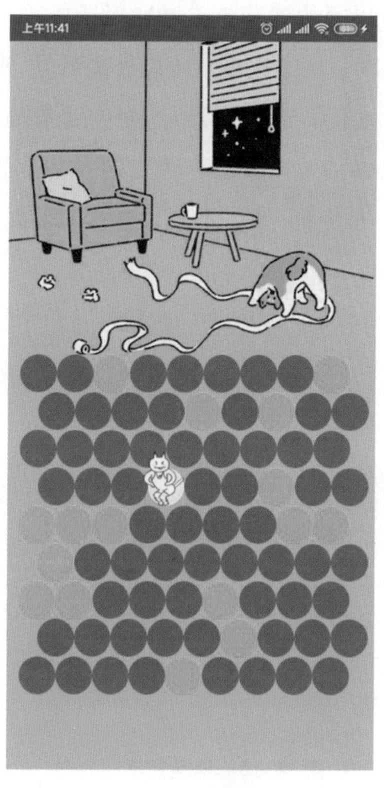

图 14－1　围住神经猫主界面

14.2 设 计 思 路

实现此游戏应用的主要技术是 RelativeLayout（相对布局管理器）、ImageView（图像视图）组件、SurfaceView。围住神经猫的具体要求设计思路如下。

(1)定义背景元素。如图 14－1 所示，背景中有许多的点，游戏中需要在背景图片上画出这些元素，所以仅通过数组下标进行区分是不合理的。为找到在画板上的位置，就要给它们添上 X、Y 坐标。根据面向对象原则，将这些点抽象成游戏中的背景元素，这里就定义一个 Paint 类代表背景元素。在 Paint 类中，有坐标 X、Y 成员属性，还有一个状态。

(2)初始化游戏。完成了背景元素的定义，就要对游戏背景进行初始化。因为需要不断更新界面并长时间频繁操作，如果将代码写在主线程中，容易造成线程阻塞，所以需要将操作放到其他线程中去。因为其他线程中不能修改 UI 元素，所以引入 SurfaceView，用于满足在其他线程绘制界面并且可以迅速更新界面的需求。创建一个 Paint 类型的二维数组，用于保存每个背景元素的位置坐标，并初始化行与列，随机生成 15 个路障。

(3)监听玩家的点击。对玩家的屏幕点击动作做出响应。当玩家点击灰色的点时，灰色变为红色的路障，猫移动一次。玩家点击点阵以外的背景，进行游戏初始化。玩家点击其他点，则不做处理。通过判断玩家点击的坐标所落得位置来做相应的颜色处理。

(4)最小路径选择。猫的移动可以分为六个方向。基本思想是贪心算法，遍历六个方向。如果猫能走到边界，则返回步长；如果不能，则返回负数，通过贪心算法选择猫向下一个方向。神经猫移动方如图 14－2 所示。

图 14－2　神经猫移动方向

14.3　设计过程

在实现此游戏时,大致需要分为搭建开发环境、准备资源、绘制界面和实现代码等四个部分,下面进行详细介绍。

14.3.1　搭建开发环境

本程序的开发环境及运行环境具体如下。

操作系统:Windows 7。

JDK 环境:Java SE Development KET(JDK) version 7。

开发工具:Android Studio 2.3。

开发语言:Java、XML。

运行平台:Windows、Linux 各版本。

14.3.2　准备资源

在实现本实例前,首先需要准备游戏中所需的图片资源,这里共包括图标、一张开始背景图片、一张主界面图、一张开始按钮图片以及 16 张神经猫不同动作的图片,并把它们放置在项目根目录下的 res/drawable - mdpi/文件夹中,放置后的效果如图 14 - 3 所示,并把它们放置在项目根目录下的 des/drawable - mbp;/文件夹中,图片资源目录如图 14 - 4 所示。

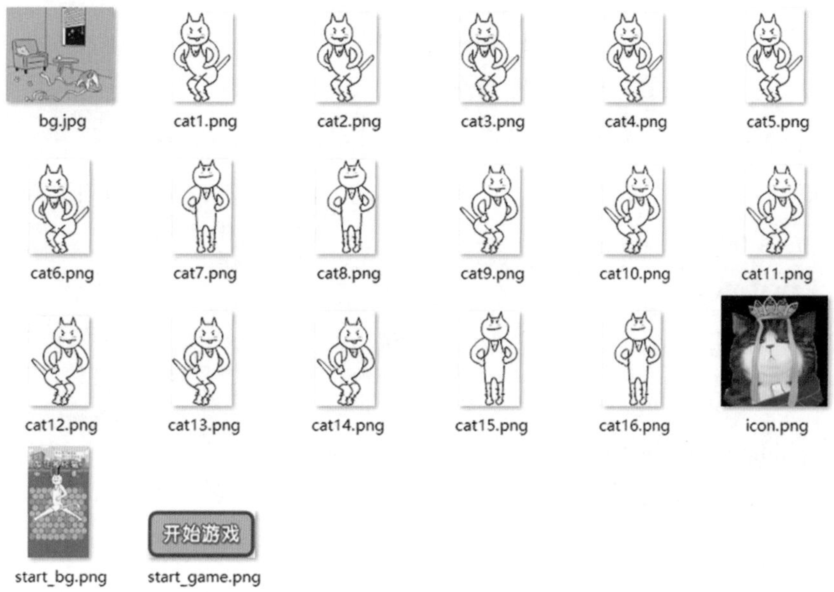

图 14 - 3　放置后的效果

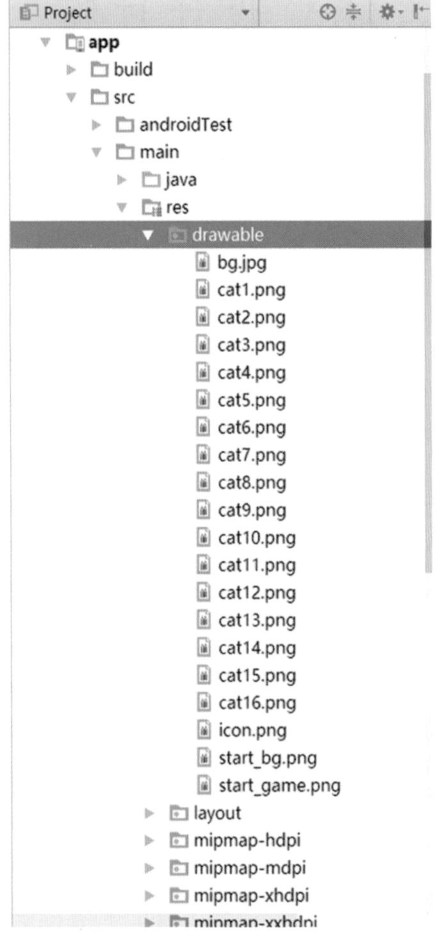

图 14－4　图片资源目录

将图片资源放置到 drawable－mdpi 文件夹后，系统将自动在 gen 目录下 com. mc 包中的 R. java 文件中添加对应的图片 id。打开 R. java 文件，可以看到下面的图片 id：

```
public static final class drawable{
        public static final int bg = 2130837579;
        public static final int cat1 = 2130837580;
        public static final int cat10 = 2130837581;
        public static final int cat11 = 2130837582;
        public static final int cat12 = 2130837583;
        public static final int cat13 = 2130837584;
        public static final int cat14 = 2130837585;
        public static final int cat15 = 2130837586;
```

```
    public static final int cat16  = 2130837587;
    public static final int cat2  = 2130837588;
    public static final int cat3  = 2130837589;
    public static final int cat4  = 2130837590;
    public static final int cat5  = 2130837591;
    public static final int cat6  = 2130837592;
    public static final int cat7  = 2130837593;
    public static final int cat8  = 2130837594;
    public static final int cat9  = 2130837595;
    public static final int icon  = 2130837596;
    public static final int login  = 2130837597;
}
```

14.3.3　绘制界面

修改新建项目的 res/layout 目录下的布局文件 activity_main. xml，首先将默认添加的相对布局管理器和 ImageView 组件删除,并为其设置背景图片。而整个游戏的背景绘制在 SurfaceView 中编写,以便在其他线程中实现 UI 界面的快速更改和计算。代码如下:

```
<? xml version = "1.0"  encoding = "utf - 8"?  >
<RelativeLayout xmlns:android = "http://schemas. android. com/apk/res/android"
    android:layout_width = "match_parent"
    android:layout_height = "match_parent"
    android:gravity = "center_horizontal"
    android:paddingBottom = "25dp"
    android:background = "@ drawable/bg" >
    <ImageView
        android:layout_alignParentBottom = "true"
        android:layout_width = "wrap_content"
        android:layout_height = "wrap_content"
        android:onClick = "onClick"
        android:src = "@ drawable/start_game"  / >
</RelativeLayout >
```

14.3.4　实现代码

(1)首先定义背景元素。创建一个 Point 类,Point 类含有坐标 X、Y 属性,还有一个状态，状态类型分为三种：

①神经猫可走的状态 STATE_ON；

②神经猫不可走的路障状态 STATE_OFF；

③神经猫所处的状态 STATE_IN。

设置对应的颜色为 ON_GRAY 灰色、OFF_RED 红色、IN_ORANGE 橙色，代码如下：

```
public class Point {
    private int x, y;
    public  enum STATUS {STATUS_OFF, STATUS_IN, STATUS_ON}
    private STATUS status;
    public Point(int x, int y) {
        this.x = x;
        this.y = y;
    }
    public int getX() {
        return x;
    }
    public int getY() {
        return y;
    }
    public STATUS getStatus() {
        return status;
    }
    public void setStatus(STATUS status) {
        this.status = status;
    }
    public void setXY(int x, int y) {
        this.x = x;
        this.y = y;
    }
}
```

(2)初始化游戏。

定义常量：

```
private static final int ROW = 9;      // 行数
```

```
private static final int COL = 9;      // 列数
private static final int BOCKS = COL * ROW / 5;      // 障碍的数量
private int SCREEN_WIDTH;      // 屏幕宽度
private int WIDTH;      // 每个通道的宽度
private int DISTANCE;      // 奇数行和偶数行通道间的位置偏差量
private int OFFSET;      // 屏幕顶端和通道最顶端间的距离
private int length;      // 整个通道与屏幕两端间的距离
private Drawable cat_drawable;      // 做成神经猫动态图效果的单张图片
private Drawable background;      // 背景图
private int index = 0;      // 神经猫动态图的索引
初始化游戏：
private void initGame() {
        steps = 0;
        //初始化背景元素数组
        for (int i = 0; i < ROW; i++) {
            for (int j = 0; j < COL; j++) {
                matrix[i][j] = new Point(j, i);
            }
        }
    //设置数组的状态,刚开始都为可通过
        for (int i = 0; i < ROW; i++) {
            for (int j = 0; j < COL; j++) {
                matrix[i][j].setStatus(Point.STATUS.STATUS_OFF);
            }
        }
    //设置神经猫的位置
        cat = new Point(COL / 2 - 1, ROW / 2 - 1);
        getDot(cat.getX(), cat.getY()).setStatus(Point.STATUS.STATUS_IN);
    //利用随机数,设置 15 个路障
        for (int i = 0; i < BOCKS; ) {
            int x = (int) ((Math.random() * 100) % COL);
            int y = (int) ((Math.random() * 100) % ROW);
            if (getDot(x, y).getStatus() == Point.STATUS.STATUS_OFF) {
```

```
                    getDot(x, y).setStatus(Point.STATUS.STATUS_ON);
                    i++;
                }
            }
        }
```

(3)绘制界面。绘制界面时,利用 surfaceView 在一个单独线程中重新绘制画面。使用 Surface 首先要继承 surfaceview 类,并实现 SurfaceView.CallBack 接口,重写 Callback 的三个方法:surfaceCreated()、surfaceChanged()和 surfaceDestroyed()。

①surfaceCreated。在 Surface 第一次创建后会立即调用该函数,可以在该函数中编写绘制界面相关的初始化代码。

②surfaceChanged。当 Surface 的状态发生变化时会调用该函数,至少会被调用一次。

③surfaceDestroyed。当 Surface 被摧毁前会调用该函数,一般在该函数中清理使用的资源。

创建 GameView 类并继承 SurfaceView,实现游戏界面的绘制和游戏逻辑的实现。具体代码如下:

```
public class GameView extends SurfaceView implements OnTouchListener {
public GameView(Context context) {
            super(context);
            getHolder().addCallback(callback);
        }
            Callback callback = new Callback() {
            public void surfaceCreated(SurfaceHolder holder) {
                redraw();  // 调用 redraw 进行重绘
            }
             public void surfaceChanged(SurfaceHolder holder,
             int format, int width, int height) {
           //屏幕宽度改变后重新绘制
                WIDTH = width / (COL + 1);
                OFFSET = height - WIDTH * ROW - 2 * WIDTH;
                length = WIDTH / 3;
                SCREEN_WIDTH = width;
            }
```

```
            public void surfaceDestroyed(SurfaceHolder holder) {
            }
        };
}
```

绘图方法是使用 Canvas 在已加载的布局中重新绘制界面，并覆盖显示。代码如下：

```
private void redraw() {
    Canvas canvas = getHolder().lockCanvas();   //先上锁
    canvas.drawColor(Color.rgb(0, 0x8c, 0xd7)); //设置主界面背景颜色
    Paint paint = new Paint();   //开始绘制到屏幕上
    paint.setFlags(Paint.ANTI_ALIAS_FLAG);   //抗锯齿
    for (int i = 0; i < ROW; i++) {
        for (int j = 0; j < COL; j++) {
            DISTANCE = 0;   //设置偏移量，奇数行和偶数行的偏移量不同
            if (i % 2 != 0) {   //i 为奇数行
                DISTANCE = WIDTH / 2;   //偏移量为元素的宽度的一半
            }
            Point dot = getDot(j, i);
            switch (dot.getStatus()) {
            //根据不同类型的状态元素，设置不同颜色
                case STATUS_IN:
                    paint.setColor(0XFFEEEEEE);
                    break;
                case STATUS_ON:
                    paint.setColor(0XFFFFAA00);
                    break;
                case STATUS_OFF:
                    paint.setColor(0X74000000);
                    break;
                default:
                    break;
            }
            //drawOval 为绘制椭圆形方法。
            // 因为绘制圆时要提供圆心，较为麻烦，所以使用椭圆
```

```
                //参数为椭圆的外接矩形,RectF 的参数为左上顶点和右下定点的坐标
                canvas.drawOval(new RectF(dot.getX() * WIDTH + DISTANCE
                        + length, dot.getY() * WIDTH + OFFSET, (dot.getX() + 1)
                        * WIDTH + DISTANCE + length, (dot.getY() + 1) * WIDTH
                        + OFFSET), paint);
            }
        }
        int left;
        int top;
        if (cat.getY() % 2 == 0) {
            left = cat.getX() * WIDTH;
            top = cat.getY() * WIDTH;
        } else {
            left = (WIDTH / 2) + cat.getX() * WIDTH;
            top = cat.getY() * WIDTH;
        }
        // 此处神经猫图片的位置是根据效果图来调整的
        cat_drawable.setBounds(left - WIDTH / 6 + length, top - WIDTH / 2
                + OFFSET, left + WIDTH + length, top + WIDTH + OFFSET);
        cat_drawable.draw(canvas);
        background.setBounds(0, 0, SCREEN_WIDTH, OFFSET);
        background.draw(canvas);
        //绘制结束后释放绘图 提交绘制的图形
        getHolder().unlockCanvasAndPost(canvas);
    }
```

(4)贪心选择路径。选择神经猫移动方法的思路为利用贪心思想,通过放回 cat 到边缘的距离或障碍物的距离来给猫一个最优的行走路径。代码如下:

```
private void move() {
    if (inEdge(cat)) {    //如果是边界,调用 failure 函数
```

```java
            failure();
            return;
        }
        Vector<Point> available = new Vector<>(); //cat 可走的 neighbor 点
        Vector<Point> direct = new Vector<>(); //返回为正数的 neighbor 点,即无路障的点
        HashMap<Point, Integer> hash = new HashMap<>();
        for (int i = 1; i < 7; i++) {
            Point n = getNeighbour(cat, i); //依次获取 cat 六个方向的相邻点
            if (n.getStatus() == Point.STATUS.STATUS_OFF) {
                available.add(n);
                hash.put(n, i);
                if (getDistance(n, i) > 0) {  //获取该方向上的距离
                    direct.add(n);
                }
            }
        }
        if (available.size() == 0) {   // 没有可用点
            win();                       // 成功围住神经猫
            canMove = false;
        } else if (available.size() == 1) {     // 只有一个可用点
            moveTo(available.get(0));
        } else {
            // 不是第一次点击,且有多条路可走
            // 根据到边界 edge 的距离 distance(包括路障等的计算)来决定走的方向
            Point best = null;  // 最终决定要移动到的元素(点)
            if (direct.size() != 0) {  // 含有可到达边界的方向
                int min = 999; // 用于记录到达边界的最小值,初始值为一个较大值
                for (int i = 0; i < direct.size(); i++) {
                    if (inEdge(direct.get(i))) {
                        best = direct.get(i);
```

```
                    break;
                } else {
                    int t = getDistance(direct.get(i),
                            hash.get(direct.get(i)));
                    if (t < min) {
                        min = t;
                        best = direct.get(i);
                    }
                }
            }
        } else {    // 所有方向都有路障
            int max = 1;    // 所有方向都有路障时,距离要么为负数,要么为0
            for (int i = 0; i < available.size(); i++) {   //负数越大越优
                int k = getDistance(available.get(i),
                        hash.get(available.get(i)));
                if (k < max) {   // 因为k是负数,所以用小于号
                    max = k;
                    best = available.get(i);
                }
            }
        }
        moveTo(best);  //移动到最合适的一点
    }
    if (inEdge(cat)) {
        failure();
    }
}
```

项目开发完成后,就可以在模拟器中或真机运行该项目了。点击运行按钮,选择创建好的模拟器,即可在Android模拟器中运行该程序。运行程序,在屏幕上将随机生成15个障碍,不断点击背景元素围住神经猫,围住后显示点击的次数。围住神经猫游戏界面如图14-5所示。

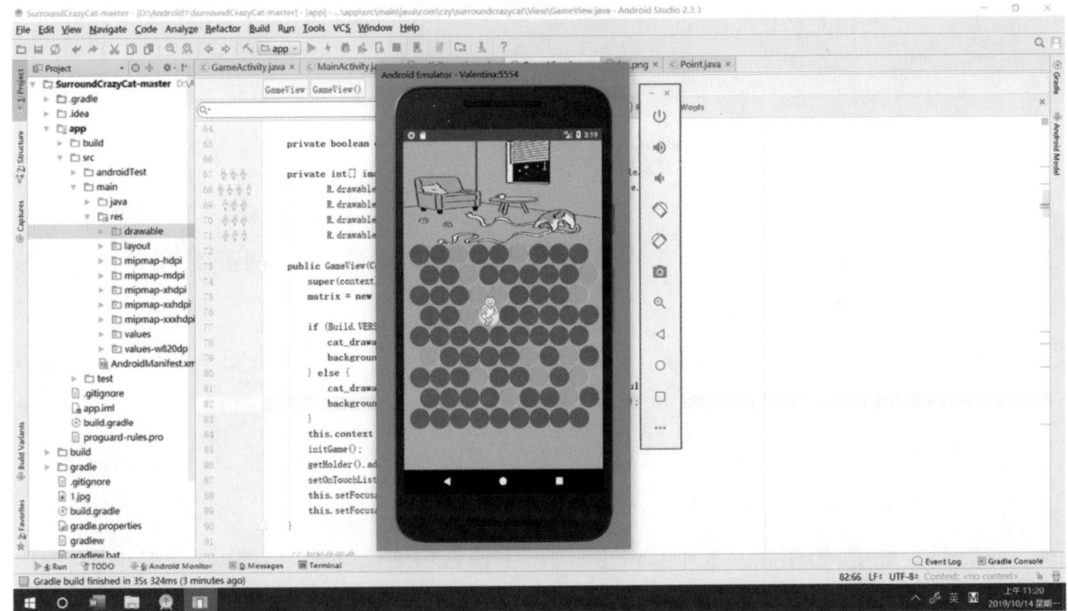

图 14－5　围住神经猫游戏界面

参 考 文 献

[1]闫祎颖,何云瑞,陈亮,等.基于CMDB的信息系统故障根因定位技术的研究[J].通信电源技术,2020(03):33-35,37.

[2]李建欣,韩杰.基于IT资产信息库的资产安全风险分析[J].网络安全和信息化,2019(11):114-116.

[3]付庆华,王贤,仉潮.数据中心运维管理系统CMDB库设计与实践[J].信息系统工程,2018(01):65.

[4]林珊珊,史莉雯,杨洋,等.基于CMDB实现IT预算精细化管理[J].中国金融电脑,2018(04):72-74.

[5]袁国泉,张明明,李叶飞,等.电力企业信息系统运行方式模型研究[J].电子科学技术,2016(03):365-368.

[6]BMCSoftare Inc. CMDB分步构建指南[M].北京:北京航空航天大学出版社,2011.

[7]刘通,周志权.ITIL与DEVOPS服务管理与案例资产详解[M].3版.哈尔滨:哈尔滨工业大学出版社,2019.